Y0-BTD-494

NUCLEAR RADIATION PHYSICS

PRENTICE-HALL INTERNATIONAL, INC., *London*
PRENTICE-HALL OF AUSTRALIA, PTY. LTD., *Sydney*
PRENTICE-HALL OF CANADA, LTD., *Toronto*
PRENTICE-HALL OF INDIA PRIVATE LIMITED, *New Delhi*
PRENTICE-HALL OF JAPAN, INC., *Tokyo*

NUCLEAR RADIATION PHYSICS

Fourth Edition

RALPH E. LAPP
Quadri-Science, Inc.

HOWARD L. ANDREWS
University of Rochester
School of Medicine and Dentistry

PRENTICE-HALL, INC.

Englewood Cliffs, New Jersey

10 9 8 7 6 5 4 3 2 1

ISBN: 0-13-625988-x

Library of Congress Catalog Card Number 75-168620

Printed in the United States of America

Preface to the Fourth Edition

Nuclear science has continued to advance since the third edition of this text in 1963. High-energy particle physics, plasmas, and solid-state physics have been the subjects of intense research efforts, and progress has been spectacular. One of the major tasks in preparing this new edition has been to select those results from the research laboratories that properly belong in a book of this scope.

Those who have used previous editions will notice a drastic change in the order of presentation. Most courses in which the book is used have had concurrent laboratory exercises. To assist in the early stages of this laboratory work, the first three chapters of the present text are devoted to nuclear instrumentation. Some questions are left unanswered until later chapters, but experience with this arrangement has shown that the early introduction to measurements is desirable.

Some of the space previously devoted to reactor economics has been used to introduce some of the new developments in space radiation and the transuranium elements. The treatment of the health and safety aspects of radiation has also been reduced, to keep the book within bounds.

The number of problems has been substantially increased, and many answers are given. Much tabular material needed for problem solving has been included, but in many cases the student must seek other sources. Although the analytical treatment is presented in the MKSA system of units, some of the tabular material has been retained in the CGS units of the original publications.

Comments and criticisms have been received from many users of earlier editions. As many as possible of these suggestions have been incorporated. For all of these expressions of interest we are deeply grateful.

RALPH E. LAPP
HOWARD L. ANDREWS

Contents

NUCLEAR RADIATION PHYSICS

1

Ionization and Ionization Chambers

1.01 Ionization and Excitation

All ponderable matter contains both positive and negative electric charges which in the normal, undisturbed state are so accurately balanced that no net charge is observed. Electric neutrality is in accord with the general principle that an undisturbed system will tend to the state of lowest energy content, and with the fact that an attractive force exists between electric charges of opposite sign. Charges of unlike sign will, therefore, tend to move together with the emission of energy. Macroscopic observation will detect no charge of either sign, although detailed probing will reveal both and will provide evidence on their spatial distribution.

When energy is added to a system that is initially neutral, one or more of the constituent electrons may be split off from the parent atom or molecule to which it was orginally attached, to become temporarily a free negative charge. The most common process involves the removal of only a single electron, which, together with the positively charged residue, forms an *ion pair*. The electron may remain free for an appreciable time, or it may promptly attach itself to some neutral structure to form a negative ion of mass comparable to that of the positive ion.

As we shall see in detail later, radiation may deliver to an atom or molecule insufficient energy to produce an ionization. The structure is then said to be in an *excited state*. This state may persist for a long period of time, or it may radiate promptly and return to the *ground state*. In some cases an excited molecule will dissociate in preference to returning to the ground state. Excited states are not directly detectable, but ions can be given directed motions by an applied electric field to produce a measurable electric current.

Each ion, positive or negative, will carry a net charge that is an exact integral multiple of the fundamental quantity of charge originally measured by Millikan in his famous oil-drop experiment. Most ions will have only a single net charge, but multiply charged ions of either sign can be produced.

By the time Millikan had determined the magnitude of the fundamental quantity of charge, scientists had established the centimeter–gram–second or CGS system of units for measuring physical quantities. There was no reason to expect that the charge unit would have any integral expression in these units, and indeed it did not. In almost all branches of physics, the meter–kilogram–second–ampere (MKSA) or SI (Système International) system of units has superseded all of the CGS variants. There are two variants of the MKSA system, depending upon the way in which the factor 4π is introduced. In this text we shall use the rationalized system in which

$$\epsilon_0 = \text{vacuum permittivity} = 8.85 \times 10^{-12} \text{ farad meter}^{-1}$$

$$\mu_0 = \text{vacuum permeability} = 1.257 \times 10^{-6} \text{ henry meter}^{-1}$$

In this system the electric force equation is

$$F = \frac{q_1 q_2}{4\pi\epsilon_0 r^2} \text{ newtons}$$

In this expression the two charges will be given in coulombs.

Many of the experimental data used in nuclear physics calculations were obtained and reported in the older system and have not been retabulated in the new. It will be necessary, therefore, to occasionally introduce quantities given in one of the CGS systems. Tables of unit equivalents will be found in the Appendix (Tables 2-5).

1.02 Ion Diffusion and Recombination

For the moment we consider only ionization in gases, where both neutral atoms and ions are free to move about. In any gas at or near room conditions there will be an enormous rate of molecular collisions because of thermal motions. Collisions between neutral molecules may be either elastic or inelastic. If the kinetic energies are below those needed to produce an excited state, the collision is necessarily elastic. Total kinetic energy and momentum will be conserved. At higher energies, inelastic collisions are probable, with part of the kinetic energy going into either ionizations or excitations.

The average distance traveled by a molecule between collisions is known as the *mean-free-path* λ. A calculation of the mean-free-path of a sphere of radius r moving in a concentration of n cm^{-3} similar spheres leads to

$$\lambda = \frac{1}{4\sqrt{2}\,\pi n r^2} \text{ cm} \tag{1-1}$$

Argon is commonly used as a filling gas in ionization chambers because of its high density. When the atomic radius of argon, 1.9×10^{-8} cm, is put into Eq. (1-1), we obtain $\lambda = 5.8 \times 10^{-6}$ cm at atmospheric pressure and 4.4×10^{-5} cm at a pressure of 10 cm of mercury. The latter pressure is commonly used in Geiger–Müller counters.

Calculations of the mean-free-path for charged particles are more complicated because of the electric forces between them. Charged-particle collisions are based on collision probabilities, usually expressed as collision cross-sections.

Positive ion–negative ion or positive ion–electron collisions can result in a *recombination*, or a return to the normal, unionized state. The cross-sections for ion–ion or ion–electron collisions depend in a complicated fashion upon the velocities of the colliding particles and upon their nature. In general, cross-sections increase with increasing energy. In some cases, notably in the noble gases, the cross-section for ion–electron recombination passes through a maximum and then drops almost to zero with increasing energy. As a result, the probability of an ion–ion recombination may be 10^4 times that for an ion–electron.

The probability of recombination of any two species is proportional to the concentration of each, and hence to the product of the two concentrations. Recombination is expressed in terms of a *recombination coefficient* α, defined as

$$\frac{dn}{dt} = -\alpha n^+ n^- \tag{1-2}$$

where n^+ and n^- are the number densities of the two combining species. If the two species have equal ion concentrations, then

$$\frac{dn}{dt} = -\alpha n^2$$

In air at normal atmospheric pressure, α has a value of about 2.2×10^{-6} $\text{cm}^{-3}\ \text{sec}^{-1}$.

When the rate of ion production is constant, as under conditions of constant radiation intensity, the ion concentration will rise to an equilibrium value n_e, where recombination will just equal production. As an example, consider a region where ion pairs are produced at a rate of n_0 $\text{cm}^{-3}\ \text{sec}^{-1}$, and where the concentration at any time t is n. Then

$$\frac{dn}{dt} = n_0 - \alpha n^2 \tag{1-3}$$

At equilibrium, $dn/dt = 0$ and we have

$$n_0 = \alpha n_e^2 \tag{1-4}$$

An example of ionic equilibrium occurs in our atmosphere, where ions are constantly being produced by cosmic rays. Conductivity measurements show an equilibrium concentration of about 1600 ion pairs cm^{-3}. Substituting values in Eq. (1-4) gives a rate of ion production of 5.4 ion pairs cm^{-3} sec^{-1}.

1.03 Energy Units

Consider the movement of ions created in a gas between two plane parallel electrodes, Fig. 1-1. Let the potential difference of V volts exist across the electrodes, which are separated by d cm. The *electric field strength* E is defined as the force acting on a unit positive charge placed at the point of interest. In general, the field strength will be equal to the negative gradient of the potential $E = dV/ds$, where ds is to be taken in the direction of the field at the point in question. In the present simple case, E will have the constant value V/d at all points between the parallel electrodes. E will be directed perpendicular to the electrode planes.

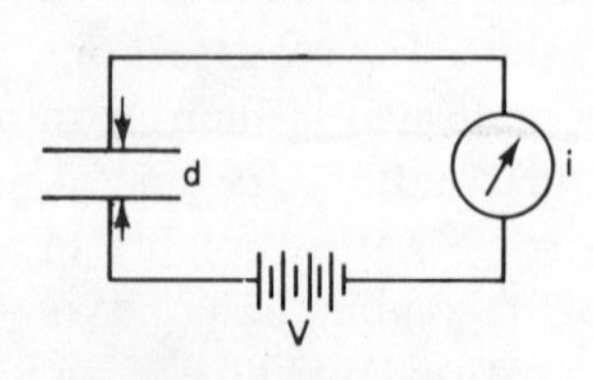

Figure 1-1. A parallel-plate ionization chamber connected in a current-measuring circuit.

The potential difference V is equal to the work done on or by a unit charge when it moves from one electrode to the other. A charge q will involve an amount of work Vq in the transport, and in the MKSA system

$$\text{volts} \times \text{coulombs} = \text{joules}$$

It is convenient in atomic and nuclear affairs to use the volt as the unit of potential and to take the fundamental quantity of electricity, the charge on the electron, as a new unit of charge. Since this charge is equal to 1.6×10^{-19} coulomb,

$$1 \text{ electron volt (eV)} = 1.6 \times 10^{-19} \text{ joule} = 1.6 \times 10^{-12} \text{ erg}$$

The electron volt is used with all of the multiple and submultiple prefixes now in common use. We shall find most useful the KeV $= 1.6 \times 10^{-16}$ joule and the MeV $= 1.6 \times 10^{-13}$ joule.

According to Eq. (1-1), an ion in a gas at normal atmospheric pressure and temperature, moving in a field of 1 volt cm^{-1}, will acquire an energy of about 2×10^{-4} eV between collisions. Statistical mechanics shows that the most probable kinetic energy of a gas molecule at a temperature of T°K is given by

$$\text{K.E.} = kT \tag{1-5}$$

where k is Boltzmann's constant, 8.62×10^{-5} eV $°K^{-1}$ or 1.38×10^{-23} joules $°K^{-1}$. Thus the most probable energy at room temperature, 0.025 eV, is considerably larger than that acquired by an ion moving in a field of moderate intensity. Obviously, diffusion processes will play a large role in the movement of ions unless they move in the presence of a strong field.

The velocity acquired by an ion in an electric field of unit strength is known as the *ionic mobility*. Ionic mobilities depend upon the mass of the ion and the other parameters that determine collision cross-sections. Mobilities are relatively independent of gas pressure but are quite sensitive to the presence of impurities. Typical values are:

air	1.4 (cm sec^{-1})/(volts cm^{-1})
argon	1.4
hydrogen	6.7

Because of its smaller mass the electron has a mobility about 10^3 times that of the heavy ions. The rationalized dimensions of a mobility are cm^2 sec^{-1} volt^{-1}, a group that conveys little of the physical significance of the parameter.

1.04 Ionization Currents

The ionization current produced in an enclosed volume, or *ionization chamber*, and measured in an external circuit, will be due to only those ions that reach the chamber electrodes. Any ions lost by recombination will be lost to the measurement. When every ion is collected, with no loss due to recombination, the chamber is said to be *saturated* and a saturation current flows in the circuit. It is obviously desirable to operate a chamber at saturation. Then the measured current will be maximal for a given radiation intensity, and will be proportional to that intensity. At saturation the ion current will be independent of small fluctuations in the source of polarizing voltage.

For simplicity, consider an ion chamber with plane parallel electrodes. At the surface of the positive electrode there will be a high concentration of negative ions, since all of this species that were produced in the entire volume, and escaped recombination, will move toward this electrode. Recombination in this region will be small because the rate of recombination is proportional to the product n^+n^-, and here n^+ is very small. All positive ions produced in this region move out promptly on their way to the cathode. An analogous situation will obtain at the cathode because of a dearth of negative ions. Recombination will be maximal at the center of the space, where the two types of ions stream past each other.

From Eq. (1-2) it is evident that the number of recombinations will be proportional to the time the ionized species exist. Ion-collection times should be short, a situation obtained by applying an adequate collection voltage. Recombination also increases with ion densities; a voltage sufficient to saturate at one intensity may be inadequate at a higher level.

The development of theoretical relations between the measured ion current i and the saturation current i_s in terms of geometrical parameters is complicated by the thermal motions of the ions. For plane parallel electrodes separated by a distance d, with a collecting voltage V, a relation of the form

$$\frac{1}{i} = \frac{1}{i_s} + k\frac{d^3}{V^2} \tag{1-6}$$

is satisfactory at high collecting efficiencies, where diffusion is relatively unimportant. The second term on the right has the analytical form that would be expected from simple considerations. Recombination should decrease with d because this decreases the distance traveled to collection and also increases the field strength. Increasing V also tends to reduce recombination by reducing the collection time. An example of the agreement between theory and measurement can be seen in Fig. 1-2. The plot of $1/i$ against $1/V^2$ is reasonably linear, as expected.

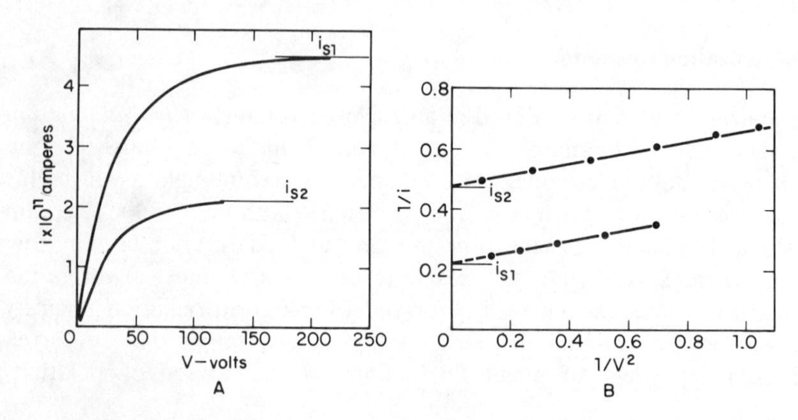

Figure 1-2. The current–voltage characteristic of an ionization chamber. (A) depicts the actual response. (B) shows a plot in the variables of Eq. (1-9) with measured points plotted along with the theoretical values.

Practically, the presence or absence of saturation is easily determined. At saturation, ion current will be independent of voltage at constant radiation intensity. If the current remains constant when the voltage is increased, the chamber is operating at saturation.

1.05 Ionization Chamber Construction

Any enclosed gas volume fitted with a pair of electrodes can function as an ionization chamber. Chamber configurations can take on many different forms and sizes, depending upon specific requirements. Probably the most common arrangement utilizes a cylinder for one electrode with a concentric cylinder or axial rod for the other. Spherical chambers may be used where an isotropic response to radiation is desired.

The analysis of a cylindrical or spherical chamber is complicated by the fact that the field strength is no longer uniform as in the case of the two parallel planes. Current–voltage relations will have the same general form as that given by Eq. (1-6) for the simpler case, but the geometrical factors will enter in a more complicated fashion.

Ion chambers were one of the earliest types of radiation detectors, and they are still in wide use, although some of their functions have been taken over by more modern methods. Ion chambers and the associated electronic measuring circuits lack the ultimate sensitivity that can be achieved with some other methods. Ion chambers are, however, the detector of choice when high radiation intensities are to be measured. Under saturation, or essentially saturation conditions, the ion current will be proportional to the amount of ionization produced in the gas volume. Thus an ion chamber will respond to the ionizing capability of the absorbed radiation. This is in contrast to some other detectors whose response is proportional to the number of ionizing events, regardless of their size.

Ion chambers may be designed so that a radioactive sample can be put directly into the gas volume. Energetic particles may be detected by introducing them into the chamber through a thin window. Very special considerations enter into the design of chambers intended to respond to photon radiation such as X or gamma rays. A discussion of these designs must be postponed until the nature of photon absorption has been considered.

Air is the gas most commonly used in ion chambers, and it is required in those chambers designed to measure the ionizing ability of photons in roentgen units (Sec. 12.10). Argon is used occasionally, perhaps under a pressure of several atmospheres in order to increase the amount of radiation absorbed in a given volume.

Ion chambers are customarily operated with the outer cylinder or sphere at ground potential and the inner electrode mounted on an insulator. This insulator must be of the highest quality to insure satisfactory performance. Leakage currents through this insulator must be kept at very low values through the use of suitable materials and through scrupulous attention to maintaining them in their original condition. Some insulators with satisfactory leakage characteristics are unsuitable because of a pseudo-leakage effect known as "soak-in." Soak-in refers to the slow migration of charge into the interior of an insulating material as long as there is a potential difference across it. When the potential is removed or reduced, the charge will reverse the direction of its movement. Both of these charge movements appear as ionization currents, in opposite directions, to an external measuring instrument.

Before the advent of modern synthetics, amber, quartz, and cast sulphur were most commonly used for ion-chamber insulators. Today there is a wide choice among the synthetic plastics. Insulator surfaces must be highly polished in order to reduce surface leakage, and so these insulators are molded in dies made with specially finished surfaces. Moisture, dust, and handling must be avoided if the initial condition of a high-grade insulator is to be maintained.

1.06 Ionization Current Measurements

Ionization currents were first measured with some form of electrometer. In one arrangement, a high resistance was placed in series with the chamber and a battery, Fig. 1-3. Ion current flowing through this resistance changed the potential of the central electrode in accordance with Ohm's law, and this change in potential was measured with the electrometer. Many electrometer designs were available using either metallized quartz fibers or light conducting vanes as the moving element. Sensitive electrometers required a good deal of attention and today they have been replaced almost entirely by specially designed vacuum tubes.

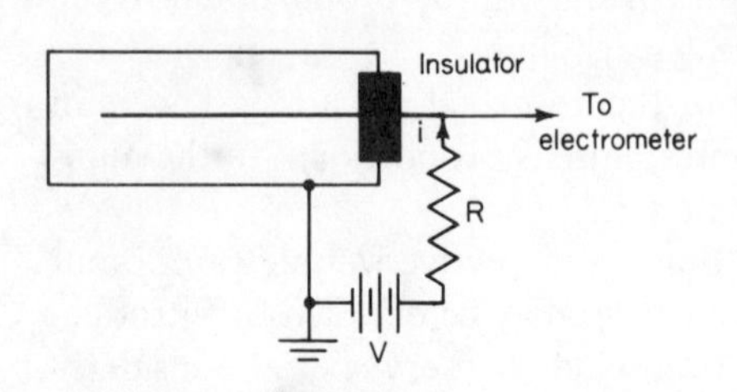

Figure 1-3. A circuit for measuring ionization currents with an electrometer.

The voltage developed across the series resistor by an ion current can be impressed on the grid of a vacuum tube, usually a triode or a tetrode specially designed and operated to have a very low grid current. The resulting current changes in the plate circuit of the tube can be either measured directly or amplified.

In most cases the series resistance will have to be very high, on the order of 10^9–10^{12} ohms, in order to attain the desired sensitivity. The plate circuit of an electrometer tube will have a resistance of only 10^6 ohms or perhaps even less, and thus the electrometer tube serves as a sort of impedance changer, taking a signal from a high-impedance circuit and feeding it out from a much lower impedance.

Because of the lower impedance, there will be no special insulator problems in the amplifiers that follow the electrometer tube. Amplifier instabilities can, however, be a serious problem. Ion-current amplifiers must be *DC-coupled*, which is to say that the plate circuit of each tube must be connected directly to the grid of the following tube. Any instability in a tube or its power supply, or any spurious signal picked up in the circuit, will be passed on and amplified by all succeeding tubes. These amplified instabilities are eliminated in *AC-coupled* amplifiers designed to pass only rapidly changing signals such as those that result from the absorption of a sudden short pulse of ionization.

The need for DC amplification has been eliminated in the *dynamic-capacitor electrometer*, which has proved to be very useful for the measurement of small ionization currents. An ion chamber can be thought of as an electrical capacitance C in which charge and potential will be related by

$$V = \frac{Q}{C} \tag{1-7}$$

With a constant chamber-circuit arrangement, a given number of ionizations will develop a constant potential that will be proportional to the amount of charge collected. Equation (1-7) shows the desirability of keeping C small, in order to develop a large signal. In the dynamic-capacitor electrometer, C is deliberately varied by cyclically changing the electrode configuration. An AC voltage will now be produced from a constant charge. This AC signal can be amplified by condenser-coupled stages that will be free of much of the long-term drift that plagues DC amplifiers.

1.07 Measurement by Charge Collection

Alternatively, an ion chamber may be connected directly to an electrometer, Fig. 1-4. Ions collected at the central electrode will change the potential of the electrometer in accordance with Eq. (1-7). Again, it is ovbiously desirable to keep C as small as possible in order to maximize the signal voltage obtained from a given charge.

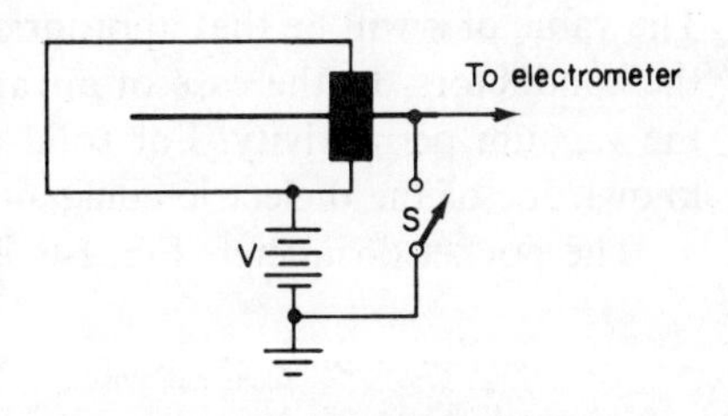

Figure 1-4. An ionization chamber connected for measuring by the rate-of-charge placed on an electrometer.

The capacitance of some simple configurations can be calculated by an application of Gauss' law. In integral form

$$\oint E_n \, dS = \frac{Q}{\epsilon} \tag{1-8}$$

In Eq. (1-8), E_n is the component of the field intensity that is directed normally outward at every point of a surface S that surrounds some charge distribution of total charge Q. The integral is to be taken over the entire surface enclosing the charge.

Let us apply this law to the concentric cylinders shown in Fig. 1-5. For simplicity, consider an axial length of 1 meter, and let there be a charge of q coulombs on each meter of length. The logical choice for the closed surface will be a cylinder of radius r, concentric with the other cylinders, and closed

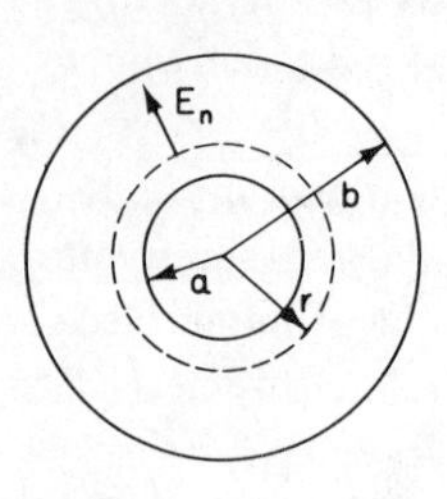

Figure 1-5. Geometrical relations in a cylindrical ion chamber.

at the ends by plane surfaces perpendicular to the axis of symmetry. Except for some fringing at the very ends of the cylinders, the electric field is radial, and hence the normal component over the ends of the closed surface will be zero. The integration needs to be carried out only over the cylindrical surface, where E_n is constant. Then

$$\oint E_n\,dS = 2\pi r E_n = \frac{q}{\epsilon} \tag{1-9}$$

and

$$E_n = \frac{q}{2\pi\epsilon r} \tag{1-10}$$

The integral of E_n from one electrode to the other must be equal to the potential difference V.

$$V = \int_a^b E_n\,dr = \frac{q}{2\pi\epsilon}\int_a^b \frac{dr}{r}$$

This gives

$$V = \frac{q}{2\pi\epsilon}\ln\frac{b}{a} \tag{1-11}$$

and

$$C = \frac{q}{V} = \frac{2\pi\epsilon}{\ln b/a}\text{ farads m}^{-1} \tag{1-12}$$

The value of ϵ will be that appropriate to the particular dielectric separating the conductors. In the case of air and most other gases, ϵ may be taken as the vacuum permittivity. For solid dielectrics, ϵ must be calculated from a knowledge of the dielectric constant $\epsilon = \kappa\epsilon_0$.

The pocket dosimeter, Fig. 1-6, is a common form of ion chamber which

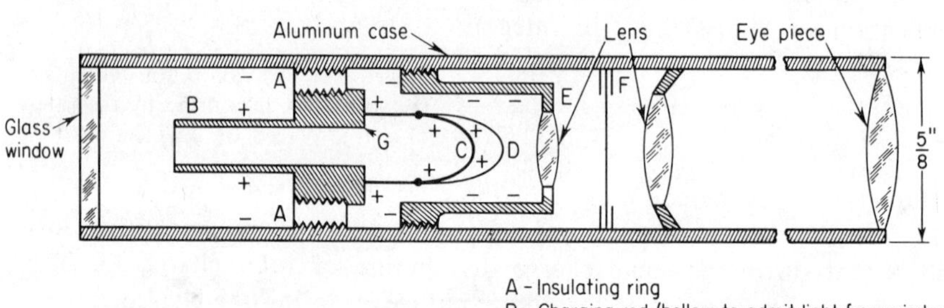

Figure 1-6. Pocket ion chamber-electroscope, or pocket dosimeter.

measures charge by direct collection. A quartz-fiber electroscope is mounted inside a small ion chamber. The indicating system consists of a fine quartz fiber, bent into the shape of a U and located close to a U-shaped wire. When a charge is placed on this system, the fiber will be deflected away from the wire. Ions formed in the chamber by radiation will neutralize some of the charges on the fiber, which will then move back toward its uncharged position. The entire unit, including the built-in microscope for reading the position of the fiber, is only slightly larger than a fountain pen.

1.08 Condenser-R-meter

As we shall see later, the basic instrument for measuring radiation exposures in roentgen units is the *standard air chamber*, Sec. 12.10. A standard air chamber is a nonportable device, suitable only for measurements in the laboratory. Field measurements are preferably made with some sort of a secondary standard instrument which can be calibrated in terms of the standard air chamber. One commonly used secondary standard is the *condenser-R-meter*, Fig. 1-7.

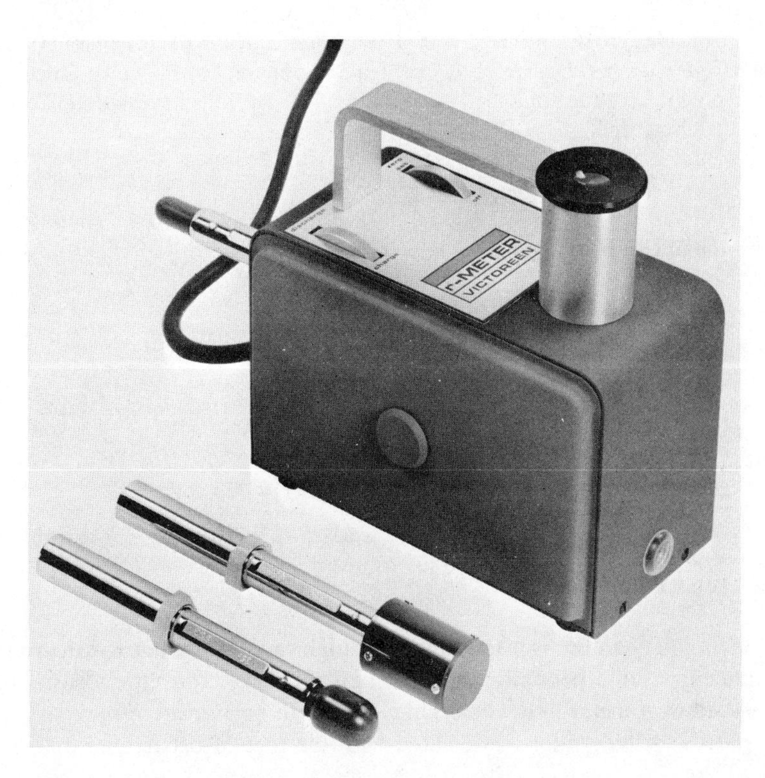

Figure 1-7. Condenser-R-meter. Courtesy of the Victoreen Instrument Co.

The condenser-R-meter consists of a quartz-fiber electrometer and a series of detachable chambers known as *thimble chambers.* In use, a chamber is connected to the electrometer and the entire system is charged until the fiber position, as read with the microscope, coincides with the zero of the internal, graduated scale. In usual designs, some 500–550 volts is required to fully charge the system. The fully charged chamber is removed from the electrometer, the charging contact covered with a protective cap, and then it is placed at the point of interest. During a radiation exposure, ions formed in the chamber will neutralize some of the charge on the electrodes, and the chamber potential will drop. After an appropriate exposure time, the partially discharged chamber is replaced in the electrometer, which has remained in a fully charged condition. When the thimble chamber is reconnected, there will be a charge sharing in accordance with the relative capacitance values and the potential as measured by the electrometer will decrease. The position of the quartz fiber will shift accordingly, and its new position can be read in terms of radiation-exposure units.

Illustrative Example

Consider an electrometer with a capacitance of 0.5 pF (picofarads) and a 6.0-pF chamber, both initially charged to a potential of 540 volts. Calculate the potential of the combined system after 7×10^9 ions have been collected in the chamber.

Initial chamber charge:

$$540 \times 6.0 \times 10^{-12} = 3.24 \times 10^{-9} \text{ coulombs}$$

Initial electrometer charge:

$$540 \times 0.5 \times 10^{-12} = 0.27$$

Chamber charge lost:

$$7 \times 10^{9} \times 1.6 \times 10^{-19} = 1.12$$

Final chamber charge:

$$3.24 - 1.12 = 2.12$$

Final electrometer charge:

$$= 0.27$$

Final system charge: 2.39

Final potential: $(2.39 \times 10^{-9})/(6.5 \times 10^{-12}) = 368$ volts

1.09 Pulse Chambers

When an ion chamber is responding to a high radiation intensity, the number of ionizing events per unit time will be very large. Ionization currents, as measured by a meter in the external circuit, will be average values, subject to the statistical fluctuations associated with the rate of occurrence of random events. When a chamber is connected to a voltage-sensitive device instead of to a current-measuring instrument, it is usually desirable to use a large circuit resistance in order to develop a large signal in accordance with Ohm's law $V = RI = RQ/t$. Here again, at high intensity, the meter will indicate the average rate of charge collection.

When the radiation intensity is low, it may be desirable to count the number of primary ionizing events, or pulses, rather than to record average values of current or voltage. Chambers designed for pulse recording must have a short ion-collection and recovery time in order to respond to individual pulses closely spaced in time.

Consider a chamber circuit, Fig. 1-8, consisting of a chamber of capacitance C and series resistance R, with an applied polarizing voltage V. At time $t = 0$, assume that a primary ionizing event has been followed by the collection of charge Q. Applying Kirchhoff's law to the series circuit,

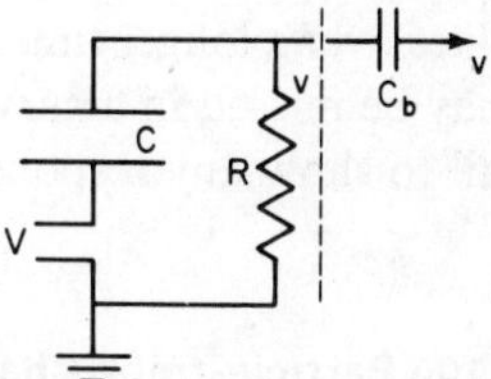

Figure 1-8. Pertinent parameters in a pulse-counting circuit.

$$V - iR - \frac{q}{C} = 0 \qquad (1\text{-}13)$$

where i and q are instantaneous and not average values of current and charge. Differentiating with respect to time,

$$-R\frac{di}{dt} - \frac{1}{C}\frac{dq}{dt} = 0 \qquad (1\text{-}14)$$

But the time rate of charge transfer is just the current, so Eq. (1-14) becomes

$$-R\frac{di}{dt} - \frac{i}{C} = 0 \quad \text{or} \quad \frac{di}{i} = -\frac{1}{RC}\,dt \qquad (1\text{-}15)$$

Integrating,

$$i = Ke^{-t/RC} \qquad (1\text{-}16)$$

The constant of integration can be evaluated from the condition that at $t = 0$, $i = Q/CR$. Then the signal voltage $-iR$ developed across R will be

$$v = -\frac{Q}{C}\,e^{-t/RC} \qquad (1\text{-}17)$$

According to Eq. (1-17), the pulse will have a polarity opposite to that applied to the electrode at which the signal is derived. The peak amplitude of the signal will be determined by Q/C, the rate of decay by the product RC.

The product RC is known as the *time constant* of the circuit. At $t = RC$, the pulse amplitude will be reduced to $1/e = 0.37$ of its peak value. When R is in ohms and C is in farads (or megohms and microfarads, respectively), the time constant will be in seconds.

For pulse counting, a time constant of 10^{-4} sec or less is usually desired. A blocking condenser C_b, Fig. 1-8, isolates the recording circuits from the DC polarizing voltage. Rapid variations in potential produced by the collection and decay of a pulse will be transmitted to an AC amplifier and a counter.

Time constants of seconds or minutes may be used in circuits designed to measure slowly varying radiation intensities. The signal voltage will then be determined by the value of R and by the value of Q averaged over several time constants. Readings will respond slowly to changes in the radiation intensity. A response to an abrupt change in radiation intensity will be 63 per cent complete in one time constant. A long time constant will produce more constant readings as the averaging of the statistical fluctuations takes place over a longer time. On the other hand, a localized area of contamination may be missed in a survey with a long-time-constant instrument, which may fail to show any response if it is passed rapidly over the area.

1.10 Particle-track Chambers

The particle-track chamber, now available in several forms, has been and continues to be one of the most important instruments used in radiation research. Originally developed by C. T. R. Wilson in 1911, the *cloud chamber* makes visible the path of an ionizing particle through the production of a trail of liquid droplets which are large enough to be seen with the naked eye.

The Wilson cloud chamber consists essentially of a cylinder closed at the upper end with a viewing window and at the lower end with a moveable piston. The space above the piston is partially filled with a liquid, usually water or a water–alcohol mixture. When the piston is moved in slowly, the gas above the liquid will be compressed and will be saturated with the liquid. If now the piston is suddenly withdrawn, the gas will expand and cool. The gas space will now be supersaturated, a condition that is favorable for the condensation of the liquid into visible droplets. Charged ions are favorite nuclei for the formation of these droplets, and thus an ionizing particle will leave behind a trail of tiny raindrops.

Tracks are usually photographed simultaneously from two directions so that the track can be reconstructed in three dimensions by projection. A skilled observer can readily identify a particle from the type of track seen in a cloud chamber. An alpha particle, or other heavy ion, will leave a thick, straight track of dense ionization. An electron, on the other hand, will leave a thin, sparsely ionized track with many turnings as it is deflected by collisions. Mesons and protons will leave trails of intermediate character.

Values of specific charge, which is the charge per unit mass, or e/m, can be obtained by applying a magnetic field to a cloud chamber. The charged particles will now move along circular arcs whose radii are determined by e/m and by the particles' velocity. If either of these factors is known, the other can be determined.

Cloud chambers of the expansion type are sensitive for only a small fraction of the time, during the short period of supersaturation. Continuously sensitive chambers can be constructed by using a gas–liquid mixture in a space whose floor and ceiling are maintained at quite different temperatures. With a steep thermal gradient across the chamber, there will be some region in a continuous state of supersaturation. Thermal-gradient chambers are useful for demonstration purposes but are little used in research. In the interest of conserving photographic film it is customary to arrange particle counters to respond only to the type of particle under investigation. When such an event occurs, the counters will trigger the expansion and the photographic equipment to record the track.

Although the cloud chamber has been invaluable in studies of high-energy particles, it has several serious shortcomings. Chief among these is that the ionization takes place in a gas of relatively low density. Because of the low density, the occurrence probability for some types of events in a chamber is relatively low, and if they do occur perhaps only a small portion of the entire track will be seen.

In 1952, Glaser constructed the first *bubble chamber*, designed to overcome some of the defects of the cloud chamber. The bubble chamber has proved to be the most important single tool used in the study of high-energy particles.

In a bubble chamber a liquid under high pressure is suddenly expanded to produce a superheated state somewhat above the boiling point at the temperature at which the chamber is operating. Theories of bubble formation show that bubbles formed below a critical size tend to shrink and disappear, whereas those above the critical size continue to grow. The presence of ions at the site of a bubble greatly enhances its tendency to grow to visible size. An ionizing-particle track is then detected by a trail of bubbles analogous to the trail of water droplets seen in the cloud chamber. Stereo photographs are taken and magnetic fields are applied in the usual way.

A variety of liquids have been used in bubble chambers, ranging from hydrogen ($0.0586\ g\ cm^{-3}$) to xenon ($2.3\ g\ cm^{-3}$). Hydrogen is the most commonly used liquid because with it all chamber reactions will be unambiguously with protons. Hydrogen chambers must be operated at very low temperatures (25°K) in order to obtain the necessary degree of superheat. As the interest in very-high-energy reactions has developed, bubble chambers of truly heroic proportions have been built or planned.

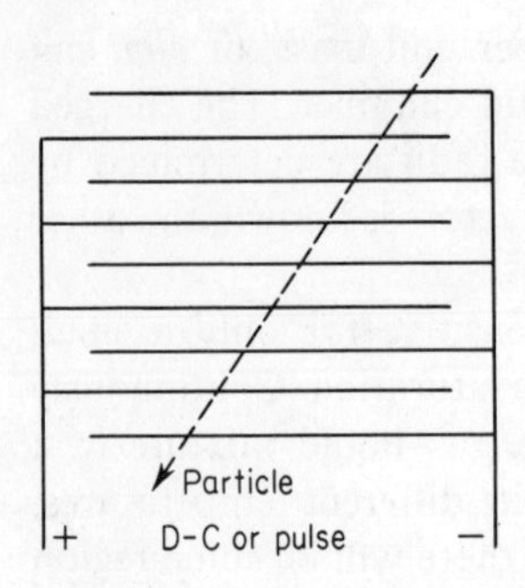

Figure 1-9. Schematic arrangement of the electrodes in a spark chamber.

Spark chambers represent a still more recent development in methods for detecting particle tracks. Alternate members of a series of parallel plates, Fig. 1-9, have a high voltage across them. An ionizing particle traversing the chamber will break down the dielectric to leave a trail of spark discharges in each gap. These sparks can be enhanced for improved photographic recording by superposing on the steady potential a high-potential pulse at the moment the desired particle is in the chamber. This pulse can be triggered by counters, as was done in the case of the cloud chamber. Figure 1-10 shows the tracks of a positron–negatron electron pair in a spark chamber. The decreasing radius of the negatron as it loses energy in the magnetic field is clearly evident.

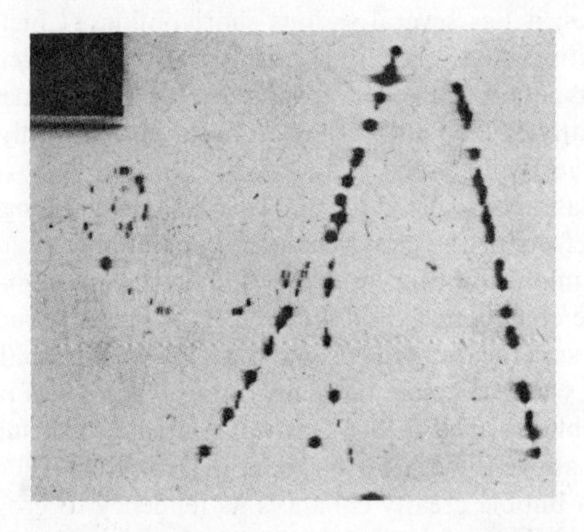

Figure 1-10. Tracks of a positron-negatron pair in a spark chamber. The slow negatron spirals under the influence of a magnetic field as it comes to rest in the chamber.

REFERENCES

Attix, F. H. and W. C. Roesch, eds., *Radiation Dosimetry*. Vols. I, II, and III. Academic Press, New York, 1966.

Cohen, E. R. and J. W. M. DuMond, "Our Knowledge of the Fundamental Constants of Physics and Chemistry in 1965." *Rev. Modern Physics*, **37**, 537, 1965.

Feynmann, R. P., R. B. Leighton, and M. Sands, *The Feynmann Lectures on Physics.* Vol. II. Addison-Wesley, Reading, Mass., 1964.

Harnwell, G. P., *Principles of Electricity and Magnetism.* International Series in Pure and Applied Physics. McGraw-Hill, New York, 1949.

Palevsky, H., R. K. Swank, and R. Grenschik, "Dynamic Capacitor Electrometer." *Rev. Sci. Instr.* **18,** 298, 1947.

Rossi, B. B. and H. S. Staub, *Ionization Chambers and Counters.* National Nuclear Energy Series. McGraw-Hill, New York, 1949.

Rutherglen, J. G., "Spark Chambers." *Progress in Nuclear Physics,* **9,** 1, 1964.

PROBLEMS

1-1. A radioactive source emits alpha particles of 5.46 MeV and is so placed in an ionization chamber that all of the energy of the particles is expended in the chamber gas. One ion pair is produced for each 35.0 eV of kinetic energy lost. What is the rate of alpha-particle emission from the source when the saturation current in the chamber is 8.70×10^{-11} ampere?

1-2. Apply Gauss' law to the case of plane parallel electrodes and prove the statement in Sec. 1.03 about the constancy of the field strength.

1-3. The "temperature" of an electron of a given energy is a quantity of considerable interest in the study of thermonuclear reactions. What is the "temperature" of a 1-eV electron? A 1-KeV electron? A 1-MeV electron?

1-4. A molecule of nitrogen has a diameter of 3.2×10^{-8} cm and can be ionized upon absorbing 14.5 eV. What potential must be applied to a parallel-plate ion chamber operating at a pressure of 50 mm of mercury, and with an electrode separation of 3.0 cm, in order to produce ionization by collision? Repeat the calculations for oxygen molecules where the pertinent constants are 2.9×10^{-8} cm and 13.5 eV, respectively.

1-5. Derive the expression for the capacitance of two concentric spheres of radii a and b.

1-6. A capacitor to be used in calibrating an electrometer is constructed of a cylinder with an inside diameter of 12 mm and an axial rod 4 mm in diameter. Neglect the contribution of the supporting insulators and calculate the capacitance for a length of 8.0 cm.

1-7. The capacitance calculated in Prob. 1-6 was used to determine the capacitance of an electrometer according to the following schedule:

Potential of combined electrometer–condenser system:	560 volts
Condenser removed and electrometer discharged to:	290
System potential with condenser reconnected:	535

Calculate the capacitance of the electrometer.

1-8. What polarizing voltage must be applied to a parallel-plate ionization chamber with an electrode area of 20 cm^2 and an electrode separation of 3 cm in order to have a collection efficiency of 99 per cent for an ion current that has a saturation value of 5.0×10^{-11} ampere? In Eq. (1-6), assume $k = 6.7 \times 10^{-11}$, d in cm.

1-9. What will be the collecting efficiency of the system of Prob. 1-7 if the radiation intensity is raised by a factor of 10? By a factor of 100? What voltages must be applied in the latter two cases to restore the collecting efficiencies to 99 per cent?

1-10. An ion chamber–electrometer circuit with a capacitance of 8.3 pF is to be used to record the pulses produced by the 4.87-MeV alpha particles from radium. A signal of at least 2 mV is desired, with a time constant of not more than 10^{-4} sec. Can these specifications be met, and if so, what is the value of the series resistor that must be used?

1-11. The integral of Eq. (1-16) from 0 to ∞ with the properly evaluated constant must give the value of the total charge collected. At what time will one-half of this charge be collected?

1-12. A dynamic-capacitor detector consists of an ion chamber whose static capacitance is 15 pF, and which can be cyclically varied by ± 0.2 pF. What will be the amplitude of the AC signal generated in response to the collection of all of the ions formed by a 5.5-MeV alpha particle?

1-13. A typical pocket dosimeter has a capacitance of 6.0 pF and is fully charged by a potential of 200 volts. What value of leakage resistance can be tolerated if the meter is to lose no more than 1 per cent of full charge in 24 hours?

1-14. Calculate the most probable velocity of an argon atom at room temperature and pressure. Compare this velocity with that produced when an argon ion moves in a field of 150 volts cm^{-1}.

1-15. An ion chamber is to respond to the pulses of ionization produced by recoil protons which expend 3.6 MeV in the chamber gas. The chamber has a capacitance of 6.5 pF and is connected to an amplifier requiring a signal of 3×10^{-4} volt. To keep the amplifier out of the radiation field, the connection is to be made with a cable which has a capacitance of 2 pF $foot^{-1}$. How long a cable can be used? What series resistor will be required to obtain a time constant of 10^{-4} sec?

1-16. How would you modify the pocket dosimeter of Prob. 1-13, retaining the present fiber sensitivity and charging voltage, to produce a meter with only $\frac{1}{50}$ of the original sensitivity, for use in radiation accidents?

1-17. The roentgen unit is defined as that amount of X or gamma radiation that will liberate 2.58×10^{-4} coulomb of charge in 1 kilogram of air. What chamber volume will be required in order to obtain an ion current of 4×10^{-12} ampere when it is placed in a radiation field of 2.0 roentgens per hour? What series resistance will be required in order to develop a signal of 10^{-3} volt?

1-18. The ion chamber of Prob. 1-17 is to be made of concentric cylinders 35 mm and 3 mm in diameter. How long must the chamber be in order to obtain the required volume? What will be the time constant of the circuit?

1-19. Repeat the calculations of Prob. 1-18 for a chamber made of cylinders 15 mm and 3 mm in diameter.

1-20. Assume an applied potential of 150 volts and calculate the field strengths at the inner and outer electrodes for the chambers in Probs. 1-18 and 1-19.

2

Gas-Filled Pulse Counters

2.01 Avalanche Ionization

Practically all gas-filled pulse counters take some advantage of the enormous increase in the number of ion pairs that can be realized through the use of secondary ionization, or *gas amplification.* An ion formed in a gas between charged electrodes will move with a constantly increasing velocity until recombination occurs, or until it loses kinetic energy through a collision. At atmospheric pressure the mean-free-path is short, and in an electric field of reasonable intensity an ion can pick up only a small amount of energy between collisions. Under these conditions all collisions will be elastic; both momentum and kinetic energy will be conserved.

If, however, the gas pressure is reduced, or if the applied voltage is greatly increased, an ion can acquire a substantial amount of energy between collisions. Inelastic collisions, in which kinetic energy is lost to molecular excitations or ionizations, now become possible. An excitation may simply lead to a return to the normal, unexcited state, or to a molecular dissociation. Each ionization resulting from a collision will, however, add an ion pair to those already present. Secondary ions may, in turn, produce more ions in subsequent collisions, thus producing very rapidly a tremendous multiplication of the number of ions formed in the primary event. This cumulative amplification process is known as *Townsend* or *avalanche ionization.* When a total of A ion pairs results from a single primary pair, the process has a *gas amplification factor* of A. A will be unity in an ion chamber where no secondary ions are formed, and may be as great as 10^{10} in a Geiger–Müller or G–M tube.

2.02 Electric Field Strength

If gas amplification is utilized, an ion must acquire, in one mean-free-path, sufficient energy to ionize a neutral molecule. Typical ionization potentials range from 10–25 eV, and so potential gradients of the order of $10/(5 \times 10^{-6}) = 2 \times 10^{6}$ volts cm^{-1} will be required at normal atmospheric pressure. Field strengths of this magnitude are impractical with plane parallel electrodes where $E = V/d$ (see Prob. 1-2).

The mean-free-path can be increased by decreasing the gas pressure, but at the same time the probability of absorbing the primary radiation in the gas decreases. The choice of an operating pressure represents a compromise between conflicting requirements. At reasonable gas pressures, the voltage needed to produce avalanche ionization is still excessive if plane parallel electrodes are used. Cylindrical configurations permit attaining high field strengths without the use of extremely high voltages. In principle, a wide variety of electrode arrangements can be used, but practical considerations favor the use of an outer cylindrical cathode with an axial wire of small diameter as the anode.

Equations (1-10) and (1-11) can be combined to give the field strength E at any radius r between two concentric cylinders of radius a (inner) and b (outer), when a potential V is applied:

$$E = \frac{V}{r \ln (b/a)} \tag{2-1}$$

With fixed values of a and b, the most intense field will be located just outside of a where the radius has the smallest possible value. Small values of a are obviously desirable in order to obtain a large value of E for a given voltage. Structural strength requirements limit the central wire to a minimum diameter of about 0.08 mm. To insure stable operation the wire must be very smooth and free of die marks or any sharp projection. A sharp point will have a very small effective radius and a correspondingly intense local field which may lead to spurious spontaneous discharges.

2.03 Gas-amplification Factor

Before discussing the mechanism of pulse amplification in detail, let us consider a cylindrical counter connected to a circuit that will measure the total charge resulting from each ionizing event. Let the counter be exposed to a radiation source that will produce a constant number of ions in each primary event. Such a source might be a radioactive nuclide emitting alpha particles, each of which will produce about 10^{5} ion pairs. Again, the source might consist of a gamma-ray emitter, with each photon capable of producing 100 primary ions.

We now consider the pulse sizes produced by these radiations as the voltage across the counter is varied. At low voltages the counter behaves exactly like an ionization chamber. As the polarizing voltage is increased the pulse sizes increase through a recombination region *A*, Fig. 2-1, to the saturation region *B*. At saturation the pulse sizes of the radiations taken as examples will differ by a factor of 10^3.

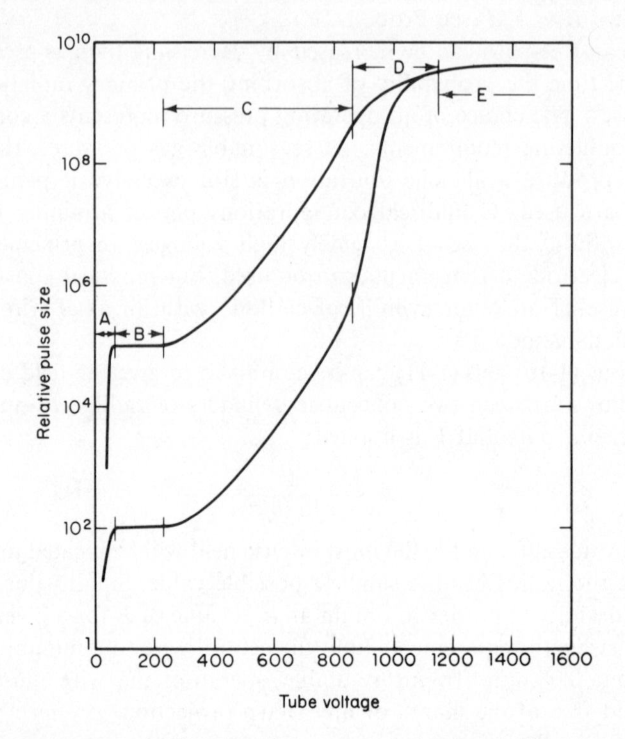

Figure 2-1. Comparative pulse sizes obtained with gas amplification from a small and a large ionizing event.

As the voltage continues to increase, secondary ions will be formed when the integral of Eq. (2-1), taken over one mean-free-path, equals the ionization potential of the filling gas. Secondary ionization will take place over the mean-free-path that lies just outside the central wire, since here the field strength is maximal. Primary ions formed outside of this mean-free-path will be accelerated, each toward the appropriate electrode, but none of these ions can acquire sufficient energy to produce secondaries.

With the onset of secondary ionization, each pulse will be larger than that produced by the primary ions alone by a factor which depends upon the voltage but not upon the number of primary ions. The size of each pulse will be related to the size of the primary initiating ionization by a constant factor. This is the *proportional region C*, Fig. 2-1. As the voltage continues to increase, secondary ions can be formed beyond one mean-free-path from the central wire. The gas amplification will increase rather sharply with voltage, but the proportionality will be maintained until A values of about 10^4 are attained.

At still higher voltages, secondaries can be formed farther and farther from the central wire, and gas amplifications of 10^5–10^7 will result. Over this range the strict proportionality between the initiating and the total ionization is gradually lost, forming the region of *limited proportionality, D*. As proportionality is lost, the two curves shown in Fig. 2-1 converge to form a single curve which represents an output pulse that depends upon the applied voltage but not upon the size of the initiating event.

Further voltage increases lead to still larger avalanches until in the *Geiger region E* as many as 10^{10} secondaries may be produced by a single ion pair. In this region, gas amplification A has no unique value since the size of the output pulse no longer bears any relation to the size of the primary event. In the Geiger region, each pulse of ionization is spread axially along the entire length of the central wire, and extends outward for many mean-free-paths. Voltage increases beyond the Geiger-region values will not produce larger pulses, which are already at the full capability of the tube.

2.04 The Proportional Region

Spherical—or more usually, hemispherical—chambers may be used when extreme values of gas amplification are not required, as in *gas-proportional counters*. In the hemispherical configuration, the central electrode, or anode, is usually in the form of a small wire loop. As in the cylindrical counter, the electric field changes rapidly with distance in the immediate vicinity of the anode; the voltage–pulse size relations are also quite similar.

In the proportional region, secondaries are formed only in the immediate vicinity of the primaries, with very little axial spread of secondary ionization. With a typical gas amplification of 10^3, the absorption of an alpha particle might lead to a pulse containing 10^8 ions, while a pulse initiated by a beta particle or a gamma ray would contain more nearly 10^5 ions. Even with this amount of gas amplification, the pulses will be too small to actuate directly most recording circuits. A *linear amplifier* can be used to raise them to the desired level while maintaining the relative pulse sizes.

Each amplified pulse can be fed into some sort of an electronic counting circuit, usually equipped with a *discriminator* which can be set to reject all pulses below some desired level. If such a circuit is set to respond only to pulses produced, say, by an avalanche of 10^7 ions or more, the counter will selectively count alpha particles in the presence of beta and gamma radiation. With very high beta or gamma intensities, several small primary events may coincide in time to produce an occasional spurious large pulse, but this is a relatively rare occurrence. Proportional counters can usually discriminate against about 10^7 small pulses per second. With stable circuit conditions, a proportional counter can be set to discriminate between pulses much closer in size than those used in the example.

With reasonable parameters of construction, voltages of 1500–4000 will be required to reach the proportional region at atmospheric pressure. Proportional counting is usually done at atmospheric pressure because the voltages present no technical problems and there are advantages to a counter that can be readily opened for the insertion of samples.

At atmospheric pressure, a sample can be introduced directly into the sensitive volume, thus eliminating losses due to window absorption and to inefficient geometrical relations. After a sample is in position, the counter is flushed rapidly with the counting gas until the room air inside has been displaced. After stable counting conditions have been established, the gas flow can be reduced to perhaps 1 cm^3 min^{-1}. Some flow is needed to provide fresh gas and to provide a slight positive pressure in the chamber to prevent the inward diffusion of room air. Very-low-energy radiations can be assayed with this arrangement, which is known as a *windowless gas-flow* counter.

Many gases and mixtures are suitable for counting in the proportional region. Argon is popular because of its high density but it has the unfortunate property of having some long-lived excited states. When these states are de-excited, at a relatively long time after formation, the energy released may trigger spurious discharges in the counter. The metastable states can be de-excited rapidly through collisions with gas molecules of a different sort. This fact has led to the extensive use of a 90 per cent argon–10 per cent methane mixture for gas-proportional counting. A mixture of 96 per cent helium and 4 per cent isobutane is commonly used when it is desired to extend the use of a proportional counter into

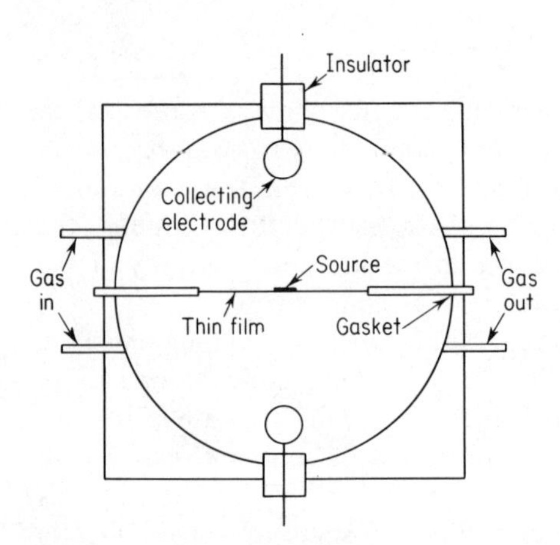

Figure 2-2. Almost 100 per cent, or 4π, geometry can be achieved with a spherical chamber and a very thin sample mount. Clamping arrangements have been omitted for simplicity.

the Geiger region. Helium–isobutane can also be used successfully in the proportional region.

With a double-chamber and a suitable sample mount, Fig. 2-2, essentially all of the particles emitted will enter the sensitive volume and will be counted. Such a counter is said to be a 4π counter, or to have a *geometry* of 4π because it responds to particles emitted through the entire solid angle of 4π steradians; 4π counters are used in *absolute* counting where it is necessary to determine the absolute disintegration rate of a radioactive sample.

Proportional counters can be used at very high counting rates because the negative ions have to move only a few mean-free-paths to be collected, and because they are moving in the most intense field in the counter. At the ion densities that exist in the proportional region, the movement of the negative ions is little influenced by the presence of the positive ions.

2.05 The Geiger–Müller Region

The remarkable sensitivity of a counter tube operating in the Geiger, or G–M, region was an important factor in the developing stages of applied radiation. As measurements became more sophisticated, the inability to distinguish between radiations proved to be a drawback, and the G–M tube has been replaced in many applications by newer devices. However, the G–M tube is by no means obsolete.

G–M tubes are used almost exclusively with voltage-activated electronic circuits. The useful signal is developed by a current flow through a resistor R, Fig. 2-3. Capacitor C blocks the high DC potential but transmits the rapid potential changes in a pulse to the recording circuits. These circuits will have a limit of sensitivity (usually adjustable) below which pulses will not be counted.

Imagine a G–M tube in such a circuit and exposed to a constant radiation intensity. Figure 2-4 shows a typical characteristic curve of count rate as a function of the applied voltage. This curve will have a threshold A, Fig. 2-4, because of the sensing limit of the circuits. In general, this threshold will be somewhere in the region of limited proportionality where the gas amplification may be 10^6–10^7. As the voltage is raised above the threshold, the count rate will rise abruptly as more and more of the smaller primary events are amplified to levels above threshold, B. Point C represents the beginning of the Geiger region CD where practically every event, no matter how small, will be amplified up to recognition levels.

The Geiger region CD is known as the *plateau*. A tube suitable for use in the Geiger region should have a long, flat plateau, since then the counting

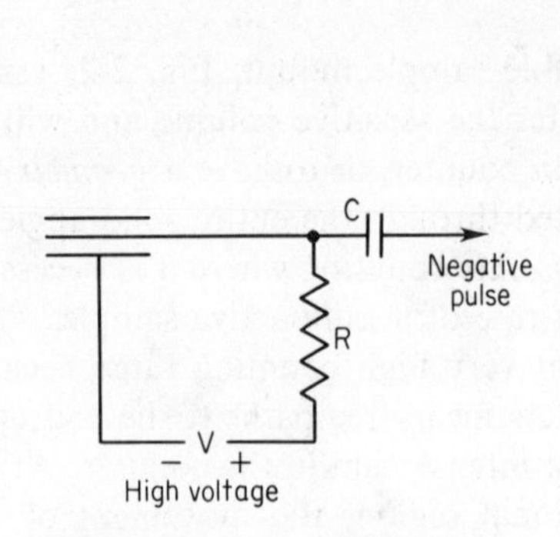

Figure 2-3. A typical G–M tube circuit. When a current pulse flows, a negative signal is developed across the resistor.

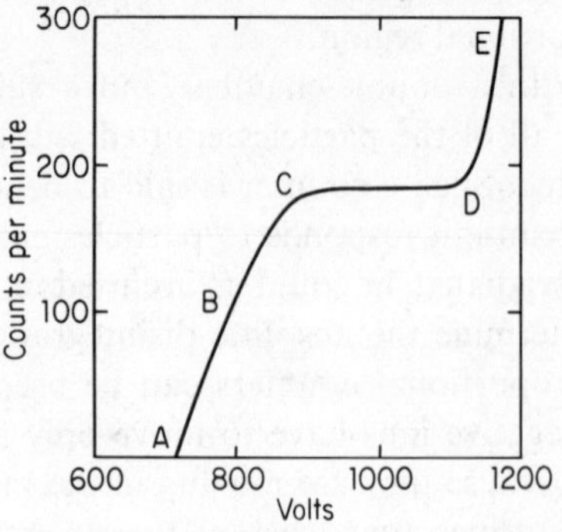

Figure 2-4. Typical variation of count rate with voltage when a G–M counter is exposed to a constant radiation intensity.

rate will not depend strongly upon the applied voltage. The plateau will have a small upward slope but it is quite possible to make tubes that show a change in count rate of only 1–2 per cent for a change of 100 volts. This is the customary unit in which plateau slopes are designated. A typical plateau may extend over a range of 200 volts. Both length and slope depend upon several variables, such as the nature of the filling gas, the condition of the central wire, and the surface properties of the cathode.

When the voltage is raised beyond the end of the plateau at *D*, the tube enters the *region of continuous discharge E*. This would be called more appropriately the multiple-discharge or spontaneous-discharge region, but custom has established the other name. In the continuous-discharge region a single pulse will trigger the tube, which then produces a series of pulses at a rate determined by the circuit constants, and independent of any radiation-initiated ionization. G–M tubes are not operated in this region. The spontaneous repetitive discharges bear no relation to the radiation being counted and in addition may shorten the counter life or damage it irreversibly. In order to understand the mechanisms that lead to continuous discharge we must consider in detail the formation and decay of an ionization avalanche.

2.06 Pulse Formation and Decay

In the first phase of an ionization pulse, a plasma made up of equal numbers of positive ions and electrons will be formed in a thin sheath just outside the central wire. The two types of ions will immediately start to separate, the electrons moving toward the center wire as the positive ions start outward toward the cathode. The electrons, with their high mobility, have to move only a short distance to be collected at the wire. Collection will be complete in about 10^{-5} sec, at which time the positive-ion cloud, or space charge, will still be close to the wire.

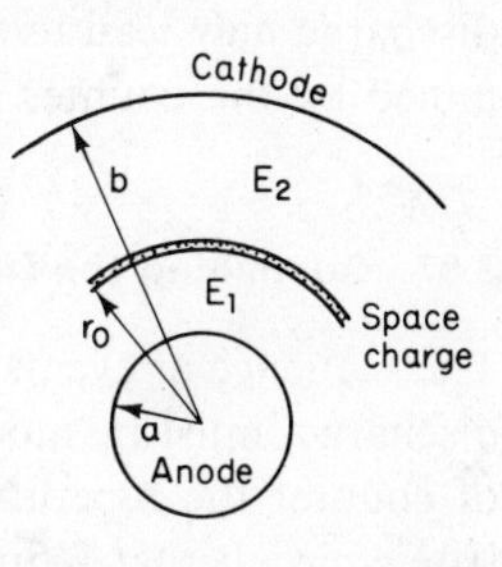

Figure 2-5. The positive ion space charge on its way to the cathode.

Consider the situation at a time when the space charge has moved outward to a radius r_0, Fig. 2-5. Let there be q charges per unit length on the wire and Q charges per unit length in the space charge. Because of the space charge, the field intensity inside the sheath will be different from that outside. From Eq. (1-10),

$$E_1 = \frac{q}{2\pi\epsilon r} \text{ inside and } E_2 = \frac{q+Q}{2\pi\epsilon r} \text{ outside}$$

Although a useful signal is developed across R, the voltage is small compared to V and we can assume that the voltage applied to the counter is constant throughout a pulse. This voltage will be the sum of the integrals of the field strengths:

$$V = \int_a^{r_0} \frac{q}{2\pi\epsilon r}\,dr + \int_{r_0}^b \frac{q+Q}{2\pi\epsilon r}\,dr \tag{2-2}$$

Integrating,

$$V = \frac{q}{2\pi\epsilon}\ln\frac{b}{a} + \frac{Q}{2\pi\epsilon}\ln\frac{b}{r_0} \tag{2-3}$$

The charge on the central wire will be

$$q = \frac{2\pi\epsilon}{\ln\dfrac{b}{a}}\left[V - \frac{Q}{2\pi\epsilon}\ln\frac{b}{r_0}\right] \tag{2-4}$$

The V term in the bracket of Eq. (2-4) refers to the charge during the quiescent period prior to ionization. The negative term in the bracket shows that q, and hence E_1, the field strength in the avalanche region, is reduced during a pulse. As the ion sheath moves outward toward the cathode, r_0 increases, the negative term decreases, and the field strength in the critical region returns toward its value in the unionized state.

In a typical counter, the positive-ion cloud will reach the cathode in about 10^{-4} sec. When a positive ion comes within perhaps 10^{-7} cm of the cathode surface it will pull an electron from the surface to become a neutral molecule. In most cases the electron will go into one of the upper energy levels of the molecule, which will then be left in an excited state. The molecule will promptly return to the ground state by radiating a series of photons of visible and ultraviolet light. Some of these photons may have sufficient energy to liberate electrons from the nearby cathode surface by the photoelectric effect. Any electron so liberated will see the high positive potential of the central wire and will be accelerated toward it. When the electron reaches the wire, it will have acquired sufficient energy to initiate a second avalanche. Thus a single discharge may lead to a self-perpetuating series of discharges at a rate determined by the counter and circuit characteristics.

2.07 Quenching the Discharge

The sequence of events described, which leads almost certainly to a repetitive discharge, must be modified if a useful device is to be obtained. In this type of counter the repetitive discharge can be interrupted by an auxiliary electronic circuit that reduces the voltage across the tube below threshold until all of the components involved in the pulse have returned to their normal state. Quenching circuits are effective but they add circuit complications, and the response rate is relatively slow.

A counter can be made *self-quenching* by adding a small amount of a quenching gas to the main counting gas, which is usually argon. Two families of quenching gases are commonly used: organic molecules such as ethyl alcohol, xylene, or isobutane, or halogen gases such as Cl_2 or Br_2.

Consider specifically a G–M counter filled with a mixture of about 90 per cent argon and 10 per cent ethyl alcohol to a total pressure of 10 cm of mercury. Both argon and alcohol molecules participate in the first phases of the avalanche, and initially the space-charge cloud will contain both species of positive ions. An argon atom requires an energy of 15.7 eV to become ionized, whereas only 11.3 eV is required to ionize the alcohol molecule. In an argon–alcohol collision it is energetically possible for the argon to become a neutral atom while ionizing the alcohol molecule. It is not energetically possible for an alcohol ion to become neutral while producing an argon ion. Each ion will make some 10^3 collisions while the space charge is on its way to the cathode and so on arrival the cloud will consist almost entirely of alcohol ions.

When an alcohol ion extracts an electron from the cathode surface, the neutral alcohol molecule will have an energy of excitation given by

$$\mathscr{E}^* = I - \phi \tag{2-5}$$

where I is the ionization potential of the alcohol and ϕ is the *work function* or the energy required to remove an electron from the cathode material. Cathode surfaces are specially chosen to have a large work function so that as little energy as possible is available for excitation. Colloidal graphite with a work function of 5 eV is commonly used as a coating on cathode surfaces.

Many organic molecules, including alcohol, prefer to de-excite by dissociation rather than by radiation, and the energies involved in dissociation are too small to trigger a second discharge. If an alcohol molecule should de-excite by radiation, most of the photons will have energies below the cathode work function, and for them electron ejection is impossible. Any photons emitted in de-excitation will most probably be reabsorbed by alcohol, which has deep, broad absorption bands in the ultraviolet. Dissociation will be the ultimate fate of almost all of the excited molecules.

Each transfer of ionization from argon to alcohol is accompanied by the emission of photons with a total energy of $15.7 - 11.3 = 4.4$ eV. This energy is too low to eject an electron from a cathode with a large work function. Again, the photons will be absorbed by alcohol molecules which will subsequently dissociate. The combination of a suitable quenching agent with a high-work-function cathode blocks almost every reaction by which a second discharge might be initiated.

There is always a small chance that an argon ion will escape neutralization in the space charge and will reach the cathode as an ion. In that case an electron will be extracted from the cathode as before, but by Eq. (2-5) there will now be 4.4 eV of energy more than in the case of alcohol. This increased energy is available for photon emission, and greatly increases the chance of electron production at the cathode, with a corresponding increase in the chance of triggering a second discharge.

The chance that an argon ion will survive to reach the cathode increases as the transit time decreases, or as the accelerating voltage increases. In addition, at higher voltages there will be a greater number of ions formed in the original plasma. Thus the chance of a spurious discharge can be expected to increase with the tube voltage. The slight upward slope of the G–M plateau is undoubtedly due to occasional spurious discharges, which become more numerous as the voltage is increased. At sufficiently high voltages the probability of a spurious discharge becomes a certainty and the tube goes into continuous discharge even in the presence of the quenching gas.

A satisfactory quenching gas must have three main properties:

1. It must have an ionization potential that is lower than that of the main counting gas in the tube. If possible, this ionization potential should be less than twice the cathode work function in order to completely suppress electron ejection.
2. It must have broad and intense ultraviolet absorption bands.
3. When in an excited state it must prefer to dissociate rather than to de-excite by radiating.

All of the organic quenching gases are vapors at room temperature, and condensation may occur upon cooling. Quenching action may become erratic with only a few degrees reduction in temperature, and may be lost entirely at the temperatures attained in some field operations.

Some of the quenching gas is dissociated at each discharge and consequently counters using an organic quench gas have a limited useful life. A typical counter may contain 10^{20} molecules of alcohol, and 10^{10} may dissociate at each pulse. The gas will be completely dissociated after

10^{10} pulses, and quenching will be erratic well before the absolute limit is reached.

Some of the halogens, notably chlorine and bromine, are used to quench the discharge in neon or in neon–argon mixtures. As little as 0.1 per cent of the halogen is needed to obtain quenching. The halogen quenching action seems to be similar to that of the organic agents in that some of the molecules dissociate at each discharge. The dissociated halogens, however, recombine at the end of the discharge, and so these tubes have an essentially unlimited life. Because of this recombination, the temporary operation of a halogen tube in the continuous-discharge region does not shorten its life appreciably.

Halogens are less effective quenching agents than the best of the organics. The plateau of a halogen tube may extend over only 150 volts, with a slope of 10 per cent per 100 volts. Comparable figures for an organic-quenched tube might be 300 volts and 2 per cent per 100 volts, Fig. 2-6.

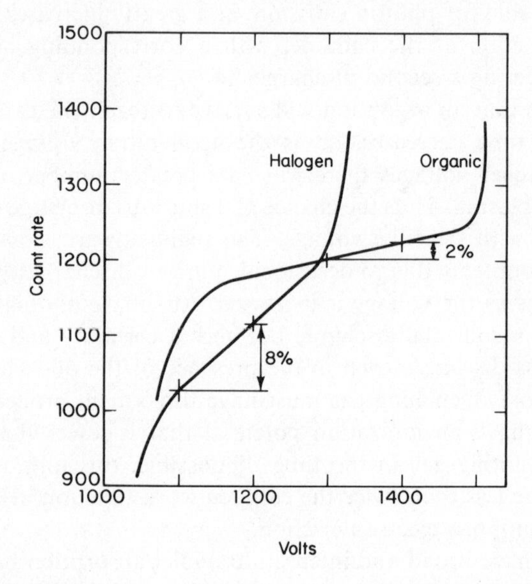

Figure 2-6. Typical plateau slopes of organic- and halogen-quenched G–M counters.

2.08 G–M Tube Construction

The conventional G–M counter is made with a cylindrical cathode 1–10 cm in diameter with a length 2–10 times as great. Counters designed for gamma-ray detection will usually have thick metal cathodes in order to increase photon absorption. Cathode surfaces are treated or coated to provide a surface with a large work function. Alkali metals have small work functions and are undesirable contaminants. In some cases a G–M tube will respond to visible light and must be completely shielded if reliable counts are to be obtained from it.

End-window counters are commonly used for counting beta particles. Windows of cleaved mica or thin glass bubbles can be made with an equivalent thickness of only 1 mg cm^{-2}. These thin windows admit all but the weakest beta particles into the sensitive volume. The central wire in an end-window counter is supported at one end, and has a glass bead on the free end to prevent spurious discharges from sharp projections.

The central wire is usually made of tungsten, but any metal that can be drawn to the desired size with a smooth surface is satisfactory. Halogen quenching gases are highly reactive and in these tubes the choice of metals is more restricted. Tungsten, tantalum, and stainless steel are satisfactory. Surface irregularities on the central wire will produce nonuniform electric fields and result in a tube with little or no plateau. Anode wires will range from about 0.07–0.25 mm in diameter. The smaller sizes are fragile and hard to handle, while the larger sizes require high voltages to obtain the required field strength.

Gases that form negative ions, such as oxygen, water, and carbon dioxide, are not well suited for use in counters. Negative ions will be attracted toward the central wire along with the electrons, but they will have the low mobilities characteristic of heavy ions. Consequently, the negative ions will arrive at the electrode long after the electrons have been collected. By this time the positive ions forming the space charge will have started to leave the vicinity of the anode, and the resulting increase in the field strength may permit the negative ions to attain energies sufficient to initiate a second discharge. Several attempts have been made to develop the use of carbon dioxide as a counting gas, in order to permit the sensitive assay of the important nuclide carbon-14. Although some of these attempts were successful, G–M counters have been superseded by other devices for most carbon-14 assays.

2.09 Pulse-counting Circuits

The pulses obtained from a gas-flow proportional counter are too small to directly actuate a counting device. A proportional counter will, therefore, be followed by a *linear amplifier* having a voltage gain of 200–5000. Frequently the required amplification is obtained in two steps. The proportional counter may be immediately followed by a preamplifier, which has little or no gain, but which produces an output signal across a relatively low impedance. This signal will then be fed into the main amplifier. This amplifier is usually carefully designed to have a gain that is strictly independent of the size of the input pulse. The pulse amplitude at the output will then be proportional to the amplitude of the pulse produced by the counter tube. Pulse-height discrimination or analysis circuits can be applied to the output signals.

G–M tube pulses are negative, rising to a maximum amplitude of perhaps 1 volt in about 10^{-5} sec and decaying in 10^{-3} sec. Pulses of this nature will actuate counting devices directly, but it is usually desirable to isolate the counter tube with at least a low-gain amplifier. This amplifier need not have a linear response because all pulse-size discrimination has already been lost in the counter.

In general, counting rates will be much too fast to be recorded directly by a mechanical counter. The pulses will be fed into some sort of a *scaling circuit*, or scaler, which will reduce the count rate by a known factor to the point where they can be recorded. One circuit in common use is the *binary scaler*, which passes on a single pulse for every two pulses received. Binary scalers lead to scaling factors of 2, 4, 8, or in general 2^n. By suitable feedback circuits, a scale of 16 can be reset to zero at the tenth pulse to produce a decade scaler, more compatible with our decimal system of numbers. Many types of scaling circuits have been devised. Detailed discussions will be found in specialized texts.

Scaling circuits record the total number of pulses that activate the detector during a known time interval. A simple division then gives the count rate, which is the quantity usually desired. Count rates can, however, be displayed and recorded by the use of a count-rate meter. In a ratemeter, appropriate circuits equalize all incoming pulses in both amplitude and duration. These equalized pulses are then fed to a vacuum tube or transistor whose output circuit contains a large capacitance. The circuit has a long time constant, and the voltage across the capacitance will be proportional to the rate at which the incoming pulses are received. The long time constant prevents a ratemeter from accurately following very rapid changes in count rate, but it will respond correctly to slow changes. Ratemeters can be designed to cover a wide range of count rates by changing the characteristics of the shaped pulses. At low counting rates the pulse duration may be lengthened to increase the sensitivity. At high counting rates the pulses may be shortened to reduce losses due to coincidences.

2.10 Resolving Time

We have already noted that the presence of a space charge effectively reduces the gas amplification to a point where secondary ionization is temporarily impossible. During this interval, known as the *dead time*, the counter is incapable of responding to a second ionizing event. As the space charge moves outward toward the cathode, its effect on the electric field near the central wire progressively decreases, and the counter sensitivity gradually returns to its original value. When all of the space charge has been neutralized, the amplification factor will have returned to the value that it had prior to the pulse.

During the period of returning sensitivity, the system may be capable of a reduced response to a second event. Figure 2-7 depicts the situation where

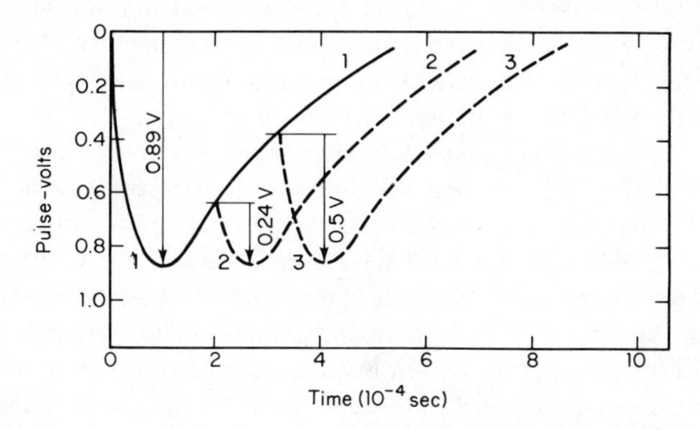

Figure 2-7. Dead time and recovery of sensitivity in a G–M tube and circuit requiring a signal of 0.5 volts.

a pulse amplitude of 0.5 volt is required to actuate the recording system. Pulse 1, starting at time zero, has a maximum amplitude of 0.89 volt, and will be recorded. Since this 0.89-volt signal represents the maximum response of the system, a second pulse, originating as at 2×10^{-4} sec, Fig. 2-7, can produce a pulse of only 0.24 volt. This response is not enough to trigger the recording circuit, and so pulse 2 will go unrecorded. If the second pulse occurs at a later time, as at 3, Fig. 2-7, it will produce a response just sufficient to trigger the counter.

The time τ between just-recordable pulses is known as the *resolving time*. Resolving time obviously depends upon the values of various circuit parameters as well as upon the characteristics of the detector itself. Resolving time is always somewhat longer than the dead time, during which the detector is incapable of producing any response.

A counting system should be capable of responding correctly to an input of *evenly spaced* pulses up to a rate of $1/\tau$. Radioactive decays are, however, random rather than evenly spaced in time, and so some pulses will be lost long before the average counting rate equals $1/\tau$. G–M counters are relatively slow ($\tau = 5 \times 10^{-4}$ sec), which leads to a loss of about 5 per cent of the incident pulses when the average rate is only 10^2 sec^{-1}.

Two types of system behaviors due to sensitivity loss and recovery are recognized. In a *paralyzable system* the recovery of sensitivity requires a certain amount of time after an ionizing event, whether or not the system responds to it. In a sense, each pulse of ions, even if not amplified to the point of being recorded, resets the recovery process back to zero. If Fig. 2-7 represents a paralyzable system, recovery will start over again with pulse 2 and will proceed along this curve rather than along curve 1. In this case, pulse 3 will produce a signal of less than 0.5 volt and will not be recorded. A paralyzable system, typified by a nonself-quenching G–M tube, has a zero response to count rates beyond its resolution capability. Such a system is potentially hazardous in a field survey instrument, since with it one may have a zero reading in an intense radiation field.

In the more common *nonparalyzable system*, recovery continues from the time that there was sufficient gas amplification to produce a recordable count, rather than being reset to zero at each primary event. In such a system pulse 2 in Fig. 2-7 does not reset the recovery, and pulse 3 will trigger the system for the second time. Most self-quenching G–M counters and gas-flow proportional counters behave as nonparalyzable systems. When these systems are exposed to very high radiation intensities, they respond at a rate determined by the resolving time rather than by the incoming pulse rate. A response of this type, although incorrect, is more informative than the zero response of a paralyzable counter.

Assume that a counting system with a resolving time τ responds at a rate n per unit time when exposed to N initiating events per unit time. In unit time the total insensitive time will be $n\tau$ and the number of counts missed will be $Nn\tau$. But the number of counts missed will also be $N - n$, whence

$$N - n = Nn\tau$$

and

$$N = \frac{n}{1 - n\tau} \tag{2-6}$$

$$n = \frac{N}{1 + N\tau} \tag{2-7}$$

If the resolving time of a system is known, an observed count rate can be corrected by applying Eq. (2-6) or by using the nomogram, Fig. 2-8, which permits a geometrical solution of the resolving-time equation.

When fast counting rates must be measured, it is desirable to make frequent determinations of the resolving time. The most usual method involves making four counts using two radioactive sources. The background-count rate is first obtained and is found to be B per unit time. One of the sources is then positioned to obtain a count rate at which resolving time loss

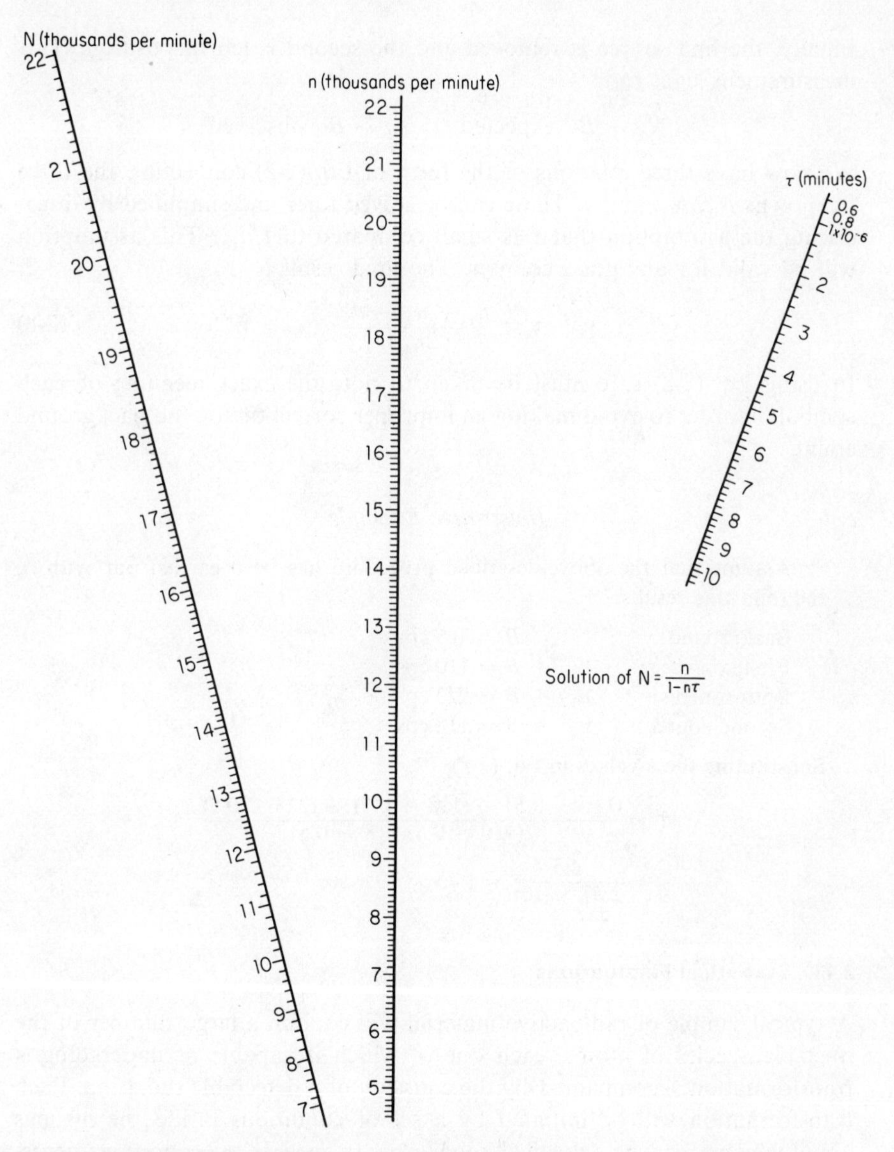

Figure 2-8. Nomogram for resolving time corrections. For solution of $N = n(1 - n\tau)$; set straightedge on proper values of n and τ and read N at the intersection.

will be appreciable. Let the expected count rate be $N_1 + B$ and that observed $n_1 + B$. With the first source still in position, the second is introduced and adjusted until the count rate is approximately doubled. We now have

$$N_1 + N_2 + B \quad \text{expected} \qquad n_{12} + B \quad \text{observed}$$

Finally, the first source is removed and the second is left in position. This measurement leads to

$$N_2 + B \quad \text{expected} \qquad n_2 + B \quad \text{observed}$$

We now have three relations of the form of Eq. (2-2) connecting the three unknowns N_1, N_2, and τ. These can be solved for τ and simplified by introducing the assumption that τ is small compared to $1/n_{12}$. This assumption will be valid for any good counter. The final result is

$$\tau = \frac{n_1 + n_2 - n_{12}}{2n_1n_2} \tag{2-8}$$

In using Eq. (2-8) care must be taken to note the exact meaning of each symbol, in order to avoid making an improper correction for the background count.

Illustrative Example

Assume that the above-described procedure has been carried out with the following results:

Background:	$B = 0.5$ cps
First source:	$n_1 + B = 110$ cps
Both sources:	$n_{12} + B = 223$ cps
Second source:	$n_2 + B = 118$ cps

Substituting these values in Eq. (2-8),

$$\tau = \frac{(110 - 0.5) + (118 - 0.5) - (223 - 0.5)}{2(110 - 0.5)(118 - 0.5)}$$

$$= \frac{4.5}{2.57 \times 10^4} = 1.75 \times 10^{-4} \text{ sec}$$

2.11 Statistical Fluctuations

A typical sample of radioactive material will contain a large number of the unstable species of atoms, each one of which is capable of undergoing a transformation accompanied by the emission of a detectable radiation. Each transformation will be initiated by a set of conditions inside the nucleus about which we have no detailed knowledge. In general, the conditions necessary for a disintegration occur only rarely; as a consequence, radioactive nuclei do not decay en masse immediately upon formation. Some will decay promptly; others will wait many years for the proper conditions to occur. Because of our lack of knowledge of nuclear conditions, radioactive decay appears to be governed by the laws of random chance rather than by precise physical laws.

Every nucleus of a given species has an equal probability of decaying in a given time interval. This probability does not vary with age, which is to say that an old nucleus has the same probability for decay that it had when it was first formed. Each disintegration is an independent event, which has no effect on the probability of decay of any other nucleus. Under these conditions, the rate of decay and the rate of emission of radiations will fluctuate in accordance with the statistical laws governing random processes.

A measurement extending over an extremely long period of time is required to establish a true value of a decay rate. Practically, measurements must be made in a much shorter time. Values thus obtained will not certainly be the true decay rate, but will have an associated uncertainty. Repeated short-term measurements of a constant rate will lead to a distribution of values around the true mean rate.

The theory of random processes shows that the frequency distribution of the short-time values will be the same as the distribution of the coefficients in the binomial expansion of $(p + q)^{N_0}$, where p is the probability that a given nucleus will decay, and q is the probability that it will not, or $(1 - p)$; N_0 is the total number of nuclei that are capable of decay. According to the binomial distribution, the probability that exactly N events will be observed is given by

$$P_N = \frac{N_0!}{N!(N_0 - N)!} p^N (1 - p)^{N_0 - N} \tag{2-9}$$

P_N will be small when N is either small or is nearly equal to N_0, and will pass through a maximum at some intermediate value of N. The N value for maximum P_N depends upon the occurrence probability p, one of the two parameters that characterize the distribution. The second parameter N_0 controls the "sharpness" of the distribution about the maximum. A measure of this "sharpness" is the *standard deviation* σ, which for small values of p is given by

$$\sigma = \sqrt{N_0 p} \tag{2-10}$$

The binomial distribution applies only to those processes in which N can take on only integral values. A plot of the frequency distribution will, therefore, be a histogram rather than a continuous curve.

2.12 Normal and Poisson Distributions

Equation (2-9) is difficult to handle when N becomes large, as it is in most practical cases. In those cases where N can take on nonintegral values, the binomial distribution can be replaced by the approximation

$$dP_N = \frac{1}{\sigma\sqrt{2\pi}} e^{\frac{-(\bar{N}_0 - N)^2}{2\sigma^2}} dN \tag{2-11}$$

In the *normal* distribution represented by Eq. (2-11) the observed variable N may take on any value from $-\infty$ to $+\infty$. The σ term in (2-11) again controls the "sharpness" of the distribution around the most probable value $\bar{N}$.

In pulse counting we shall always be dealing with integral numbers of counts and here the binomial distribution can be approximated by the *Poisson* distribution:

$$P_N = \frac{(\bar{N}_0)^N e^{-\bar{N}_0}}{N!} \tag{2-12}$$

where $\bar{N}_0$ = true mean number of events occurring in time t
P_N = probability of observing exactly N events in t

The Poisson distribution is expressed in terms of only one parameter, $\bar{N}$. Inherent in the derivation of the expression is the relation

$$\sigma_{N_0} = \sqrt{\bar{N}_0} \tag{2-13}$$

Unlike the binomial distribution, the Poisson distribution will be unsymmetrical about $\bar{N}_0$ for extreme values of N. In almost every case we are concerned with values of N close to $\bar{N}_0$ and the Poisson distribution is a satisfactory approximation.

2.13 Standard Deviation of a Pulse Count

An evaluation of Eq. (2-12) for various values of N leads to the probability distribution shown in Fig. 2-9. According to this plot there is a probability of 0.317 (31.7 per cent) that a single determination will differ from the true value by 1σ or more. This probability drops to 4.55 per cent for 2σ and to 0.27 per cent for 3σ. This is equivalent to saying that there is a probability of 0.683 that the single determination lies within $\pm 1\sigma$ of the true value, and so on.

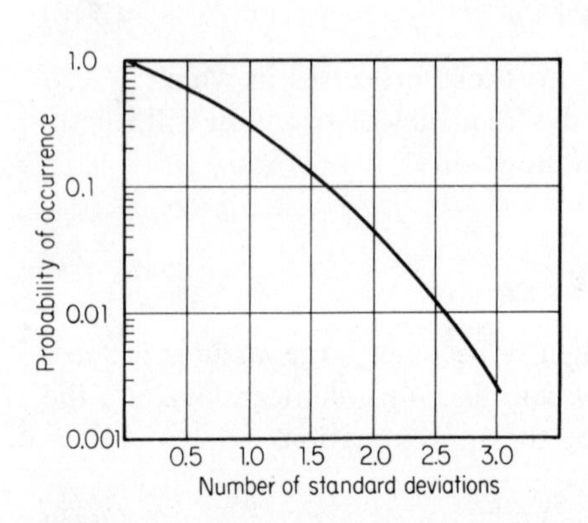

Figure 2-9. Probability, in a Poisson distribution, that a measured value will differ from the mean by more than the given number of σs.

Now, the true value can only be known from a measurement made over a very long time, and hence it can never be known with certainty from a single, or even from a series, of short-time measurements. Accordingly, Eq. (2-9) cannot be evaluated exactly. A series of measurements of N will, however, have a mean value $\bar{N}$ which has a high probability of being very nearly equal to $\bar{N}_0$. We may, then, replace $\bar{N}_0$ by $\bar{N}$ to a good approximation.

It is usually more convenient to

express the results of radioactive assays in terms of rates rather than in the total numbers of pulses recorded. Since any rate R is given by $R = N/t$, Eq. (2-12) becomes

$$P_R = P_N = \frac{(\bar{R}t)^{Rt} e^{-\bar{R}t}}{(Rt)!} \tag{2-14}$$

In any properly operating system the counting time t will be known precisely and so the standard deviation will be based only on the fluctuations in N. That is,

$$\sigma_R = \frac{\sqrt{N}}{t} \tag{2-15}$$

A series of n observations designed to measure either $\bar{N}_0$ or $\bar{R}_0$ will produce a series of N or R values distributed according to the Poisson relation. The standard deviation of the mean of this series of observations can be calculated from

$$\sigma_N^2 = \frac{1}{n-1} \sum_1^n (N_n - \bar{N})^2 \tag{2-16}$$

The value of σ_N thus obtained will agree with that calculated from Eq. (2-13) if the counting system is working properly and if the series of observations is sufficiently long.

It is sometimes desirable to express the standard deviation as a fraction of the quantity being measured. The *coefficient of variation* V is then given by

$$V_N = \frac{\sqrt{N}}{N} = \frac{1}{\sqrt{N}} \tag{2-17}$$

and

$$V_R = \frac{\sigma_R}{R} = \frac{\sqrt{N}/t}{N/t} = \frac{1}{\sqrt{N}} \tag{2-18}$$

Illustrative Example

The following count rates were obtained with a G–M counter and a constant source radiation:

Count no.	*Counts min*$^{-1}$	*Count no.*	*Counts min*$^{-1}$
1	1739	11	1782
2	1824	12	1869
3	1826	13	1845
4	1768	14	1812
5	1810	15	1778
6	1733	16	1833
7	1767	17	1808
8	1814	18	1772
9	1784	19	1867
10	1801	20	1743

$$\text{mean } R = 1799 \text{ cpm}$$

$$\sigma = \sqrt{1799} = 42.4 \text{ cpm} \qquad V = \frac{1}{42.4} = 0.0236$$

Five of the measured values fall more than 1σ from the mean, in satisfactory agreement with the six expected. One value more than 2σ from the mean would be expected; none is found, but nos. 12 and 19 approach it.

2.14 Counting Schedules

If it were not for the presence of background radiation, a desired value of V_S could be chosen from the demands of the assay, and the required number of counts calculated from Eq. (2-18). In practice, account must be taken of the count rate due to background, unless the count rate of the sample is relatively very large. When a sample is being counted, the count rate actually observed, R_T, will be

$$R_T = R_S + R_B \tag{2-19}$$

where the subscripts refer to the total, sample, and background, respectively. The sample activity will be obtained as a difference.

The standard deviation of a sum or difference is given by

$$\sigma_S = \sqrt{\sigma_T^2 + \sigma_B^2} \tag{2-20}$$

whence

$$\sigma_S = \sqrt{\frac{N_T}{t_T^2} + \frac{N_B}{t_B^2}} = \sqrt{\frac{R_T}{t_T} + \frac{R_B}{t_B}} \tag{2-21}$$

where t_T and t_B are the counting times with and without the sample, respectively. The corresponding coefficient of variation is

$$V_S = \frac{\sigma_S}{R_S} = \frac{\sigma_S}{R_T - R_B} \tag{2-22}$$

A desired value of V_S and, hence, of σ_S is set by the particular requirements of the assay. If an approximate value for R_S is known, Eq. (2-21) can be used to set up a schedule in terms of either counting times or total counts required to obtain the desired precision.

The exact significance of V and σ must be kept in mind when choosing values of V_S. When V_S is set equal to 0.01, there is only a 68 per cent chance that a single pair of measurements, R_T and R_B, will yield a value of R_S within 1 per cent of the true value. If V_S is taken as 0.005, the chance that a single pair will yield a value within 1 per cent rises to 95 per cent.

Another application arises when one desires to determine the presence or absence of radioactivity in a very weak sample. Again an absolute statement cannot be made. If the net sample activity is twice the standard deviation of the measurement, there is a 95 per cent probability that the net count is within 1 per cent of the true value, and the presence of activity can be considered established.

An examination of Eq. (2-21) shows that a given value of σ_S can be obtained with a variety of combinations of t_T and t_B. To obtain a unique solution, some other requirement must be introduced. Three cases are of practical interest.

1. To minimize the total time required for an assay $(t_T + t_B)$, we set the differential $d(t_T + t_B)$ equal to zero. This requires that

$$dt_T = -dt_B \tag{2-23}$$

Squaring Eq. (2-21) and taking differentials,

$$2\sigma_s\, d\sigma_s = -\frac{R_T}{t_T^2}\, dt_T - \frac{R_B}{t_B^2}\, dt_B \tag{2-24}$$

Since σ_s is held constant, $d\sigma_s = 0$ and, from Eqs. (2-23) and (2-24),

$$t_B = t_T \sqrt{\frac{R_B}{R_T}} \tag{2-25}$$

Combining this result with Eqs. (2-21) and (2-22), we obtain

$$t_T = \frac{R_T + \sqrt{R_T R_B}}{V_S^2 R_S^2} \tag{2-26}$$

With a chosen value of V_S and a value of R_S obtained by estimate or a rough measurement, Eq. (2-26) can be used to calculate t_T and t_B.

2. Many scaling circuits are arranged so that they will count unattended for predetermined counting times. It may then be convenient to use an assay schedule where $t_T = t_B$. This requirement leads to

$$t_T = \frac{R_T + R_B}{V_S^2 R_S^2} \tag{2-27}$$

Again, the schedule can be determined from a choice of V_S and an estimate of R_S.

3. Alternatively, the circuit is set to count equal predetermined numbers of total counts. That is, $N_T = N_B$. Now,

$$N_T = \frac{R_T^2 + R_B^2}{V_S^2 R_S^2} \tag{2-28}$$

and the schedule is calculated as before.

Consider now case 2, where the counting system is set to count for equal preset times. For any value of V_S, the time required for an assay will be proportional to $(R_T + R_B)/R_S^2$, and this factor may be taken as a sort of figure of merit, M, of the system. Thus we may write

$$M = \frac{R_S + 2R_B}{R_S^2} \tag{2-29}$$

The smaller the value of M, the more advantageous the system, from this limited viewpoint. The most useful application of Eq. (2-29) occurs with very weak samples, when $R_S \ll R_B$ and the required counting times become long. Under this condition, Eq. (2-29) becomes approximately

$$M = \frac{2R_B}{R_S^2} \tag{2-30}$$

and we see that the counting time will be determined by the relative response of the system to background and to the sample. Any increase in R_S relative to R_B will decrease the assay time and even proportional increases in both rates will help, because of the presence of the squared term in the denominator.

REFERENCES

Campion, P. J., "A Study of Proportional Counter Mechanisms." *Int. J. Applied Rad. and Isotopes* **19**, 219, 1968.

Lowenthal, G. C., "Secondary Standard Instruments for the Activity Measurement of Pure β Emitters. A Review." *Int. J. Applied Rad. and Isotopes* **20**, 559, 1969.

Montgomery, C. G. and D. D. Montgomery, "The Discharge Mechanism of Geiger–Mueller Counters. *Physical Review* **57**, 1030, 1940.

Porges, K. G., "Note on the S^2/B Criterion of Quality for Counting Equipment." *Int. J. Applied Rad. and Isotopes* **19**, 711, 1968.

Present, R. D., "On Self-quenching Halogen Counters." *Physical Review* **72**, 243, 1947.

Price, W. J., *Nuclear Radiation Detection.* McGraw-Hill, New York, 1964.

Stever, H. G., "The Discharge Mechanism of Fast G–M Counters from the Deadtime Experiment." *Physical Review* **61**, 38, 1942.

PROBLEMS

2-1. An ionization chamber is used with an electrometer capable of measuring 5×10^{-11} ampere to assay a source of 0.835-MeV beta particles. Assume saturation conditions, and that all of the particle energy is expended in the chamber gas. Calculate the rate at which the beta particles must enter the chamber to just produce a measureable response.

2-2. The beta particles in Prob. 2-1 produce on the average 80 ion pairs in a counter operating in the proportional region with a gas-amplification factor of 6×10^4. What rate of particle incidence will be required to produce an average current of 5×10^{-11} ampere?

2-3. The initial avalanche in a G–M tube contains 2×10^8 ion pairs in a volume of 0.8 cm^3 close to the central wire. Calculate the loss due to recombination during the 10^{-6} sec before the electrons are collected.

2-4. A G–M counter with a cathode 3.6 cm in diameter is filled to a pressure of 9 cm of mercury with argon, with an additional pressure of 1.4 cm of ethyl alcohol. How many collisions will an average ion make as it travels toward the cathode? How many of these collisions will be with alcohol molecules? Assume that the molecules of alcohol and argon have equal radii and that the ions behave like neutral molecules.

2-5. A G–M tube with a cathode 5.0 cm in diameter and a wire diameter of 0.012 cm is filled with argon to a pressure such that the mean-free-path is 7.8×10^{-4} cm. Calculate the value of the voltage that must be applied to just produce an avalanche.

2-6. A G–M tube with a cathode diameter of 4.0 cm and a wire diameter of 0.016 cm is filled with argon and alcohol to a pressure such that the mean-free-path is 4.6×10^{-3} cm. Calculate the maximum radius at which secondary ions will be formed when 1200 volts is applied to the tube.

2-7. Use the summation relation to compute the standard deviation of an observation in the data series given in the illustrative example of Sec. 2.13. How does this value compare with that calculated from the average count? What do you infer about the operation of the system?

2-8. A nonself-quenching G–M tube has a cathode 3.0 cm in diameter, a central wire 0.025 cm in diameter, and an effective length of 8.0 cm. What is the time constant of this circuit when a series resistance of 5×10^9 ohms is required to insure quenching? The dead time of this circuit extends until the voltage pulse due to a discharge has decreased to 5 per cent of its original amplitude. What is the maximum rate of response of the system to evenly spaced events?

2-9. Assume that a series of assays are to be made using "equal-time" counting to obtain a coefficient of variation of 0.02. Let R_S take on a series of values from $0.2R_B$ to $10R_B$, calculate the required counting times, and make an appropriate plot of these values as a function of the R_S/R_B ratio.

2-10. A G–M counting system with a background count of 20 cpm is to be used in the assay of a series of samples which are expected to have count rates of about $3.0R_B$. A commercial laboratory has contracted to carry out the counts to a coefficient of variation of 0.01 for $8 per hour of counting time. What are the relative costs of the three counting schedules? One background count will be made for every ten active samples.

2-11. The assay system of Prob. 2-10 was used in a study that generated 100 samples, and required a rare radioactive compound that cost $1200. Assume "equal-time" assays with a background count for every ten unknowns, and calculate the economic advantage or disadvantage of doubling the amount of radioactive material.

2-12. A G–M counter with a volume of 35 cm^3 is filled with argon to a pressure of 9 cm of mercury, and ethyl alcohol is added to a total pressure of 10 cm. What will be the useful life of this counter in a space probe operating where the average count rate is estimated to be 280 sec^{-1}? Assume that each pulse contains 2×10^9 ion pairs and that the useful life will end when 25 per cent of the quenching gas has been dissociated.

2-13. Develop an expression that can be used as a figure of merit for a counting system operating in the "equal-count" mode.

2-14. A G–M counter with a background count rate of 26 cpm is used to assay samples which average 90 cpm. What will be the coefficient of variation and the standard deviation of an "equal-count" assay of 10,000 counts each? What time will be required for one assay plus background?

2-15. The counter in Prob. 2-14 is replaced with a scintillator which increases both background and sample counts by a factor of 10, to 260 and 900 cpm, respectively. What will be the coefficient of variation, and the standard deviation of an "equal-count" assay of 10,000 counts each? What time will be required for each complete determination?

2-16. By introducing an energy selective component into the circuit (single-channel pulse-height analyzer), the sample count rate in Prob. 2-15 was reduced to 450 cpm while the background rate was cut to 36 cpm. Repeat the calculations of Prob. 2-15 with the new conditions.

2-17. A human whole-body counter (Sec. 20.06) has a background count of 8000 cpm. Assume an "equal-time" counting mode of 10 min each and calculate the minimum count rate that is needed to achieve a coefficient of variation of 0.03.

2-18. The radiation intensity of a gamma-ray source will vary as $1/d^2$ when the distance to the detector is greater than 20 times the size of the source or the detector. This fact can be used to determine the resolving time of a pulse counter, using a single source, whose activity need not be known. Let the expected count rates from a source be N_1 and kN_1, where k is known, and let the observed count rates be n_1 and n_2, respectively. Develop an expression for the resolving time.

2-19. The output circuit of a ratemeter consists of a resistance of 50,000 ohms shunted by a condenser. The output pulses from the ratemeter have a constant amplitude of 5.0 milliamperes, and a variable width. What pulse width is required to produce a voltage of 5.0 volts across the "tank" circuit at a count rate of 200 cpm? What pulse length will be required to obtain the same voltage at a count rate of 50,000 cpm? What capacitance will be required to obtain a circuit time constant of 2 sec?

2-20. Calculate the resolving time of a pulse counter from the following schedule of count rates: background 65 cpm, first source in position 8350 cpm, both sources in position 16,832 cpm, second source in position 9416 cpm.

3

Solid and Liquid Radiation Detectors

3.01 Conduction and Fluorescence

We have seen that it is possible to collect all of the ions formed, with little loss due to recombination, when radiation is absorbed in a gas. Because of their low density, gas-phase detectors are poor absorbers of penetrating radiations, and consequently have relatively low counting efficiencies. The higher densities of liquids increase the amount of radiation absorbed in a given volume, but the greatly increased molecular concentrations increase recombination to unacceptable levels. Liquid detectors based on the direct collection of ions are rarely, if ever, used.

Solid-phase detectors have the advantage of high density, and several types have been developed in which recombination is not a serious drawback. Diamonds and a few other crystals have been used as *photoconductors*, in which the measurement is made by the direct collection of the electrons released by the radiation. Solids suitable for detectors by photoconduction must be almost perfect insulators. Any conducting crystal will contain relatively large numbers of free electrons, whose movement under a field constitutes the electric current. Detection of any small increase in this number that might be produced by radiation absorption in a conducting solid is impossible. Practical solid-phase detectors must, therefore, operate on quite different principles.

Roentgen's original discovery of X rays in 1895 was made through the observation that radiation absorption causes crystals of zinc sulphide, ZnS, to emit visible light. Later, Rutherford made use of this phenomenon in a classic series of measurements on the scattering of alpha particles by atomic nuclei. The Rutherford experiments were exceedingly tedious. A carefully prepared ZnS screen was viewed with a microscope after the observer's eye had been thoroughly dark-adapted. Tiny flashes of light, or *scintillations*, could then be seen where the alpha particles struck the screen, and the rate of arrival of these particles could be determined. Statistical considerations require that large numbers of scintillations be counted, but the human recording system is capable of operating only at very low speeds. Observation times thus became very long. With the development of high-speed electric circuits and photoelectric cells, scintillation counting has become one of the most powerful methods of radiation detection.

There is an ill-defined time boundary between two classes of light emission following the absorption of radiation. When visible or ultraviolet light is emitted within 10^{-8} sec or less after the radiation absorption, the emission is called *fluorescence*. *Phosphorescence* refers to delayed light emission, which may follow the radiation absorption by minutes, days, or even years. Each of the two classes has important applications. For the moment we shall confine our attention to the prompt fluorescent response.

3.02 Photomultiplier Tubes

Scintillation counting has been removed from the limitations of the human recording system by the development of the photomultiplier tube, Fig. 3-1.

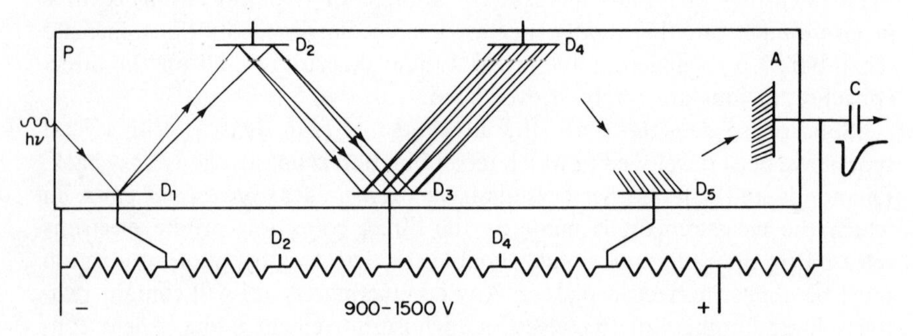

Figure 3-1. Schematic diagram of a photomultiplier tube and the voltage divider network.

Light incident upon a photocathode *P* will eject one or more electrons by the well-known photoelectric effect. These electrons will be attracted to the first of a series of *dynodes*, each of which is maintained at a succesively higher positive potential by a voltage source and a potential divider. Typically, each dynode surface consists of a carefully prepared composite layer of cesium and antimony, or other metal combination. These surfaces will emit several electrons when struck with a single electron of sufficient energy. The dynodes are so shaped and arranged that each secondary electron will move toward the next stage, accelerated by the higher positive potential. Secondary electron multiplication will take place at each of the 10–12 dynodes that are incorporated in the usual photomultiplier tube.

Each dynode surface will produce 4–8 secondaries instead of the two shown schematically in Fig. 3-1. Thus a 10-stage tube with a gain of 6 per stage will have an overall gain of 6^{10} or 6×10^7. The gain per stage is a sensitive function of the accelerating voltage. To insure constant gain, the supply voltage of 900–1500 volts, or about 100 volts per stage, must be carefully stabilized. Electron trajectories are also influenced by magnetic fields, and for some applications care must be taken to keep magnetic conditions constant.

The secondary electrons emitted by the last dynode will be attracted to an anode A, connected to the positive voltage supply through a series resistance. A flash of light on the photocathode will result in the appearance of a pulse of electrons at the anode, which will produce a negative output pulse because of the voltage drop across the load resistor. This negative pulse can be transmitted to the recording circuits through capacitance C.

At room temperature there will be an appreciable emission of thermionic electrons from the cathode to produce an undesirable *dark current*. If the initial light pulse is sufficiently large, the useful signal will be much greater than the thermionic "noise" and no interference will result. When very weak light pulses must be detected, it may be necessary to reduce the dark current by operating the phototube at a low temperature, even down to that of liquid nitrogen.

3.03 The Perfect Crystal Lattice

Consider a perfect crystal, specifically an alkali halide such as sodium iodide, in which the atoms are precisely arranged in a three-dimensional lattice. Only one plane of the lattice is shown in Fig. 3-2. Such a perfect crystal is electrically neutral, both on a macroscopic and a microscopic scale. At low temperatures there are no free electrons to carry an electric current and so the crystal is essentially a nonconductor.

When radiation is absorbed by the crystal, an electron may be removed from its original bound position, as at B, Fig. 3-2. A negative charge is now free to move through the crystal lattice and a net positive charge exists at A, where the electron was ejected. In the absence of an external electric field the electron may move away by diffusion, but recombination will be prompt, and electrical neutrality will be quickly restored.

When an external field is applied, the freed electron will move toward the anode. On the way, the original electron may exchange with a bound

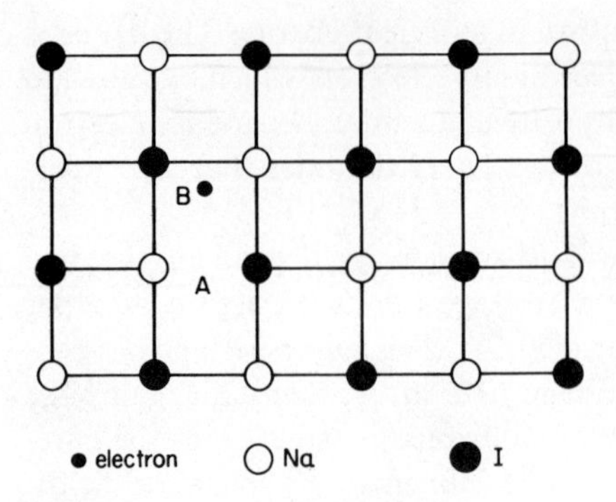

Figure 3-2. A perfect crystal lattice of NaCl with an electron released at *A* by radiation absorption and bond breakage.

electron which, in turn, may exchange with another bound electron, and so on. These exchanges are of no consequence, since the net effect is that of a single electron moving toward the positive electrode.

The position where the electron is missing is known as a *hole*, and this too can effectively move under the influence of the field. A neighboring electron may move in to replace the one originally released, but this only creates a new hole, which then acquires a neutralizing electron, and so on. The net effect is a movement of the hole toward the cathode. The combined effect of electron and hole movements is an electric current that may be detected in the external circuit. There will be no charge multiplication during the charge movements; the situation is analogous to that in a gas-filled ion chamber. As in an ion chamber, there is a small probability of an electron–hole recombination before collection.

Energy was put into the system to free the electron from its original bound state; this energy will be released at recombination. Some of the energy releases will be in the form of visible or ultraviolet light. Part of the emitted light will be absorbed in the crystal itself, to be gradually degraded and dissipated as heat. The rest will escape from the crystal and will then be available for actuating recording equipment. Any recombination energy not emitted as light will be given to the lattice structure in the form of increased vibrations, to be eventually dissipated as heat.

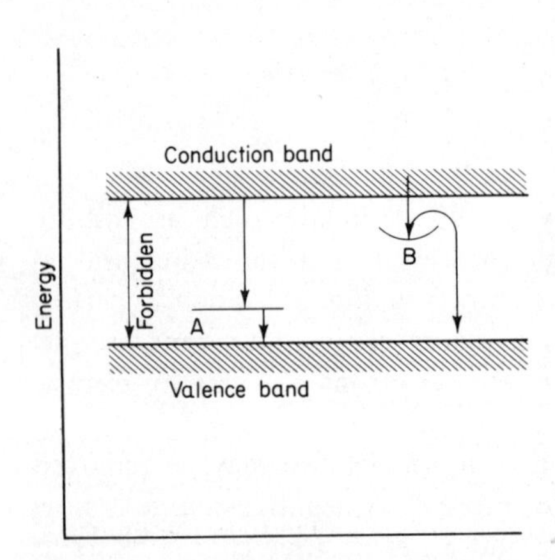

Figure 3-3. Energy relations in a crystal. Electrons in a perfect crystal cannot exist in the forbidden band. Impurities permit trapping centers in the forbidden band near either edge, *A* or *B*.

Perfect crystals exist only in principle. Actual crystals will contain lattice imperfections and impurities that act to produce some free conduction electrons in the absence of radiation. The energy relations in a crystal lattice may be pictured as in Fig. 3-3. Those electrons forming the valence bonds will have energies in the region known as the valence band. A definite amount of energy is required to break a valence bond and free an electron. When this is done, the electron is said to be in the conduction band. An electron absorbing less than the required energy will remain bound and will give up its excess energy to lattice vibrations. Thus there is a *forbidden energy band* which cannot be occupied by any electron in a perfect crystal. The width of the forbidden band is about 7 eV in the case of diamond.

3.04 Crystal Impurities

An atom of an impurity, incorporated into an otherwise perfect crystal, may have a valence differing from those of the main crystal constituents. The impurity may have an extra unpaired valence electron, or one may be lacking. Even with equal valences, an impurity atom will have energy levels distinct from those of the atoms making up the body of the crystal. For example, an atom of an impurity may have allowed energy levels within the band that are forbidden to the atoms in the regular lattice structure.

Even minute quantities of an impurity can profoundly affect the behavior of a crystal. An impurity may permit the presence of some electrons in the conduction band at all times. Such a crystal will have a relatively large electrical conductivity on which a small conductivity due to radiation is to be superposed. Thus in principle perfect diamonds should be useful as radiation detectors, but in practice few are sufficiently free of impurities. Light emission rather than electrical conduction has been the most widely exploited mechanism for adapting crystals to radiation detection.

A crystal *scintillator* is produced by deliberately adding a carefully controlled amount of a chosen impurity to the basic crystal material. A large number of impurity centers will then be created throughout the lattice when the crystal is grown, either from solution or from a melt. A free electron or a hole, moving simply by diffusion through the lattice, will quickly encounter one of the impurities and may become attached there, or *trapped*. The trapping atom will be in an excited state and will attempt to relieve this excitation very rapidly. Energy may be given directly to the lattice bonds in the form of energy packets known as *phonons*. Other atoms will favor removing the excitation through the emission of visible light. A desirable impurity is one which will favor light emission over phonon transfer.

Since the energy levels of the excited impurity will be quite different from those in the main body of the lattice, there is a good chance that the emitted light will escape from the crystal rather than be lost inside by resonance absorption. Energy loss inside a crystal, either by absorption or by energy degradation, is known as *quenching*. Note that this use of "quench" has nothing in common with its use in connection with the G–M counter, Sec. 2.07.

A crystal scintillator is an inefficient energy converter. A good sodium iodide crystal *activated* with thallium, commonly denoted by NaI(Tl), may produce one electron–hole pair for every 50 eV of radiation absorbed. The electron–hole energy will be of the order of 2eV, and so the energy-conversion efficiency of the crystal will be about 4 per cent. Only the enormous amplifying capabilities of the photomultiplier make practical the use of crystal scintillators.

3.05 Scintillation Detectors

A variety of crystal–activator combinations have come into general use as radiation detectors. NaI(Tl) is commonly used for the detection of X and gamma rays, for reasons that will appear in our discussion of photon absorption, Chapter 12. One of the oldest combinations, ZnS(Ag), is still favored for use with alpha particles. Organics such as anthracene can be used effectively for beta-particle measurements in spite of somewhat unfavorable quenching characteristics, because in this application the crystals can be made rather thin. Other organic molecules may be dissolved in a solvent to produce a *liquid scintillator*, which is something of a misnomer since the light emission will still come from the molecules of the dissolved solute. Scintillators may be combined with monomers which are then polymerized to produce plastic scintillation detectors in sizes and shapes not obtainable with naturally grown crystals. Table 3-1 lists pertinent properties of some commonly used scintillators.

TABLE 3-1
PROPERTIES OF SOLID SCINTILLATORS

Scintillator	*Effective atomic no. (Z)*	*Density ($g\ cm^{-3}$)*	*Emission max. (Å)*	*Decay time (nsec)*
Anthracene	5.8	1.24	4400	26
p-Terphenyl	5.7	1.17	4100	10
Stilbene	5.7	1.19	4100	8
NaI(Tl)	50	3.67	4100	0.25
ZnS(Ag)	27	4.1	4500	9

For maximum response, the wavelengths of the light emitted by the scintillator should match the spectral response of the photocathode of the multiplier tube. Most phototubes used with scintillators have the so-called S-11 spectral response shown in Fig. 3-4. A good many scintillators emit light at wavelengths somewhat shorter than is desirable for the most effective energy transfer to the photocathode. In some liquid scintillator formulations this mismatch is partially corrected by adding a secondary scintillator or

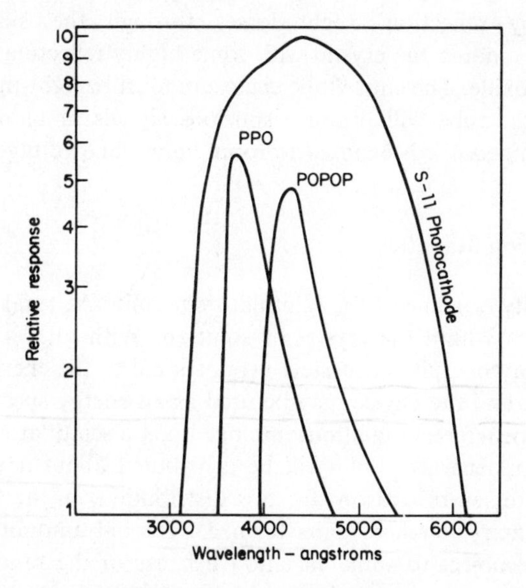

Figure 3-4. Spectral response of a typical photocathode, and emission spectra of some common scintillators.

"wave shifter." The secondary scintillator absorbs the primary scintillators and emits light at somewhat longer wavelengths, better suited to the response of the photomultiplier. A typical liquid formulation suitable for detecting the very-low-energy beta particles from carbon-14 or hydrogen-3 is:

Primary scintillator:	2, 5-diphenyloxazole (PPO)	5 g
Secondary scintillator:	1, 4-bis-(5-phenyloxazole) benzene (POPOP)	0.1 g
Solvent:	Toluene	1 liter

Close coupling must be maintained between the crystal and the phototube so that the latter will "see" the emitted light as effectively as possible. Sodium iodide is highly hygroscopic, and so this crystal must be hermetically sealed to prevent the entrance of moisture. The crystal is customarily "canned" in a thin-wall aluminum or stainless steel container with an optical glass window on one face, Fig. 3-5. This window is coupled to the face of the photomultiplier tube with an optical grease whose index of refraction is chosen to minimize

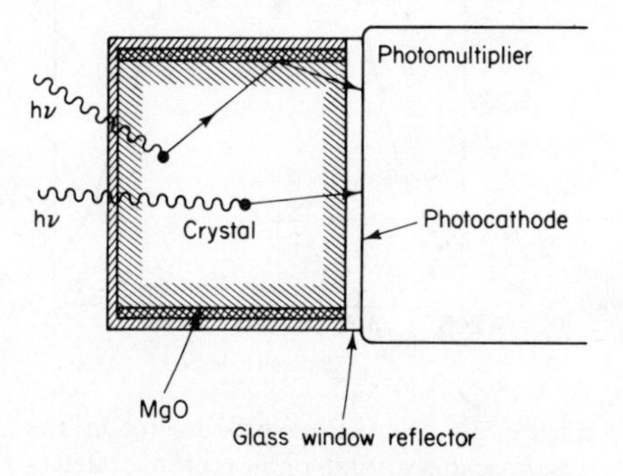

Figure 3-5. Coupling between a solid crystal scintillator and a photomultiplier tube.

light losses by reflection. Light losses through the sides are minimized by surrounding the crystal with some highly reflecting material such as magnesium oxide. The entire tube enclosure must be light-tight. Any weak light striking the tube will produce spurious signals; a phototube can be permanently ruined if it is exposed to room light while voltage is applied.

3.06 Scintillation Statistics

The high density of some of the scintillators permits the total absorption of many radiations within the crystal or solution. With total absorption, the light output may be quite accurately proportional to the energy of the incident radiation, and the device may be used as an energy spectrometer.

When monoenergetic radiations impinge upon a scintillator, the response will not be monoenergetic, but will be distributed about a most probable value. There are several reasons for this distribution. In the first place, the number of primary ion–hole pairs formed by equal amounts of absorbed energy will be subject to some variation because of the random nature of the interactions that lead to ionization. Additional variations will arise from variations in the energy division between emitted light and phonons. Light originating in various parts of the scintillator will not all reach the phototube with equal efficiency. Finally, there will be some variations in output current because of sensitivity variations at the photocathode and the dynode surfaces.

All of the processes are essentially independent, and so the overall response of a scintillator system should obey Poisson statistics. When the number of primary events is large, the discontinuous Poisson distribution may be replaced by the continuous normal distribution. Figure 3-6 shows the response of a sodium iodide-phototube system to the monoenergetic gamma rays (662 KeV) emitted by radioactive cesium-137.

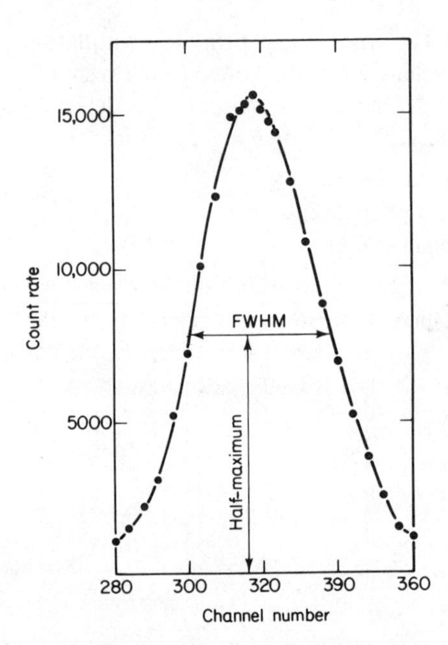

Figure 3-6. Statistical broadening in the response of a crystal–photomultiplier detector to monoenergetic photons.

The width of a statistical distribution of events is usually expressed by the standard deviation σ, given according to Eq. (2-13) by $\sqrt{N}$, where N is the number of events counted. In scintillation spectrometry it is more usual to characterize the distribution by the width of the distribution curve at one-half of maximum height. Two abbreviations are used, FWHM and HWHM, meaning, respectively, full-width at half-maximum and half-width at half-maximum. If the distribution is normal, these values can be related to σ if desired (see Prob. 3-1). HWHM, expressed as a percentage of the energy value at maximum response, is commonly used to specify the energy resolution of a detector system.

3.07 Phosphorescence

All of the previous discussion has been based on the assumption that any energy trapped at an impurity will be emitted almost instantaneously as light. For reliable pulse counting, only very small delays between trapping and emission can be tolerated. Certain crystals, or crystal–impurity combinations, will hold charge carriers, either electrons or holes, for long periods of time. Such devices are obviously not suited to recording individual pulses, but are extremely useful for measuring radiation intensities integrated over some extended time period.

In the case of prompt fluorescence, an energy level in the forbidden band will appear as at A, Fig. 3-3. If the transition down into the valence band is *allowed* by certain *selection rules*, the electron will not linger at the trap, but will drop within perhaps 10^{-10} sec to one of the lower energy states. A particular transition into the valence band may be *forbidden* by the selection rules, or the energy level at the trap may have a shallow minimum, as at B, Fig. 3-3. In either case, the trapped carrier cannot escape until a small amount of extra energy is added.

Certain glasses, typified by silver-phosphates or cobalt-borosilicates, have the types of trapping centers that delay the release of the charge carriers. In some of these glasses, the carriers will be trapped at *color centers*, and the color density as seen with visible light can be used as a measure of the radiation absorbed by the glass during the exposure period. Care must be taken in using these glasses as radiation dosimeters, because the light used to study the color may release the trapped carriers, thus causing either partial or total fading.

Other glasses will remain colorless after a radiation exposure, even though they may contain a large number of trapped carriers. In some of these glasses, the trapped carriers can be released by flooding the glass with ultraviolet light. The light released by the untrapping can be measured by a photocell equipped with filters to prevent its response to the stimulating ultraviolet. In most of the photoluminescent glasses, some of the trapped carriers will drop out of the traps spontaneously during the first few hours following a radiation exposure. The resultant loss of sensitivity has been one of the drawbacks to the widespread use of these glasses as radiation detectors.

3.08 Thermoluminescent Dosimeters

A good many substances, when exposed to radiation, will leave carriers in traps that can be relieved by heating. Practical developments of thermoluminescent dosimeters, or TLD's, have centered on CaF_2(Mn) and LiF. The former is the more sensitive; the latter has radiation-absorption characteristics more nearly resembling those of living tissue.

In use, a few milligrams of the TLD material, either in the form of powdered crystals or incorporated in a plastic, is exposed to the radiation. Carrier trapping will occur in traps that are relatively stable at room temperature. After exposure the TLD is put through a rapid heat cycle to a maximum temperature of 200–300°C. The light emitted as the trapped carriers are released is picked up by a photomultiplier tube and recorded as a *glow curve*, which displays the intensity of the emitted light as a function of either the heating time or the sample temperature. The maximum value of the glow curve can be used as a measure of the total amount of radiation received by the TLD, provided that care is taken to standardize the rate and limits of the heating cycle. Alternatively, the light output emitted during the entire heating cycle can be integrated and displayed as a single number representing the radiation dose.

In good TLD material the total light liberated during the heating cycle will be strictly proportional to the energy absorbed over an exposure range of $10^5 : 1$. If the radiation exposure is not close to the upper limit of usefulness, the TLD material can be put through an annealing cycle by heating up to 400°C and reused with no loss of sensitivity. Normal LiF will contain two isotopes of lithium, ^{6}Li and ^{7}Li. Each species exhibits thermoluminescence, but the ^{6}Li has an enhanced sensitivity to neutrons. By using TLD's made up of the separated isotopes, it is possible to make measurements in mixed radiation fields.

3.09 Detection by Charge Collection

We have already pointed out the relative inefficiency of the scintillation detector. A charged, ionizing particle from the incident beam of radiation moves through a crystal to leave a trail of free electrons and holes. A few of these return to the configuration of an undisturbed lattice through processes which lead to the emission of visible light, some of which can be picked up by a photomultiplier. A considerable increase in efficiency could be achieved if each electron–hole pair could be utilized directly to produce an electrical signal. The requirement of an extremely low intrinsic conductivity in the crystal has already been mentioned.

Leaving aside for the moment the problem of intrinsic conductivity, let us consider the characteristics that are desirable in a solid-state detector operating by direct charge collection. Charge carriers produced in the solid by radiation absorption will move under the influence of an externally applied field toward the appropriate electrodes. In the general case, the electrodes need not be parallel planes as shown in Fig. 3-7.

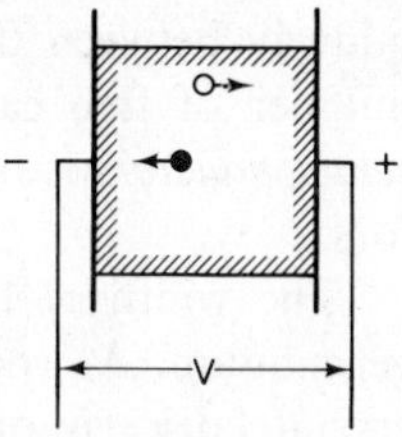

The force acting on a carrier will be Ee, where E is the electric field intensity. In a short interval of time, dt, the work done on the moving charge will be

$$dW = Ee\,dx = Eev\,dt \tag{3-1}$$

o electron
● hole

Figure 3-7. Movements of charge carriers in a crystal with an externally applied field.

where v is the carrier velocity. The work done on the charge must be obtained from the external field. The two electrodes form a condenser of capacitance C whose energy is $W = Q^2/2C$, whence

$$dW = \frac{Q}{C}\,dQ = Vi\,dt \tag{3-2}$$

where i is the current flow at the electrodes. Combining Eqs. (3-1) and (3-2),

$$i = \frac{Eev}{V} = \frac{Ee}{V}\frac{dx}{dt} = \frac{e}{V}\frac{dV}{dt} \tag{3-3}$$

The signal generated will be proportional to the effective charge collected, $q = idt$, so finally,

$$q = e\frac{dV}{V} \tag{3-4}$$

The signal will, from Eq. (3-4), be proportional to the fraction dV/V of the total potential traversed by the carrier before it is lost at an electrode, at a metastable trap, or by recombination. Preferably each carrier should journey all the way to its attracting electrode. In the general case the fraction dV/V depends upon the carrier mobility (in cm^2 $volt^{-1}$ sec^{-1}) and the average lifetime of the carrier before it is immobilized. The product of these two factors must be large in any acceptable detector. Equation (3-4) also suggests that a charge-collection detector should be relatively small. The smaller the detector the greater the chance of collecting a carrier before it is lost.

3.10 Semiconductors

The requirements of the mobility–lifetime product point to the elements silicon and germanium as outstanding candidates for charge-collection detectors. Unfortunately, each of these elements has an intrinsic conductance midway between the values for good insulators and metal conductors. The number of free carriers available at room temperature in either of these *semiconductors* is far too high to permit their use directly as radiation detectors.

The width of the conduction band is 1.2 eV in silicon and 0.78 eV in germanium. At room temperature the most probable phonon energy in a crystal lattice is only 0.025 eV, but the actual energies will be distributed above and below this value according to the Maxwell–Boltzmann distribution. The number of carriers in a small energy increment dW at energy W will be

$$dN = N_0 e^{-W/kT}\, dW \tag{3-5}$$

where k = Boltzmann's constant = 8.62×10^{-5} eV°K^{-1}
T = absolute temperature

The number of carriers with energies greater than 1.2 eV is far too great for the practical use of silicon, and the situation for germanium is much worse.

The energy distribution is a strong function of the exponent W/kT in Eq. (3-5). A change of W from 1.2 to 1.3 eV will decrease the number of conducting carriers by a factor of 50. Cooling a crystal to liquid-nitrogen temperature will sharply reduce kT to the point where intrinsic conductivity is no longer troublesome. In the case of silicon, modifications can be introduced which will permit effective operation at room temperature.

3.11 Doped Crystals

In a perfect crystal of either Si or Ge, adjoining atoms will be held together by *covalent bonds* formed by the sharing of the four valence electrons of each atom. Figure 3-8A depicts the situation for Si with its 1.2-eV forbidden band, at room temperature. Many of the covalent bonds are incomplete where electrons have escaped to the conduction band upon the acquisition of more that 1.2 eV from the energy of the lattice vibrations.

Figure 3-8B shows the situation when a pentavalent impurity such as phosphorous is introduced into a silicon crystal. After all of the covalent bonds are satisfied, there will be an unpaired electron very loosely bound to the phosphorous atom. In silicon the energy of this bond will be only 0.050 eV below the energy of the conduction band. This level is only twice the thermal energy at room temperature, and so almost every electron that was originally attached to a phosphorous atom will be free to take part in electrical conduction. The impurity or *donor* atom has contributed negatively charged carriers and the crystal is said to be *n-type* or *n*-doped.

Similarly, a trivalent impurity such as boron, Fig. 3-8C, will be unable

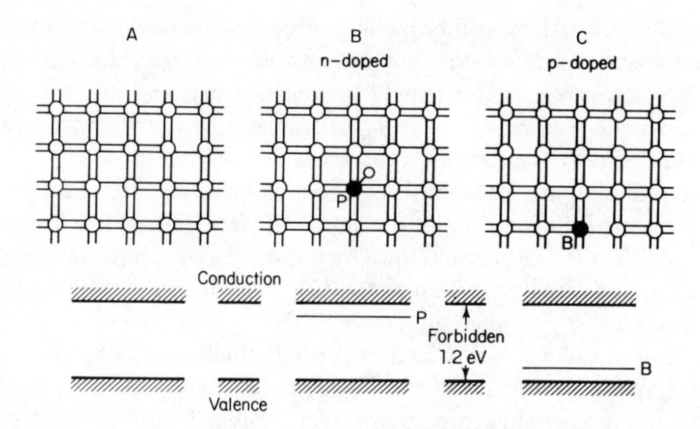

Figure 3-8. Energy levels in a perfect silicon crystal, (A), with *n*-doping, (B), and *p*-doping, (C).

to form a full set of covalent bonds with its neighbors, and so a hole will be created. The hole can accept an electron from a neighboring position and so this type of impurity is an *acceptor*. Electrical conductivity will again be enhanced if the energy level of the hole lies in the forbidden band of silicon, but this time by the movement of positive charges. The silicon is now *p-type* or *p*-doped.

Neither a *p*- nor an *n*-type crystal makes a suitable detector, because of the increased conduction resulting from the impurities. Consider now a silicon wafer that has been diffused partially with donors and partially with acceptors, Fig. 3-9A. There is now a *p–n junction* on one side of which is a preponderance of holes, and on the other side, of electrons. A polarizing field, Fig. 3-9B, will pull most of the holes toward the negative electrode while most of the electrons move toward the anode. When the wafer has

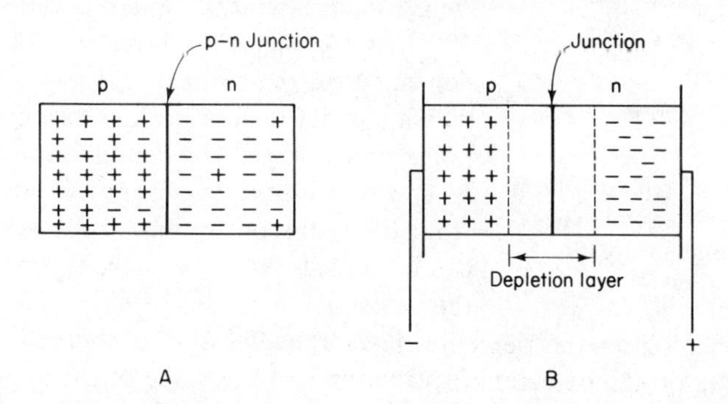

Figure 3-9. (A) In an unpolarized *p–n* crystal there are many free charges of each sign. (B) The application of a polarizing field creates a depletion layer, devoid of charge carriers.

been thus polarized, there will be a charge-free or *depletion layer*, from which all carriers have been removed. Conduction will cease after the initial polarization since the *n*-region can supply no positive carriers, nor the *p*-region electrons. The depletion layer acts as an insulator, and the crystal now has the low conductivity required in a radiation detector.

When radiation is absorbed in the depletion layer, free electron–hole pairs will be formed. These will move under the influence of the field to produce a signal in the external circuit. Once these carriers have migrated into the proper regions, charge movement will cease until another pulse of carrier pairs is produced by another primary event.

Depletion layers can be formed in germanium by the process described, but in germanium the forbidden band is so narrow that there is still unacceptable conduction at room temperature. Germanium detectors must be kept at the temperature of liquid nitrogen at all times in order to maintain the low conductivity and to prevent the diffusion of the two dopants throughout the entire body of the crystal. As with silicon, the highest degree of purity must be maintained throughout the manufacturing process. The starting materials must be extraordinarily free of impurities and the penetration of the dopants very carefully controlled.

3.12 Surface-barrier Detectors

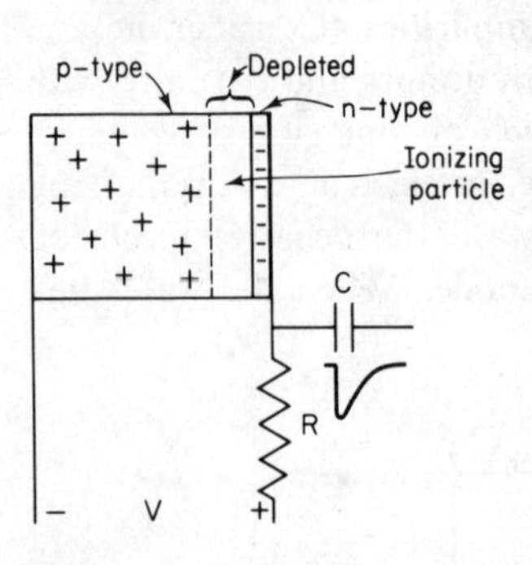

Figure 3-10. A typical silicon depletion-layer detector. A negative pulse appears across R.

Figure 3-10 shows the main features of a practical silicon depletion-layer detector. The main body of the detector is *p*-type silicon with an electrical connection to the negative pole of the voltage source made on the back facc. The *n* type region is made extremely thin, usually about 0.1 micron, which is only 10^{-5} cm. A very thin layer of gold is then deposited on the surface of the *n*-layer to provide a contact for the positive terminal of the polarizing voltage. The combined thickness of *n*-layer and electrode absorbs so little energy from the radiation beam that the *p*–*n* junction can be considered to be at the front surface of the detector.

The thickness of the depletion layer ranges from 10 microns to 5 mm, depending upon the detector construction and the value of the applied voltage. A typical surface barrier detector for alpha-particle measurements may have a depletion layer of 100 microns with an applied potential of 100 volts. Sensitive areas of 2–3 cm² are available and larger ones will surely be developed.

Several outstanding features have made surface-barrier detectors important tools in nuclear physics.

1. *High conversion efficiency.* A charged particle will produce, in silicon, one carrier pair for each 3.5 eV absorbed. This is an order of magnitude greater than the production of ion pairs in a gas. Energy conversion in a scintillation crystal is even more unfavorable, but they can be made in large volumes and this keeps them in competition with the much smaller surface-barrier detectors.
2. *High energy resolution.* The size of the output pulse from a surface-barrier detector is strictly proportional to the amount of energy absorbed in the depletion layer. If this layer is thick enough, the particle will come to rest in it, and the pulse size will then be a measure of the initial energy of the particle.

   The high conversion efficiency also contributes to the high energy resolution. A high efficiency means that there will be a large number of carrier pairs for a given energy deposition, and this in turn means a narrow distribution of pulse sizes around the mean, in accordance with standard statistical considerations. Figure 3-11 shows the energy resolution of a surface-barrier detector recording the alpha particles from plutonium-238 (5.499 MeV) and from polonium-210 (5.305 MeV).

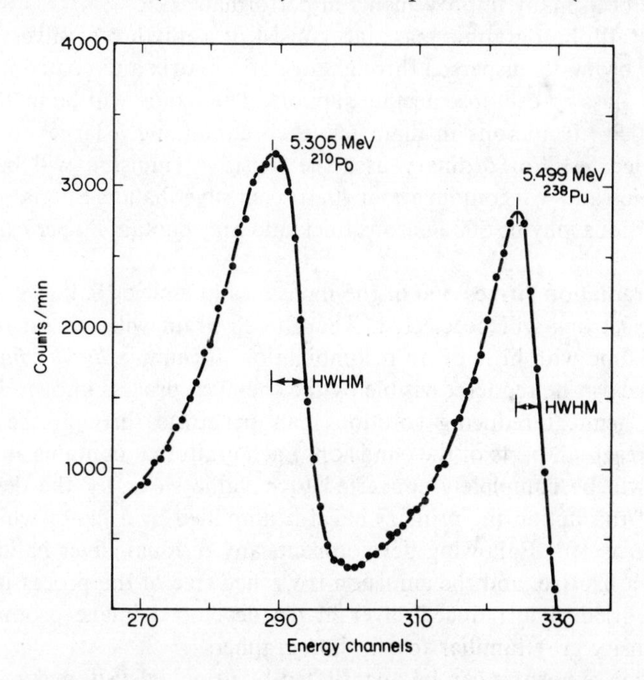

Figure 3-11. Alpha particle spectrum of a composite source obtained with a silicon barrier layer detector and a pulse height analyzer. The HWHM is measured on the high sides of the peaks to minimize the effects of the finite source thickness.

3. *High speed of response.* Carrier mobilities are high, about 1500 cm^2 $volt^{-1}$ sec^{-1} for electrons and 500 cm^2 $volt^{-1}$ sec^{-1} for holes. Carriers have to move only short distances before collection, and pulse widths are in the order of nanoseconds.
4. *Small size.* For some applications, the small sizes are a drawback. In other cases the small size, together with a clearly defined sensitive volume, are important assets.
5. *Differential sensitivity.* Surface-barrier detectors are relatively insensitive to neutrons or photons. This feature permits them to detect charged particles in the presence of other radiations with little or no interference. At the same time the response to charged particles depends only upon the energy and not upon the type of particle involved.

3.13 Photographic Emulsions

Photographic emulsions are one of the oldest and, today, one of the most common detectors of radiation. During the time emulsions have been used for this purpose there has been little change in the basic process, although there have been many improvements in performance.

Almost all photographic materials consist of a mixture of silver halides, mostly the bromide, dispersed through a gelatin matrix and coated as a thin layer on a glass or cellulose acetate support. The halide will be in the form of grains 0.3–1.0 microns in diameter, each containing a large number of halide molecules. For ordinary uses the sensitive emulsion will be 10–20 microns thick and will contain about 40 per cent silver halide. Special nuclear-track emulsions may be 500 microns thick and may contain 80 per cent silver halide.

When radiation strikes one of the molecules in a silver halide grain, one or more electrons will be ejected. The altered grain will remain for long periods of time with little or no recombination, forming a *latent image.* The latent image can be rendered visible by the chemical process known as development. Chemical reducing solutions can penetrate through the gelatin matrix to reach all parts of the emulsion. Each grain that contains an altered molecule will be completely converted to metallic silver by the developer. Because of this action the primary effect is amplified by a factor which may be as large as 10^9. Following development, any residual silver halide is removed by a fixative, and the emulsion is washed free of the processing solutions and dried. The reduced silver in the developed image produces the neutral-density gray familiar to all photographers.

Radiation exposures can be quantitated by using a densitometer to measure the amount of a visible light beam absorbed by the silver image. A focused light beam, Fig. 3-12A, with an intensity I_0 is passed through a small spot on the processed emulsion. A photocell measures the fraction of the light transmitted. Optical density of any absorber is defined by

$$\text{O.D.} = \log_{10}\frac{I_0}{I} \qquad (3\text{-}6)$$

Nearly all photographic materials have an optical density–exposure dependence typified by the curve in Fig. 3-12B. There will be a threshold of sensitivity below which no detectable darkening of the emulsion can be measured. Just beyond the threshold a curved portion (1) of the characteristic leads into a linear portion (2). At still higher exposures all available silver has been reduced and no further darkening will be observed (3).

Uses of photographic emulsions as radiation detectors fall into four broad categories:

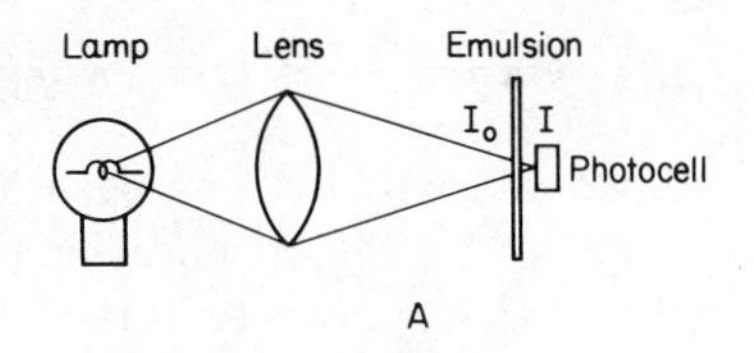

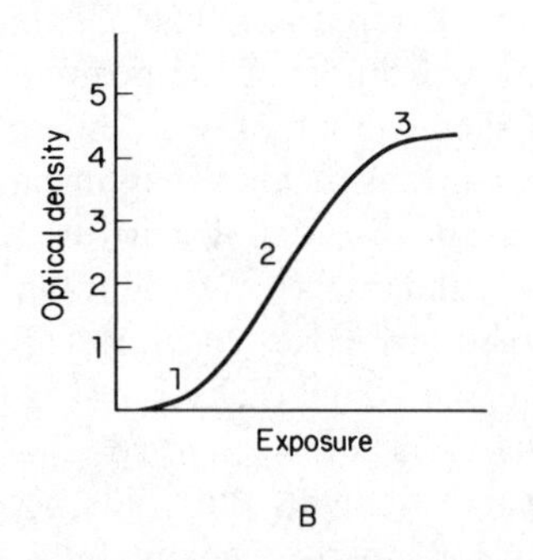

Figure 3-12. (A) Schematic of photoelectric densitometer for measuring emulsion densities. (B) Characteristic response curve of a photographic emulsion.

1. Modified versions of ordinary emulsions are used by physicians and dentists in radiography. Differential absorption of an X-ray beam as it passes through parts of the body produces corresponding variations in density on an exposed film. Industrial radiography uses the same process to search for flaws in manufactured items.
2. In film-badge dosimetry, small film packets provide measures of the integrated radiation exposures of personnel or areas. An important feature of the film is that it provides the only permanent record of the exposure.
3. A thin slice of tissue or other material containing radioactive material can be brought into close contact with an emulsion to produce an autoradiograph. With a low-energy beta-particle emitter such as hydrogen-3, the source of the radioactivity can be located to the limit of the grain size of the emulsion.
4. Special formulations, known as nuclear-track emulsions, are used to record the tracks of individual ionizing particles. The length of a track is a measure of the energy lost in the emulsion. Particle charge and mass can be inferred from a study of the spacings between the developed grains. Nuclear-track emulsions have been one of the powerful tools in nuclear physics and in the study of radiation in space.

3.14 Particle-track Etch Patterns

When a heavily ionizing particle moves through a dielectric, it leaves behind a trail of broken chemical bonds and a generally disordered structure. The main action appears to be the formation of dense ionization with a subsequent forcing of the freed positive ions into the surrounding material by mutual electric repulsion. The phenomenon is not seen in conductors, where prompt neutralization of the positive ions is to be expected. The damaged area has a diameter of only a few angstroms, so it is visible only under favorable conditions at high magnification in an electron microscope.

The disordered structure along the track is more chemically reactive than the undamaged surrounding material. If the energy deposition along the track has exceeded a critical value and if the track extends outward to the surface of the dielectric, it is possible to enlarge the track by chemical etching. When the enlarged track diameter becomes comparable to the wavelength of light, the voids become scattering centers that can be seen in an optical microscope. Etching solutions attack undamaged material as well, but less vigorously. Optimum etching solutions and etching conditions are determined by trial and error.

Track formation in solid dielectrics appears to be a quite general phenomenon, although a few materials have particularly favorable properties. Typical materials are:

Material	*Etchant*	*Critical energy* (*MeV* mg^{-1} cm^2)
Phosphate glass	48% HF	15
Mica	48% HF	15
Mylar	6N NaOH	5
Cellulose nitrate	6N NaOH	2

Practically any dielectric will respond to the massive ionization densities produced by fission products. Cellulose nitrate will produce etchable tracks from alpha particles and low-energy deuterons. These track detectors are being used to study the highly ionized components of cosmic radiation, as well as the heavily ionizing particle beams produced in accelerators. Even under extreme environmental conditions the latent tracks do not heal, and are available for etching centuries after they were produced. Ancient samples are now being etched and studied to learn about radiation conditions long ago. Quantitative track counts can be used to date minerals bearing fissionable materials. Visual track scanning and counting is a tedious process but some mechanical and electrical counting aids promise to make this task much easier in the future.

REFERENCES

Fleischer, R. L., P. B. Price, and R. M. Walker, "Solid State Track Detectors: Applications to Nuclear Science and Geophysics." *Ann. Rev. Nuc. Sci.* **15,** 1, 1965.

Heath, R. L., *Scintillation Spectrometry and Gamma-Ray Spectrum Catalog.* 2nd ed., Vols. I and II. Clearinghouse for Federal Scientific and Technical Information, Springfield, Va., 1964.

Lovett, D. B., "Track Etch Detectors for Alpha Exposure Estimation." *Health Physics* **16,** 623, 1969.

Miller, G. L., W. M. Gibson, and P. F. Donovan, "Semiconductor Particle Detectors." *Ann. Rev. Nuc. Sci.* **12,** 189, 1962.

Parmentier, J. H. and F. E. L. TenHaaf, "Developments in Scintillation Counting Since 1963." *Int. J. Applied Rad. and Isotopes* **20,** 305, 1969.

Tavendale, A. J., "Semiconductor Nuclear Radiation Detectors." *Ann. Rev. Nuc. Sci.* **17,** 73, 1967.

PROBLEMS

3-1. The shape of the normal probability distribution, given by Eq. (2-11), is controlled by the exponential factor, the coefficient serving merely to normalize the integral to unit probability. Show that for a normal distribution the half-width at half-maximum is HWHM $= 1.177\sigma$.

3-2. The peak response to the 662-KeV gamma ray from ^{137}Cs, shown in Fig. 3-6, occurs in energy channel 318, with half-maximum points in channels 301 and 336. Assume the pulse-height analyzer response to be linear with energy and calculate the standard deviation and the coefficient of variation of the energy determination.

3-3. Calculate the standard deviation and the coefficient of variation for each of the alpha-particle peaks shown in Fig. 3.11. Note that the shape of each of these peaks depends upon the sample thickness as well as the response of the detector.

3-4. A depletion-layer detector has an electrical capacitance determined by the thickness of the insulating dielectric, which is to say the depletion layer. Calculate the capacitance of a silicon detector with the following characteristics: area 1.5 cm^2, dielectric constant 12, depletion layer 50 microns. What potential will be developed across this capacitance by the absorption of a 4.5-MeV alpha particle which produces one ion pair for each 3.5 eV expended?

3-5. Surface-barrier detectors require special preamplifiers designed with strong negative feedback to cancel a large fraction of the detector capacitance. Calculate the time constant of an uncompensated detector circuit which has a capacitance of 87 pF, when the signal is generated across a resistance of 100 megohms.

3-6. A silicon surface-barrier detector has a depletion layer of 50 microns when polarized with 75 volts. What time will be required to collect all of the electrons, and all of the holes, from an ionizing event?

3-7. A silicon detector has an area of 3.0 cm^2 and a depletion layer of 60 microns when polarized with 50 volts. Calculate the time constant of the circuit when used with a 100-megohm resistance, the time required to collect all of the holes, and the amplitude of the voltage pulse produced by the absorption of a 3.78-MeV alpha particle.

3-8. The depletion depth of a surface-barrier detector varies as the square root of the applied voltage. Repeat the calculations of Prob. 3-7 for a polarizing voltage of 100 volts.

3-9. Assume normal distributions of ionizing events and calculate the expected energy resolution for a 5.5-MeV alpha particle detected by a gas-proportional counter, a crystal scintillator, and a surface-barrier detector. Repeat the calculations for a 100-KeV beta particle.

4

Classical Mechanics, Relativity, and Quantum Theory

4.01 The Nuclear Radiations

Soon after Henri Becquerel discovered radioactivity in 1896, many scientists attacked the problem of determining the nature of the emitted radiations. Physicists recognized that energy could be transmitted in one of two distinct ways: through the motion of particles, each having a certain kinetic energy, or through a wave motion where the transmitted energy depends upon the amplitude of the wave. The early experiments on the nature of radioactive emissions appeared to yield unequivocal results in line with this division into particles and waves.

Alpha particles (or alpha rays, as they were first called) were identified as helium atoms carrying a positive electric charge. Beta particles (or rays) were shown to be negatively charged particles moving with high velocity. The identification of the gamma rays presented some technical difficulties, but they were eventually shown to be electromagnetic waves of very short wavelength, similar to the X rays discovered by Roentgen in 1895.

As research proceeded, other nuclear radiations—the proton, positron, neutron, and several kinds of mesons—were discovered and their properties determined. All of these appeared to be particles rather than waves.

It soon became evident that the classical division between waves and particles was not so distinct as it had at first seemed. Indeed, in some experiments the electromagnetic waves appeared to behave like particles; in others, particles unquestionably showed wavelike properties. It now appears that both wave and particle properties must be invoked to describe the behavior of entities whose exact nature we do not understand. Before discussing the details of this duality, it will be well to review briefly some of the concepts of classical mechanics.

4.02 Some Classical Mechanics

The classical concept of a particle is a relatively concentrated structure which has inertia, or a resistance to any change in its velocity. The measure of this inertia is the *mass* m of the particle, the unit in the MKSA system being the kilogram. Ingenious experiments to be described later have permitted mass measurements of particles as light as the electron, 9.11×10^{-31} Kg.

Classical mechanics associates with a moving mass a kinetic energy $\frac{1}{2}mv^2$ and a momentum mv, where v is the velocity of the particle. The velocity may have any value from zero to c, the velocity of light, and it may change rapidly as the particle gains or loses energy.

Size and shape are important characteristics of macroscopic particles, but these characteristics cannot be uniquely determined for the small particles making up atoms and atomic nuclei. For many purposes, the electron, for example, can be considered as a dimensionless, mathematical point. Other considerations lead to a spherical electron of radius about 2×10^{-13} cm. In this domain, size appears to have no unequivocal definition but depends upon the type of measurement made to determine it. Most probable values can be given, but it must be realized that these do not have the precise meaning to be ascribed to values of charge or mass.

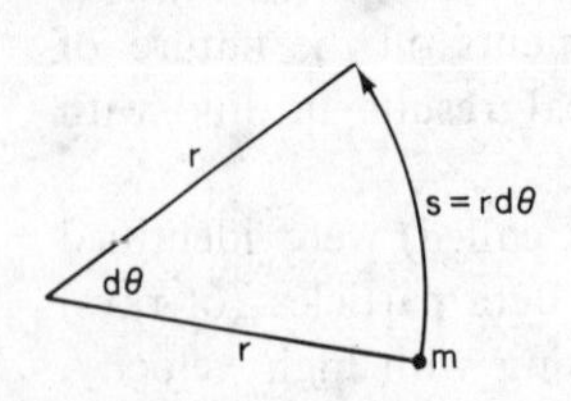

Figure 4-1. Linear and angular relations for a point-particle revolving about a fixed center.

We shall find that rotations play an important role in atomic and nuclear affairs. Consider a point particle of mass m rotating at a radius r about a fixed center, Fig. 4-1. Let the rotation be at a constant angular velocity $\omega = d\theta/dt$ radians per second. The kinetic energy T of the rotating mass will be

$$T = \tfrac{1}{2}mv^2 = \tfrac{1}{2}(mr^2)\omega^2 \tag{4-1}$$

where ω is the rotational analog of linear velocity v, and $(mr^2) = I$ is the *moment of inertia*, the analog of mass. Similarly, the angular momentum p_θ is

$$p_\theta = (mr^2)\omega \tag{4-2}$$

If the rotating body is so large that it can no longer be considered a point at constant radius, the moment of inertia will be given by

$$I = \int (mr^2)\, dr \tag{4-3}$$

The development of classical mechanics has relied heavily on conservation laws, which are not susceptible to absolute proof, but are considered, from long experience, to be inviolate. For example, the principle of conservation of energy states that in a closed system energy cannot be created or destroyed, but can be altered in form, as from potential energy to kinetic. The development of the relativity theory by Einstein required that mass be recognized as a form of energy. With this extension, energy conservation still appears to hold. Again, the requirements of momentum conservation, both linear and angular, have been important factors in the study of mechanical systems.

In macroscopic systems, the gram or the kilogram are convenient units of mass. Either of these units is inconveniently large for present purposes. A smaller unit, the *atomic mass unit* (amu), was defined in terms of the mass of that form of oxygen whose nucleus contains 16 nucleons. Chemical considerations led to a preference for a unit based on carbon, and now the physical and chemical scales have been brought into conformity with the definition of the *universal mass unit*, the u. One u is exactly equal to $\frac{1}{12}$ of the mass of one atom of that form of carbon whose nucleus contains 12 nucleons. This is equivalent to saying that 1 gram-atom of carbon of that form shall have a mass of exactly 12.000 grams. Avogadro's number N_a now becomes 6.02252×10^{23} mole^{-1}, and the unified (universal) mass unit is

$$1 \text{ u} = \frac{1}{N_a} = 1.66043 \times 10^{-27} \text{ Kg} \tag{4-4}$$

4.03 Relativity Theory

Classical physics had long been using a principle known as Galilean relativity, which states that it is impossible, by means of any experiment in mechanics, to determine the absolute motion of the laboratory in which the experiments are conducted. In other words, an experiment on, say, freely falling bodies conducted in a fixed laboratory will yield results identical with those conducted in a train moving with a constant velocity. Einstein extended this principle and stated that any experiment, and, in particular, measurements of the velocity of light, will yield identical results in laboratories that are moving with respect to each other at constant velocity.

Some of the predictions of the *special theory of relativity* were so startling that they were not immediately accepted by all physicists. Later, Einstein extended his ideas in the *general theory*, which deals primarily with gravitational forces and their interactions. Some of the predictions of each of the theories are susceptible to experimental test. In every case measurements have not negated any of the predictions.

One of the results of the theory is that it is impossible to accelerate a body up to or beyond the velocity of light, *c*. Note carefully the exact wording of this statement. As we shall see presently, some objects, notably photons and neutrinos, travel with exactly the velocity of light, but they were originated with this velocity rather than being accelerated up to it. Particles moving faster than the velocity of light (tachyons) have been postulated, but have not been observed. If they exist, they will have been given their high velocity at the moment that they were created, and they will be unable to slow down to the velocity of light or below it.

The velocity limitation requires a revision of the classical formulas for the addition of velocities. Assume a fixed observer and a system moving toward him with velocity v. Further, assume an object having a velocity u' relative to the moving system in its direction of motion. Classical mechanics predicts that the velocity relative to the fixed observer u will be $u' + v$. This postulate would violate the results of the theory if u' happened to be a light beam of velocity c. According to the relativity theory the observed velocity u is given by

$$u = \frac{u' + v}{1 + u'v/c^2}$$

whence, if $u' = c$,

$$u = \frac{c + v}{1 + cv/c^2} = c$$

A most important conclusion drawn from the relativity theory is that the mass of a body is not constant but will vary with its velocity. Rest mass m_0 is related to the mass m exhibited by the same body when moving with velocity v by

$$m = \frac{m_0}{\sqrt{1 - v^2/c^2}} \tag{4-5}$$

For small values of v, m is practically equal to m_0, but as v approaches c, m increases without limit. Appendix Table 10 gives values of m/m_0, v, and $\beta = v/c$ as functions of electron energy.

A third important prediction of the Einstein theory states that mass and energy are interchangeable, being two different manifestations of a single entity. They are related by

$$E = mc^2 \tag{4-6}$$

where E is the total energy of the body in ergs and m is the mass in grams. In our preferred units of u and MeV, the conversion becomes

$$1 \text{ MeV} = 1.074 \times 10^{-3} \text{ u} \qquad 1 \text{ u} = 931.48 \text{ MeV} \tag{4-7}$$

Relativistic relations account accurately for the behavior of charged particles accelerated to high energies by electric fields. The energy derived from the field, Ve, will be equal to the kinetic energy of the particle which, by classical mechanics, will be $m_0v^2/2$. This relation permits the velocity to increase without bound as V increases, a result not permitted by relativity.

Combining Eqs. (4-5) and (4-6),

$$E = \frac{m_0c^2}{\sqrt{1 - v^2/c^2}} \tag{4-7}$$

If now we set $v = 0$, Eq. (4-7) becomes

$$E_0 = m_0 c^2 \tag{4-8}$$

where E_0 must represent the energy inherent in mass itself. The term E_0 has no counterpart in Newtonian mechanics. Many experiments have demonstrated the validity of Eq. (4-8). A nuclear detonation, in which mass is converted to energy on a stupendous scale, furnishes the most spectacular verification of the mass–energy relation.

From Eq. (4-7), the relativistic expression for kinetic energy becomes

$$T = E - E_0 = m_0 c^2 \left(\frac{1}{\sqrt{1 - v^2/c^2}} - 1 \right) \tag{4-9}$$

The first term inside the parentheses can be expanded in a power series in v^2/c^2. This leads to

$$T = m_0 c^2 \left(1 + \frac{1}{2}\frac{v^2}{c^2} + \frac{3}{8}\frac{v^4}{c^4} + \cdots - 1 \right) \tag{4-10}$$

When v/c is small, the leading term in Eq. (4-10) becomes just the classical expression for kinetic energy. As v/c approaches unity, more terms in the expansion must be used.

Squaring Eq. (4-7) and rearranging leads to

$$E^2(c^2 - v^2) = m_0^2 c^6 \tag{4-11}$$

For $v = c$, Eq. (4-11) can be satisfied only by requiring that $m_0 = 0$. Thus an object traveling with the velocity of light must have a zero rest mass.

At velocities where relativistic mechanics must be used, the classical relation between kinetic energy and momentum, $T = p^2/2m_0$, must be replaced by

$$T = m_0 c^2 \sqrt{1 + \left(\frac{p}{m_0 c} \right)^2} \tag{4-12}$$

This relation can be obtained from Eq. (4-9) and the definition of momentum, which remains as $p = mv$ for relativistic velocities. The inverse relation is frequently useful:

$$p = \frac{1}{c} \sqrt{T(T + 2m_0 c^2)} \tag{4-13}$$

4.04 Charged Particles in a Magnetic Field

We have seen that a charged particle in an electric field moves in the direction of the field and acquires energy from it. When a charged particle enters a magnetic field, the situation is somewhat more complicated. A particle of charge e moving with velocity v perpendicular to a magnetic field B will be acted on by a force Bev. This force will act in a direction perpendicular to both B and v, which is the condition necessary to produce circular motion.

Elementary mechanics shows that for circular motion in a path of radius ρ there must be a force mv^2/ρ in a direction always perpendicular to v. Then,

$$Bev = \frac{mv^2}{\rho} \tag{4-14}$$

and the particle will move along a circular arc of radius ρ or in a complete circle if the field is sufficiently strong.

Equation (4-14) was used many years ago in the determination of the *specific charge* e/m of the electron and a variety of positive ions. Today, improved techniques yield values with uncertainties of only a few parts per million. In most cases the particle velocities are so high that relativistic relations must be used in (4-14). When this is done, we have a useful relation between the *magnetic stiffness* $B\rho$ and the kinetic energy:

$$B\rho = \frac{\sqrt{T^2 + 2m_0c^2T}}{ce} \tag{4-15}$$

The inverse relationship can be obtained from Eq. (4-12):

$$T = m_0c^2\left[\sqrt{1 + \left(\frac{B\rho e}{m_0 c}\right)^2} - 1\right] \tag{4-16}$$

Magnetic deflection methods have played an important role in establishing the properties of charged particles and in charged-particle spectrometry. Equations (4-15) and (4-16) are basic in relating measured values of $B\rho$ to momentum and energy. Experimental values of $B\rho$ are usually given in gauss-cm. If these values are used directly, all of the other quantities in Eq. (4-15) must be taken in the CGS system, with e in electromagnetic units. The same requirements apply to the quantities within the square brackets of Eq. (4-16). All values will be in the MKSA system when $B\rho$ is expressed in webers m^{-2}. For routine calculations it is customary to convert the equations so that the energy will appear directly in MeV (see Probs. 4-6 and 4-7).

4.05 Electromagnetic Separation of Ions

The method of magnetic separation of masses outlined in the previous section has been highly refined to obtain e/m values with extraordinary precision. Since e is always exactly a small integral multiple of the fundamental charge, the separation depends essentially upon the mass values. Neutral ions can be ionized by some suitable means such as electron bombardment, high temperature, or by an electric-spark discharge. Positive ions thus formed are accelerated by a negative potential of a few KV and passed through a defining slit. Many geometrical combinations of electric and magnetic fields have been devised to optimize the collection of ions with a particular value of e/m at a suitable measuring device.

Instruments in which the focused ion beam impinges upon a photographic plate are known as mass spectrographs. In a *mass spectrometer* the ion beam is caught in a Faraday cup, which is an electrode designed to minimize charge loss by secondary emission, and the resulting current is measured with an electrometer.

There are many design variations. For accurate mass determinations, high resolving power is needed and beam intensity can be sacrificed. For the large-scale production of separated ions, beam intensity is all-important and considerations of resolving power become secondary.

The mass of a particle that travels through a mass spectrometer will be that of the neutral atom less the mass of the one or more electrons that were removed at ionization. Three components make up the mass of the ion: the remaining atomic electrons, the nuclear protons, and the nuclear neutrons. Two sets of mass values are commonly tabulated and care must be taken to keep in mind which is being used. *Atomic mass values* are the actually measured values with the masses of the missing electrons added; these are the masses of the neutral atoms. *Nuclear mass values* are tabulated after subtracting all contributions of atomic electrons to the measured values. Table 4-1 lists the masses of some of the elementary particles, expressed in several different units. More complete values will be found in Table 14-2.

TABLE 4-1
MASSES OF SOME ELEMENTARY PARTICLES

Name	*Symbol*	*umu*	*MeV*	m_e	*Charge*	*Comments*
Electron	e^-	0.000549	0.511	1	-1	stable
Positron	e^+	0.000549	0.511	1	$+1$	life 10^{-6} sec
Muon	μ	0.11320	105.659	206.4	-1	unstable
Pion	$\pm\pi$	0.14990	139.578	273.2	± 1	
Proton	p	1.007276	938.256	1836.1	$+1$	stable
Neutron	n	1.008665	939.550	1838.6	0	life 12 min

Before proceeding it will be well to introduce a set of symbols and definitions:

A = mass number, an integer equal to the sum of the number of protons and neutrons in a nucleus

Z = atomic number, an integer equal to the number of protons in the nucleus; Z is also equal to the number of electrons attached to the nucleus in a neutral, unionized state

N = neutron number, an integer equal to the number of neutrons in the nucleus

Nucleon = either nuclear constituent, a proton or a neutron

Nuclide = any grouping of nucleons capable of more than a transient existence

Isotopes = nuclides having equal numbers of protons but different numbers of neutrons
Isotones = nuclides having equal numbers of neutrons but different numbers of protons
Isobars = nuclides having equal numbers of nucleons (equal mass numbers) but with different combinations of protons and neutrons

The following forms will be used in specifying a particular nuclide

$$^{A}\text{chemical symbol} \quad \text{or} \quad ^{A}_{Z}\text{chemical symbol}$$

Thus the form will be $^{12}_{6}C$ for the carbon isotope adopted as the standard of the u system; $^{13}_{6}C$ for the heavier isotope; and so on. The numbering scheme given here has only recently been adopted. In the older literature the mass number will be found as a superscript on the right instead of on the left. Strictly, we do not need the subscript, since its value is fixed when the chemical symbol is given. Both numbers are useful in balancing nuclear equations, however, and will be given frequently.

Electromagnetic focusing conditions depend upon the ratio e/m and not upon either value alone, and so doubly charged ions will appear at the same focal point as singly charged ions of exactly half the mass. Thus, singly charged oxygen-16, O^+, and doubly charged sulphur-32, S^{++}, will fall close together as a *mass doublet*. These relative masses can then be determined with a very small error associated with the dispersion of the instrument.

An extension of the doublet method utilizes appropriate hydrocarbon molecules to compare with the mass being measured. Since C and H form a wide variety of chemical combinations, a close doublet can be formed with almost any desired nuclide.

Figure 4-2, a spectrogram taken at moderately high resolution, shows doubly ionized oxygen appearing at the place where singly ionized mass-8 would be expected. Note also the relative intensities of the ^{12}C and ^{13}C lines.

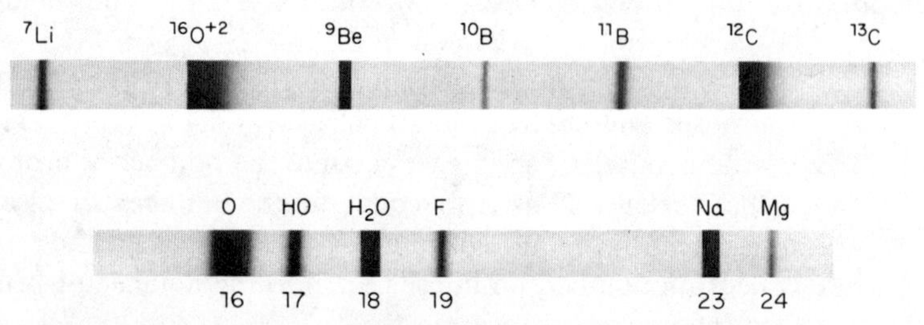

Figure 4-2. Mass spectra showing impurities in a metal sample. Doubly ionized ^{16}O appears in the position that would be occupied by a singly ionized atom $A = 8$.

4.06 Natural Isotopic Abundance

Figure 4-3 shows some typical mass spectra obtained from purified samples of elements as they are found in nature. The height of the curve at each mass number is proportional to the number of positive ions and hence to the number of atoms of that mass in the original sample. These numbers may

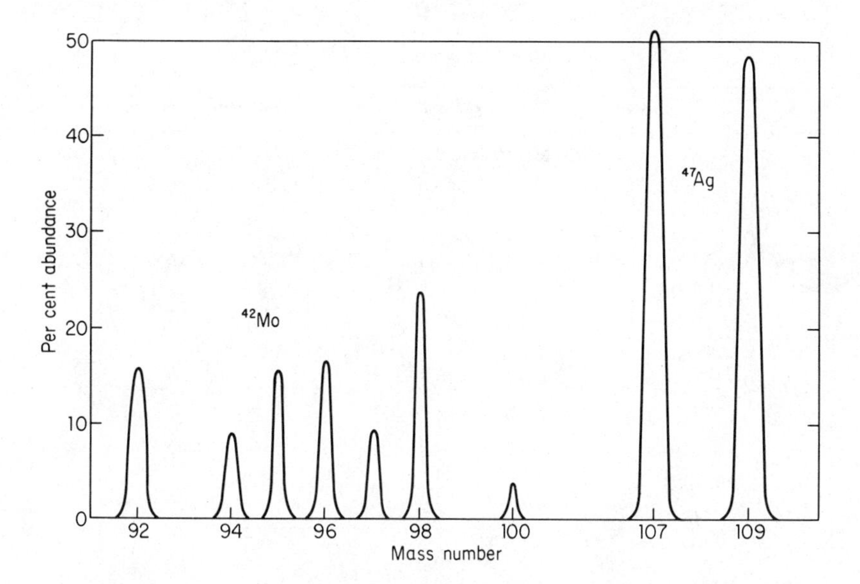

Figure 4-3. Mass spectra of molybdenum (even Z) and silver (odd Z).

be expressed as *per cent abundance*, where each is given as a fraction or a percentage of the total. *Relative abundance* is calculated by expressing the amount of each constituent as a percentage of the abundance of the most common isotope of the element. Table 4-2 gives both sets of values for the elements shown in Fig. 4-3. Other abundance values are tabulated in the Appendix, Table 9.

A study of the mass spectra of the elements as they occur in nature shows widely varying patterns of abundance. For example, gold consists of a single isotope, whereas ten isotopes of tin ranging in mass number from 112 to 124 have been identified. Elements of odd atomic number usually consist of one or at most two isotopes, whereas elements of even atomic number rarely have only one isotope.

An enormous amount of experimental data shows that with a few noteworthy exceptions the isotopic composition of a given element is independent of the location from which the sample is taken. Meteoric iron, for example, has the same isotopic composition as iron taken from the earth. We do not fully understand the laws governing isotopic abundances. Our empirical knowledge can be summarized:

1. Nuclides having an even atomic number and an even neutron number are most abundant in nature. These "even–even" nuclides constitute the majority of the elements in the earth's crust.

TABLE 4-2
ISOTOPIC ABUNDANCES

Mass number		*Relative abundance*	*Per cent abundance*
		Molybdenum ($Z = 42$)	
92		66.6	15.86
94		38.4	9.12
95		66.1	15.70
96		69.4	16.50
97		39.8	9.45
98		100.0	23.75
100		40.5	9.62
	Total:	420.8	100.00
		Silver ($Z = 47$)	
107		100.0	51.35
109		94.9	48.65
	Total:	194.9	100.00

2. "Even–odd" and "odd–even" nuclides having respectively even Z and odd N, or odd Z and even N, are moderately abundant in nature.
3. "Odd–odd" nuclides are extremely rare and only four are known to exist in stable form. These are, ${}^{2}_{1}H$, ${}^{6}_{3}Li$, ${}^{10}_{5}B$, and ${}^{14}_{7}N$. It will be noted that these are light elements with equal numbers of protons and neutrons.

Stable nuclides found in nature are distributed as follows:

even Z–even N: 163   odd Z–even N: 50
even Z–odd N: 57   odd Z–odd N: 4

Lead is one of the few elements whose isotopic composition depends upon the origin of the sample. The stable isotopes ${}^{206}Pb$, ${}^{207}Pb$, and ${}^{208}Pb$ are known to be formed from the radioactive decay of ${}^{238}U$, ${}^{235}U$, and ${}^{232}Th$, respectively, while ${}^{204}Pb$ was presumably formed at the time when all other stable elements became differentiated. Because of the different origins of the isotopes, the isotopic abundance of lead as measured today depends upon the relative amounts of lead, uranium, and thorium present when the ore-bearing material was laid down, and also upon the age of the deposit. Very careful measurements of the isotopic abundances in lead fix the time of formation of the elements at 5.5 billion years ago, with the solidification of the earth's crust 3.5 billion years ago. Recent corrections have brought age estimates from astronomical data in good agreement with the values determined from isotopic abundances.

4.07 Isotope Separation

It is frequently desirable to separate an element into its component isotopes; in general this is a difficult job. In principle any property which depends upon atomic mass can be used to effect a separation, but only a few have proved practically useful.

Electromagnetic deflection methods (mass spectrometry) give the most complete separations, but the yield of products is extremely small unless specially designed equipment is used. The separated ions are condensed onto suitable targets which are removed at intervals for collection. Many of the stable isotopes supplied by the Oak Ridge National Laboratory are obtained by electromagnetic separation.

Gaseous diffusion through small openings can be utilized for isotope separation, since by Graham's law the diffusion rate is a function of the molecular weight

$$n = \frac{\text{constant}}{\sqrt{M}}$$

where n = number of molecules diffusing in unit time
M = molecular weight of the gas

Obviously, the degree of separation is small for gases of high molecular weight. Porous barriers of large area can be used and the gas passed through a series of stages so that eventually a high degree of separation can be achieved with large volumes of gas. Gaseous diffusion is used extensively by the Atomic Energy Commission for the separation of ^{235}U and ^{238}U.

Thermal diffusion provides a relatively simple and effective but expensive method for isotope separation. In the thermal-diffusion method, the gas is contained between two concentric, vertical cylinders maintained at different temperatures. The heavier isotope will concentrate near the cold surface and the lighter isotope near the hot surface. Convection currents will carry the cold heavy isotope to the bottom of the volume and the hot light isotope will rise to the top, where it can be collected. The gas molecules transfer heat continuously from the hot cylinder to the cold, and the power consumption is high.

Some isotopic separation can be achieved by high-speed centrifugation, and plants for quantity production have been designed. Some of the mechanical design problems are formidable, and it is not certain that centrifugation can compete economically with some of the other methods.

Separation by vaporization is based on the fact that at constant temperature lighter molecules will have a slightly greater velocity and, consequently, will tend to leave the liquid phase more readily. By condensing the vapor phase, a fraction slightly enriched in the lighter isotope is obtained, and a considerable separation can be achieved by repeated fractionations.

Separation of the stable isotopes of hydrogen can be effected by electrolysis. The hydrogen gas evolved in the electrolysis of water is slightly enriched in the light isotope, and deuterium is slightly concentrated in the residual. A high degree of separation can be achieved by repeated electrolyses of the deuterium-rich fractions.

Although all isotopes of a given element enter into the same chemical reactions, the rates of reaction will differ slightly and each will have a slightly different equilibrium concentration. In a typical exchange reaction such as

$$\underset{\text{gas}}{^{15}NH_3} + \underset{\text{solid}}{^{14}NH_4Cl} \rightleftharpoons \underset{\text{gas}}{^{14}NH_3} + \underset{\text{solid}}{^{15}NH_4Cl}$$

there will be a small difference between the isotopic concentrations of the gas and solid phases at equilibrium, and some separation can be obtained. Separation is enhanced at low temperatures; this makes the method practical for light elements such as C, N, or S.

Although reaction rates and equilibrium constants differ only slightly with isotopic mass, these differences are great enough to be of concern in some cases. Any differences will be greatest in the case of hydrogen, where the masses are in the ratio of 2:1. Mice receiving high deuterium concentrations in food and water die when about 30 per cent of the light hydrogen in their bodies has been replaced by the heavier isotope.

4.08 Common Properties of Waves

It is easy to visualize the transmission of energy by moving particles that can give up their kinetic energy upon collision. Less obvious, but equally important, is the transfer of energy by wave motions. Most of the sun's energy reaching the earth is transmitted by electromagnetic waves; all audible sounds are transmitted by sound waves; some of the greatest radiation hazards arise from short-wavelength electromagnetic waves.

The general concept of wave transmission is simple. Some type of transmitter disturbs and sets up stresses in the surrounding medium. The medium, in returning to its original unstressed state, propagates the stresses, which travel out from the source and act on some suitable receiver. Thus when human vocal cords vibrate, a succession of pressure impulses is applied to the surrounding air. Pressures above and below normal exist and are transmitted by each volume of air to that immediately beyond. Pressure variations travel out from the source and upon striking an eardrum set up vibrations that one interprets as sound.

The most obvious property of a wave is its *frequency*, or number of vibrations per second, denoted by the letter ν. Frequency is really to be associated with the source of the disturbance, since the wave is merely the transmission of the disturbance. It is sometimes more convenient to use the time of one vibration τ, known as the *period*, instead of the frequency. The two are related by

$$\nu = \frac{1}{\tau} \tag{4-17}$$

Since a wave is propagated, there will be a velocity associated with the wave motion. The velocity is a property of the medium and not of the source, for once the disturbance has been transmitted to the medium, it is independent of the behavior of the source. In a given medium a wave motion will be propagated only with its characteristic velocity. This behavior is in sharp contrast to the motion of particles whose velocities depend upon their energies.

If a source is steadily emitting a frequency ν and the velocity of wave propagation is c, at the end of one second there will be ν waves spread over a distance c in space. One wave will occupy a distance λ, where

$$\lambda = \frac{c}{\nu} \tag{4-18}$$

The distance λ, or *wavelength*, is the shortest distance between consecutive similar points on the wave train. Thus it is the distance between consecutive crests or peaks, or consecutive troughs. Spectroscopists frequently find it more convenient to use the *wave number* $\tilde{\nu}$, or the number of waves per unit length. Obviously,

$$\tilde{\nu} = \frac{1}{\lambda} \tag{4-19}$$

Some phenomena, such as *reflection*, are common to both particle motion and wave motion, but *interference* and *diffraction* are unique properties of waves. If two waves of equal wavelength and opposite phase are superimposed, crest will fall on trough; the combined effect will be a complete cancellation, and the two waves *interfere destructively*. If the two waves are in phase, crest will fall on crest, an increased amplitude will result, and there is *constructive interference*. It is extremely difficult to conceive of particles showing interference.

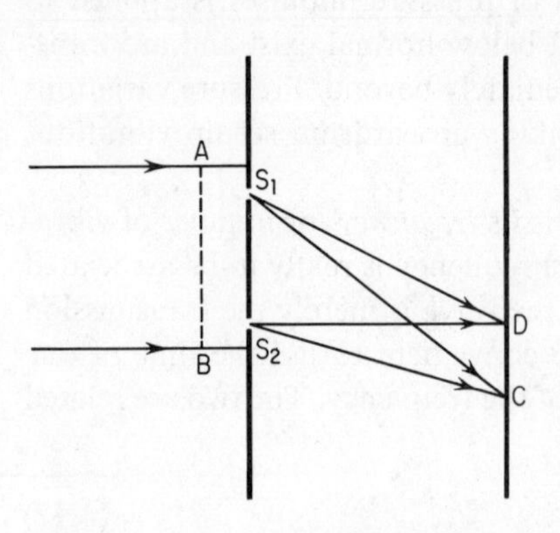

Figure 4-4. Wave interference, constructive at C, destructive at D, from a wavefront diffracted at two slits. The diffraction pattern will be symmetrical about the center line of the two slits.

Interference is most easily demonstrated through the diffraction or scattering of waves as they pass through small openings. Figure 4-4 shows a wave front AB of visible light striking a diaphragm with two small slits, S_1 and S_2. Because of diffraction, each portion of the wavefront passing through a slit will spread out in a hemispherical wave. At a point C on the screen, so chosen that the distance S_1C is exactly one wavelength greater than S_2C, there will be constructive interference and light will be seen. If S_1D is one-half wavelength greater than S_2D, there will be destructive interference and no light will be seen. It is obvious from the figure that the angle of diffraction will be appreciable only if the slit separation is comparable to the wavelength.

4.09 Electromagnetic Waves

Visible light, X rays, and gamma rays are electromagnetic waves, since they consist of oscillating electric and magnetic fields. A *field* is defined as a region in which a particular kind of force is exerted. For example, masses near the surface of the earth experience a gravitational pull and hence are in the gravitational field of the earth. If an electric charge is brought near a second charge, there will be forces acting on the charges because each is in the field of the other. Similarly, a magnet has an associated magnetic field because a second magnet brought near will be acted upon by a force. The strength of a field at any point is equal to the force exerted on a unit (of mass, charge, or magnetic-pole strength) at that point.

Consider the production of a radio wave at the antenna of a broadcasting station. In the antenna, electric charges are accelerated back and forth along the wire at a frequency determined by the generating equipment, and these charges will produce an oscillating electric field in the vicinity of the wire. The theory of electrodynamics shows that an accelerated electric charge produces a changing magnetic field, and this, in turn, produces an electric field. Thus an electromagnetic disturbance originates at the antenna and is propagated outward from it.

An electromagnetic wave consists, then, of an oscillating electric field and a similar magnetic field, inextricably connected, with each depending on the other for its existence. Each component vibrates with the same frequency, and in free space the two components are in equal phase. The two component fields oscillate at right angles to each other and at right angles to the direction of propagation.

In 1865, Maxwell predicted that visible light is an electromagnetic wave, and later experiments have completely verified his basic ideas. All electromagnetic waves, regardless of wavelength or mode of generation, consist of linked electric and magnetic fields. All the waves travel with the same velocity c in empty space. Very accurate measurements of c have been made, yielding a value 2.997925×10^8 m sec^{-1}. For all but the most exact calculations the value of 3×10^8 m sec^{-1} may be used. When electromagnetic waves travel in a ponderable medium, their velocity may be substantially less than c.

4.10 Particulate Nature of Electromagnetic Radiation

Experiments have shown that all electromagnetic radiations have wave properties; they show interference and diffraction. Other types of studies indicate that the energy carried by electromagnetic waves is discontinuous or atomic in nature rather than the continuously graded energy expected from a wave motion. The photoelectric effect will serve as an example of the discontinuous or *quantum* nature of these radiations.

A photoelectric cell consists of a metallic surface (*cathode*) sealed into an evacuated glass bulb together with a positively charged collecting electrode, called an *anode*, Fig. 4-5. When light falls on the cathode surface, electrons may be ejected from the metal. These electrons will be attracted to the anode and an electric current will flow.

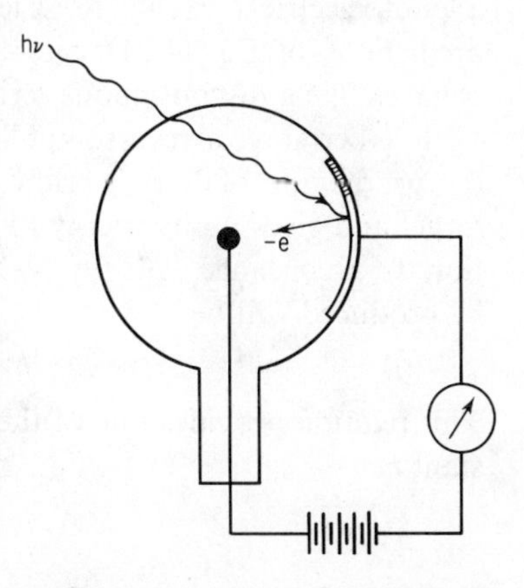

Figure 4-5. An energetic photon may eject a photoelectron from the cathode of the photocell.

According to wave theory, the vibrations of the incoming electromagnetic waves set the electrons of the metallic atoms into vibration with increasing amplitude until they have acquired sufficient energy to break loose from the metal. It would seem, then, that a weak light would require more time to liberate an electron than a strong light and that there might be a light wave so weak that the electrons could never attain the energy necessary for escape. Such a prediction is at complete variance with the facts.

The color or frequency of the light plays

an important role in producing photoelectrons. Thus the most intense beam of red light (low frequency) will not yield a single photoelectron from most metals; whereas the feeblest blue light (high frequency) will instantly produce a few. This circumstance suggested to Einstein, who reasoned, from earlier work by Planck, that radiation is not a smooth, continuous flow of energy, as pictured by the wave theory, but is, rather, a series of discontinuous packages of energy. The energy in each package, which is known as a *photon* or *quantum*, increases with the frequency of the light, being given by

$$E = h\nu \tag{4-20}$$

where h is Planck's constant ($h = 6.62509 \times 10^{-34}$ joule-sec) and E is the energy of the quantum in joules.

A photon is apparently indivisible and either exists or disappears completely upon giving up its energy in some process such as the photoelectric effect. The energy relations in the photoelectric effect are given by the Einstein equation,

$$h\nu = \phi + \tfrac{1}{2}mv^2 \tag{4-21}$$

The left-hand side of this equation is the energy available in the incoming photon. The *work function* ϕ is the energy required to remove an electron from the surface of the metal. If the photon has an energy greater than ϕ, the electron will be ejected with a kinetic energy given by the last term in Eq. (4-21).

This result immediately explains the photoelectric behavior of light of different frequencies. Red light has a low frequency, and the associated quantum energy, by Eq. (4-20), may be less than ϕ, and hence no amount will liberate an electron. Blue light has a higher frequency, and the associated quantum energy may be larger than ϕ and hence may be sufficient to produce a photoelectric current. A series of experiments by Millikan verified the predictions of Eq. (4-21), and there can be no doubt that here radiation behaves like a discontinuous series of energy packages.

Furthermore, a reverse application of Eq. (4-21) quantitatively explains the production of X rays. Here an electron with kinetic energy T strikes a target and gives up its energy to produce a photon of electromagnetic radiation. In accordance with Eq. (4-21) the maximum frequency photon that can be produced will be

$$h\nu_{\max} = T - \phi \tag{4-22}$$

This relation provides one of the best methods of determining Planck's constant h.

4.11 The Electromagnetic Spectrum

The known electromagnetic radiations can be arranged on a scale of wavelength, frequency, or energy content, Fig. 4-6. The spectrum is divided into

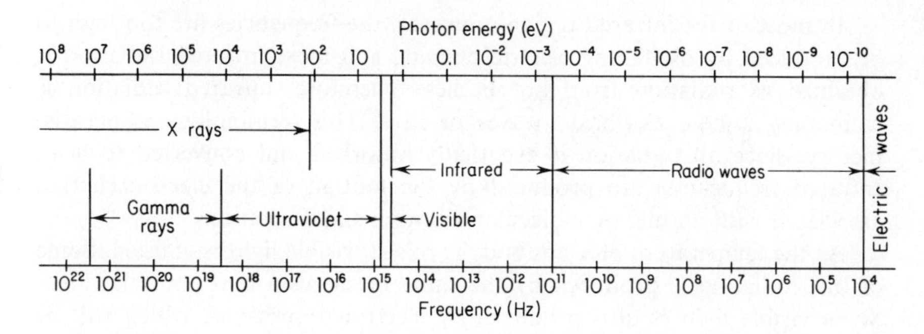

Figure 4-6. The known electromagnetic spectrum on a log (energy) scale.

several regions, but one must realize that these are arbitrary and not at all rigid. In the main, the divisions are based upon the methods used to produce the radiation, and it is possible to produce a given frequency by two or more methods. The properties of the radiations are independent of the method of production; only the frequency is important in determining the properties.

The range of electromagnetic frequencies already known is truly enormous. Electric power lines radiate electromagnetic waves at the generator frequency, which is usually 60 Hz, but it is not difficult to generate much lower frequencies. Thus the lower-frequency limit is essentially zero. Of the whole spectrum, the part taken up by visible light is very small. All sense of color, and indeed all vision, is conveyed in a frequency range of somewhat less than one octave. The limits of the visible range, as determined by the human eye, are more sharply defined than any of the others. Man is not equipped with senses that allow him to investigate anything except the visible spectrum directly. In all other regions he must rely upon an extension of his senses through the use of instruments. Even in the parts of the spectrum closely adjoining the visible region, man's sensitivity to the electromagnetic radiation (ultraviolet and infrared) is not a direct one.

The range of the electromagnetic spectrum is so great that several units are used in measuring wavelength. Wavelengths in the radio-frequency region are commonly measured in meters. Visible-light wavelengths are usually expressed in centimeters or in *angstroms* (Å), equal to 10^{-8} cm. The angstrom is also used almost exclusively in measuring wavelengths in the X-ray and gamma-ray region. The X unit (X.U.), equal to 10^{-11} cm, is also used to some extent for specifying X and gamma rays.

Methods of producing the various wavelengths vary tremendously. All wavelengths from long electric and radio waves down to the infrared can be produced by purely electrical means. One can make oscillating circuits of a natural frequency so high that there is actually an overlap with the infrared part of the spectrum.

In most of the infrared region, however, the frequencies are too high to be produced by oscillating electric circuits, and most infrared radiation is obtained as radiation from hot bodies. Therefore, infrared radiation is commonly known as "heat" waves or rays. This terminology is unsatisfactory, since all radiation is eventually absorbed and converted to heat. Infrared frequencies are produced by the motion of the electric charges associated with atomic or molecular vibrations and rotations.

As the temperature of a hot body is raised, visible light is emitted. Some of this radiation is produced by the same mechanism that emits infrared. Some visible light is also produced by electron transitions, which will be described in detail in Chapter 6.

When an electric arc is struck between metal electrodes or when an electric current passes through a gas, visible and ultraviolet light result. The ultraviolet frequencies are due entirely to electronic excitations. As one increases the energy of excitation, high ultraviolet frequencies will appear, some of which will overlap the lower frequency limit of the X-ray region.

X rays are produced by bombarding a target, usually made of a heavy metal, with high-speed electrons. As the energy of the bombarding electrons is increased, the radiated frequencies increase, overlap, and even extend beyond the gamma-ray region. Gamma rays, which after emission are identical with X rays of the same wavelength, are emitted from nuclei excited in radioactive or other high-energy processes.

4.12 The Transmitting Medium

It is difficult to conceive of a wave without reference to some medium of propagation, and in most cases the medium is easily recognized. A sound wave, for example, is transmitted by air, and in a vacuum no sound waves can be propagated. The medium of propagation for electromagnetic waves has been called the *luminiferous ether*, or simply *ether*, and it is evident that this ether must have some very unusual properties. Light waves readily pass through the best laboratory vacuum and through the vacuum of interstellar space, so that the ether must be an all pervading medium that cannot be removed from a vessel. All bodies must move through the ether without appreciable friction, or the solar system would have "run down" long ago. The alternative hypothesis, that each body drags its own ether along with it, is untenable on a variety of grounds.

If there is an all-pervading ether, it would be of the greatest interest to measure the velocity of the earth relative to it. In principle the experiment is relatively simple, as can be seen by the analogous experiment with sound waves. In still air, sound travels with a velocity of about 330 m sec^{-1}. If the wind is blowing steadily from north to south at 10 m sec^{-1}, an observer due south from a source will measure a velocity of 340 m sec^{-1}, and at due north the velocity will appear to be 320 m sec^{-1}. Thus from a series of measurements the velocity of the propagating medium, relative to the observer, can be determined.

The velocity of the earth in its orbit is about 3×10^4 m sec^{-1} and, during the course of a year, has many different directions in space. This velocity is small compared with the velocity of light, but optical instruments capable of detecting the expected differences have long been available. A determination of the relative velocity of the earth and the ether appears, therefore, quite possible. In 1887, Michelson and Morley started a long series of experiments designed to measure this relative velocity, or ether drift. In every case their results showed the velocity of light to be a constant, independent of the direction of propagation.

Several explanations were advanced for the failure of the Michelson–Morley experiment, but none were completely satisfactory until Einstein proposed his theory of special relativity in 1905.

4.13 Black-body Radiation

In their attempts to account for the electromagnetic radiation emitted by hot objects, physicists developed the concept of an ideal black body, capable of emitting and absorbing all frequencies without preference. Such an ideal radiator can be approximated by a furnace, enclosed except for a small opening through which the radiations can be viewed. These radiations will depend only upon the temperature of the furnace and not upon any unique characteristics of the device itself.

During the 1890s, experiments had determined the spectral distribution of black-body radiation. Typically, there was a maximum energy emission at a wavelength λ_m, related to the absolute temperature by the Wien displacement law.

$$\lambda_{\max} T = \text{const} = 2.8979 \text{ m°K} \qquad (4\text{-}23)$$

Emission at wavelengths above and below the maximum dropped to very low values, Fig. 4-7.

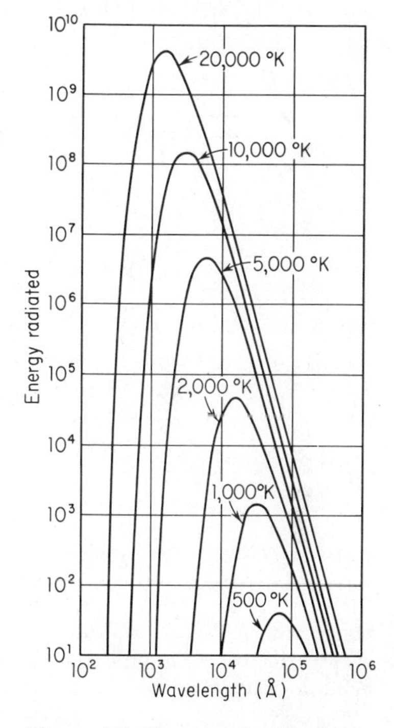

Figure 4-7. The wavelength dependence of the energy emitted by a black body.

The radiation was assumed to be emitted by myriads of small linear oscillators, each capable of vibrating at an amplitude and frequency determined by its energy content. In thermal equilibrium, the oscillator energies were supposed to be distributed according to the Maxwell–Boltzmann law, Eq. (3-5). These assumptions led to a spectral distribution with far too much energy in the shorter wavelengths, a result that has been called the "ultraviolet catastrophe." All attempts to reconcile theory and experiment failed until Planck introduced, in 1901, the idea of discrete energy states and energy quanta in place of the continuous distributions previously assumed. Although the original quantum theory has been found to be inadequate, it pioneered a revolutionary approach to the study of physical phenomena, and paved the way for the more successful quantum mechanics.

4.14 The Quantized Linear Oscillator

Consider a small mass m oscillating about an equilibrium position under a restoring force whose magnitude is proportional to the displacement of the particle, Fig. 4-8A. Then, $F = -kx$ and, by Newton's second law of motion,

$$\frac{d^2x}{dt^2} = -\frac{kx}{m} \tag{4-24}$$

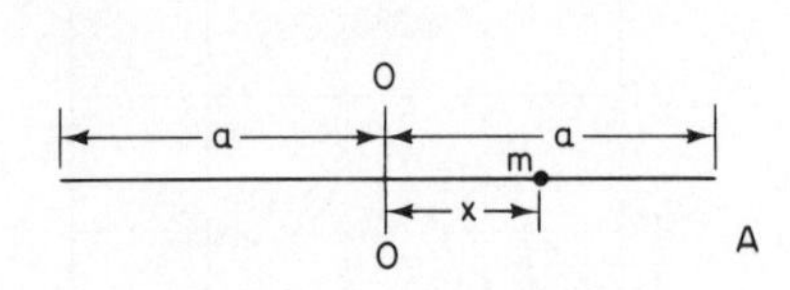

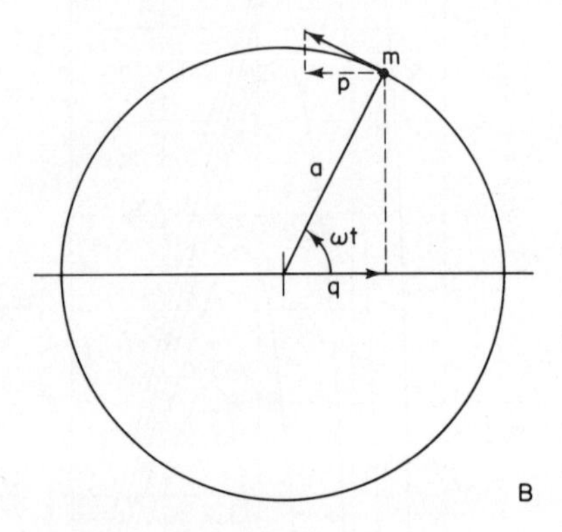

Figure 4-8. A linear oscillator, (A), may be treated as the projection of a uniform circular motion, (B).

There is no damping term in Eq. (4-24) and so the solution will be cyclic with a constant maximum amplitude. The solution of Eq. (4-24) will have the form

$$x = ae^{i\sqrt{k/m}t}$$
$$= a\left[\cos\sqrt{\frac{k}{m}}t + i\sin\sqrt{\frac{k}{m}}t\right] \tag{4-25}$$

where a is a constant depending upon the energy and $i = \sqrt{-1}$. We may drop the sin term in Eq. (4-25) with no loss of generality, and then the motion may be considered as the projection of a uniform circular motion of angular velocity $\omega = \sqrt{k/m}$ onto a linear coordinate axis, Fig. 4-8B.

It is convenient to describe the motion in terms of two *conjugate variables*, displacement q and momentum p. The exact definition of conjugate need not concern us here. It is enough to note that the two variables suffice to describe the motion completely, and that they are essentially independent. From Fig. 4-8B,

$$q = a \cos \omega t \qquad p = -ma\omega \sin \omega t$$

$$\text{or} \quad \cos \omega t = \frac{q}{a} \qquad \sin \omega t = -\frac{p}{ma\omega} \tag{4-26}$$

Squaring and adding,

$$\frac{p^2}{(ma\omega)^2} + \frac{q^2}{a^2} = 1 \tag{4-27}$$

Equation (4-27) is recognized as the equation of an ellipse in the variables p and q, with semiaxes $ma\omega$ and a. By analogy with the variables of plane geometry we can think of p and q as defining a two-dimensional fictitious or conceptual *phase space*. Phase space is not restricted to the three mutually perpendicular coordinates of Euclidian geometry. Our one-dimensional motion led to a two-dimensional space; translation in three dimensions leads to a six-dimensional phase space; and so on.

No restrictions have been placed on values of the amplitude a and not, consequently, on the values of the oscillator energy. According to Eq. (4-27) there will be an ellipse in phase space for every value of a and k/m. Planck *quantized*, or restricted, the motions to those which satisfy a relation that was later generalized to

$$\oint p \, dq = nh \tag{4-28}$$

where the integration is to be taken over one complete cycle of the motion. The universal constant h has the dimensions of action, (erg-sec), and n is a running integer 1, 2, 3, ...

When the quantum condition is applied to the linear oscillator, we have

$$\pi \times a \times ma\omega = nh \tag{4-29}$$

The left-hand side of Eq. (4-29) is just the area of an ellipse in phase space. Thus the quantization has restricted the motions to those in which the areas of the phase-space ellipses are integral multiples of Planck's constant h, Fig. 4-9.

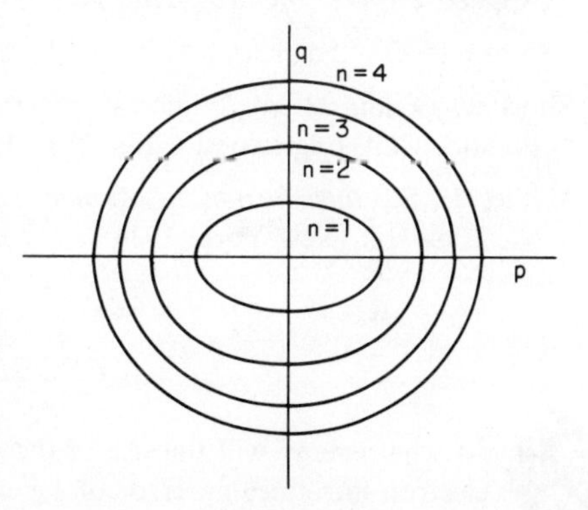

Figure 4-9. The ellipses in phase space that are allowed by the quantum theory. The area of each elliptical area will be an integral multiple of h.

The total energy of the oscillator will be made up of the sum of its potential and kinetic components. The total energy can be evaluated most easily at $\omega t = \pi/2$, when the particle is passing through the equilibrium point, and all of the energy is kinetic. At this point,

$$W = T = \frac{p^2}{2m} = \frac{ma^2\omega^2}{2} \tag{4-30}$$

When the quantum condition is introduced,

$$W = \frac{nh\omega}{2\pi} = nh\nu = n\hbar\omega \tag{4-31}$$

The factor $h/2\pi$ appears so frequently that it is given a special symbol $\hbar$.

When the linear oscillator changes its motion from one quantum state to another it will either absorb or radiate energy in integral multiples of $h\nu$. This is in sharp contrast to the nonquantized radiating capabilities of the classical oscillator.

The introduction of quantum conditions into the linear oscillator led to the proper law for the spectral distribution of black-body radiation. There was, however, no satisfactory explanation as to why a quantum condition was needed, nor why the condition took the specified form.

REFERENCES

Bergmann, P. G. *Introduction to the Theory of Relativity*. Prentice-Hall, Englewood Cliffs, N.J., 1942.

Feynmann, R. P., R. B. Leighton, and M. Sands, *The Feynmann Lectures on Physics.* Vols I and III. Addison-Wesley, Reading, Mass., 1965.

Leighton, R. B., *Principles of Modern Physics.* International Series in Pure and Applied Physics. McGraw-Hill, New York, 1959.

Newton, R. G., "Particles that Travel Faster then Light?" *Science* **167,** 1569, 1970.

Shortley, G. and D. W. Williams, *Elements of Physics.* 5th ed., Vols. I and II. Prentice-Hall, Englewood Cliffs, N.J., 1971.

White, H. E., *Introduction to Atomic Spectra.* International Series in Physics, McGraw-Hill, New York, 1934.

PROBLEMS

4-1. At what energy will the use of the classical energy–velocity relation for an electron introduce an error of 1 per cent in the velocity calculation?

4-2. Calculate the wavelength of the electromagnetic radiation which is just capable of ejecting an electron from a cesium photocathode which has a work function of 1.52 eV. Calculate the velocity of the photoelectron ejected from this cathode by light having a wavelength of 4047 Å.

4-3. Calculate the loss in mass associated with a nuclear detonation having an explosive equivalent of 50 megatons of TNT. The energy release in a TNT detonation is 1000 cal g^{-1}.

4-4. Assume that the maximum temperature in the detonation in Prob. 4-3 was (9.3×10^6)°K and that the fireball radiates like an ideal black body. Calculate the wavelength of the maximum energy emission and the quantum energy of this emission. In what region of the spectrum is this located?

4-5. A series of measurements indicates that space is filled with black-body radiation at a characteristic temperature of 3°K, the residual from the radiation emitted at the "big bang" of creation some 5×10^9 years ago. What is the quantum energy of the emission maximum? In what region of the spectrum is this located? What kind of equipment would you choose to measure this radiation?

4-6. Show that for electrons $B\rho = 3330\sqrt{T^2 + 1.02T}$, where T is in MeV and $B\rho$ is in gauss-cm. Deduce the corresponding expression for protons.

4-7. Adjust Eq. (4-16) so that the energy will be given directly in MeV.

4-8. Derive Eqs. (4-15) and (4-16).

4-9. An electron is accelerated in an electron synchrotron by the repetitive application of 100 KV. Calculate the change in mass and velocity produced in going from an energy of 100 KeV to 200 KeV; from 10 MeV to 10.1 MeV.

4-10. Singly charged copper ions are accelerated by a potential of 18,000 volts and then enter a magnetic field of 8000 gauss. Calculate the linear separation of the ^{63}Cu and ^{65}Cu ions after they have been deflected 180° by the field.

4-11. A mass doublet is formed in a precision mass spectrometer between doubly ionized dihydrobenzene, $^{12}C_6{}^1H_8$, and singly ionized ^{40}Ar. Calculate the atomic mass of the argon, measured to be 0.06896 u lighter than the doublet.

4-12. An isotope-separation mass spectrometer with a beam current of 650 μA is used to separate the naturally occurring isotopes of silver. Calculate the time required to collect 1 gram of ^{107}Ag.

4-13. The electromagnetic separation of ^{235}U and ^{238}U has been carried out in a production unit known as a calutron. What time will be required to obtain 100 g of ^{235}U in a unit operating with a beam current of 2.5 mA?

4-14. Uranium isotopes can be separated by gaseous diffusion using uranium hexafluoride, UF_6. Calculate the degree of enrichment expected after a passage through one porous barrier. How many barriers will be required to attain an enrichment factor of 1.1?

4-15. Quantize a system consisting of an electron constrained to move at constant velocity v in a circle of radius a about a fixed center.

4-16. Assume that an electron, acting as a linear oscillator, has a frequency equal to that of green light of wavelength 5400 Å. What will be the restoring force on the electron when it is displaced 10^{-9} cm from its equilibrium position? Could this force be produced by a single electric charge located at the equilibrium position?

4-17. The diatomic structure O–H has a vibrational frequency which produces radiations with a wave number ($1/\lambda$) of 3735 cm^{-1}. What is the force constant of this motion? What is the amplitude of vibration when the motion is in the lowest quantum state? Assume the O atom to be stationary.

4-18. The C–H vibration produces radiation with a wave number of 2860 cm^{-1}. How many of these quanta, arriving simultaneously and in phase, will be required to liberate a photoelectron from a photocathode which has a work function of 1.25 eV? Assume the C atom to remain stationary and calculate the amplitude of vibration required to produce an emission with sufficient energy to actuate the photocell.

5

Particle–Wave Relations

5.01 Interference Patterns

The double-slit experiment mentioned in Sec. 4.08 is technically very simple, but it leads to some most profound conclusions about the ultimate nature of radiation and matter. It is time to consider this experiment in more detail.

Consider a barrier pierced with two narrow slits 1 and 2, Fig. 5-1, mounted vertically above a collecting screen. Let a large number of small, identical particles such as fine, uniform grains of sand fall upon the barrier. Further imagine that each grain is coated with an adhesive so that it will adhere to any surface which it strikes.

Most of the grains will fall upon the barrier and are of no further interest to us. Some will pass through slit 1, to produce a statistical distribution as shown at $A1$ in the figure. Others will pass through slit 2 to produce the distribution $A2$. If both slits are open, the accumulation on the screen will be the arithmetical sum of the individual accumulations, shown as $A(1 + 2)$. This distribution will be obtained whether the accumulation through 1 and 2 occurred simultaneously or consecutively. What is happening at one slit has no effect on the grains passing through the other.

We now replace the sand with a monochromatic light source which will flood the barrier with quanta or photons. According to Eq. (4-20), each of these quanta will have an energy $h\nu$. If slit 2 is closed, a photoelectric cell will measure a distribution of light $A1$, similar to that obtained with the sand when only slit 1 was open. As with the sand, distribution $A2$ will be observed when slit 2 is open and 1 is closed. If slits 1 and 2 are opened *consecutively*, distribution $A(1 + 2)$ will again be observed. If, on the other

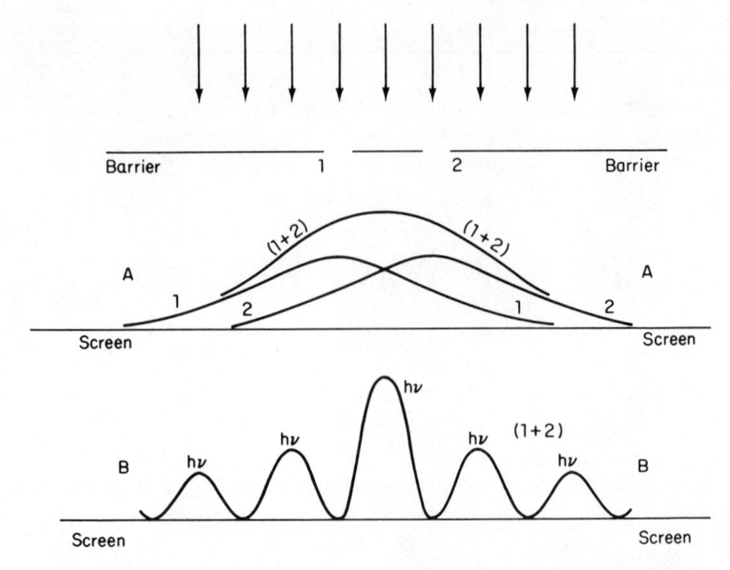

Figure 5-1. Passage of particles through a pair of slits. (A) The pattern formed when the two streams of particles combine by simple addition. (B) The pattern due to the interference of waves whose wavelength is comparable to the separation of the two slits.

hand, the two slits are opened *simultaneously*, the distribution will be given by the complex curve shown in $B(1 + 2)$ rather than that representing simple summation. Some points, well within the previously illuminated area, will now be dark. Other points, originally outside the lighted area, will now be bright, although the intensity will be less than that at the central spot.

It seems impossible to explain an interference pattern of type *B* by postulating that electromagnetic radiation consists of a series of particles. However, the individual photon energies are indivisible and individually non-additive, as would be expected of wave amplitudes. The photocell will record that every photon in pattern *B* will have exactly the energy of those in the initial beam.

According to a wave model, an in-phase wavefront strikes the two slits and from these slits two spherical wavefronts spread out behind the barrier. The central maximum in *B* is formed when the two path lengths from 1 and 2 are equal, and the two spherical wavefronts meet in phase. The first two minima occur when the path lengths differ by $\lambda/2$, and so on. This explanation does serve to locate correctly the maxima and minima, but it leaves some pertinent questions unanswered.

For example, it is not clear how the energy $h\nu$, presumably spread over a hemispherical wavefront, can suddenly be collected at one point when it meets in phase with another, similarly extended energy distribution. If, on the other hand, we assume that a photon with all of its energy passes through one slit, it is not clear how this photon knows whether the second slit is open or closed. Somehow this must be so, since we observe pattern *B* and not *A*. Again, we can leave both slits open and reduce the intensity of the incident light to the point where there will usually be only a single photon at a time traveling between the barrier and the screen. Mutual interference between the two slits seems impossible, yet pattern *B* will be observed with both slits open simultaneously. Since we must regard pattern *B* as an interference pattern, we can only conclude that the photon is in some way capable of interfering with itself. Even this explanation does not satisfy the requirements of the reduced-intensity experiment, where almost all photon passages will be consecutive rather than simultaneous.

Our troubles are compounded when we replace the light beam with a beam of monoenergetic electrons. Innumerable measurements have identified electrons as particles, but with two slits we observe pattern *B* and not *A*12. It is obvious that we are dealing with entities whose exact nature we do not understand, but which exhibit some of the properties of waves and also those of particles. There now seems good reason to include the photon in any list of "fundamental particles" rather than to treat it as something apart.

5.02 X-ray Diffraction

The penetrating radiation discovered by Roentgen in 1895 was named "X" because for some time its nature was not known. Electromagnetic waves were suspected, but many experiments designed to demonstrate wave properties failed, because of the very short wavelengths involved. As we shall see, wave interferences can be detected only when the separation of the slits, or other diffracting centers, is comparable to the incident wavelengths. Once the short wavelengths of X rays were recognized, interference phenomena were quickly detected.

Diffraction of X rays can be demonstrated by replacing the diffraction grating of visible-light optics with a diffracting crystal. Each atom in the crystal will serve as a scattering or diffracting center, as at A_1 and B_1, Fig. 5-2. When the difference in path length CB_1D is equal to an integral number of wavelengths, constructive interference will result and a diffracted beam will be formed. From the figure, the requirement for constructive interference is

$$n\lambda = 2d \sin \theta \tag{5-1}$$

Many crystals have interatomic spacings comparable to X-ray wavelengths and serve to produce diffracted beams at easily measured angles.

X-ray diffraction was first demonstrated by von Laue, who placed a

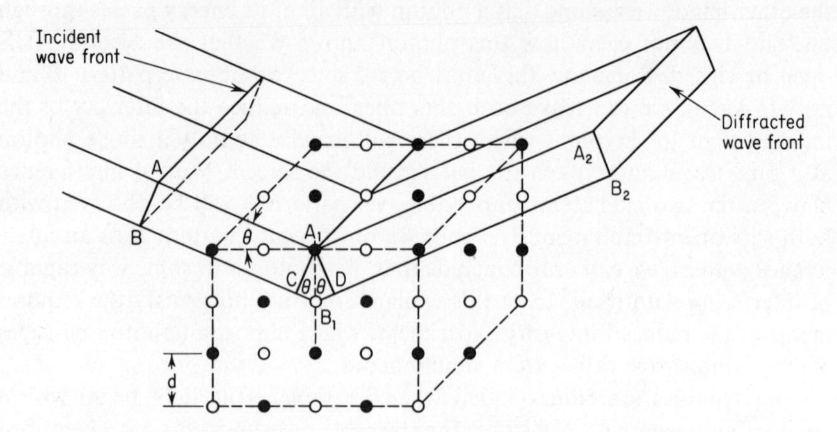

Figure 5-2. Interference of X-ray photons diffracted at the atoms in a crystal lattice.

Figure 5-3. A Laue-type interference pattern produced by X rays.

photographic plate behind a crystal carefully aligned in the beam. The three-dimensional crystal lattice produces a two-dimensional diffraction pattern, Fig. 5-3. Later, W. H. Bragg mounted a crystal on a spectrometer and arranged an ionization chamber to maintain the angular relations, shown in Fig. 5-4, as the angle of incidence was varied.

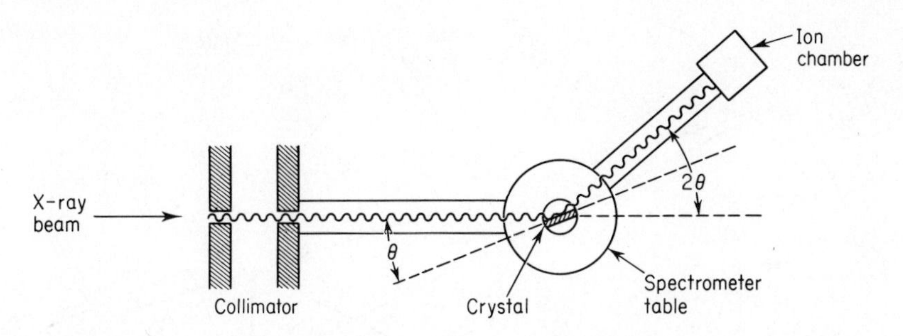

Figure 5-4. Angular relations in an X-ray spectrometer for constructive interference.

Interatomic spacings for simple structures such as cubic NaCl or KCl can be calculated from the crystal densities and from these X-ray wavelengths can be determined. With the wavelengths known, X-ray diffraction can be used to determine the interatomic spacings in more complex structures. In 1930, DuMond and Kirkpatrick introduced the use of *bent crystals*, an arrangement now used universally for the high-precision measurement of X- and gamma-ray wavelengths. A bent crystal is usually a thin slab of quartz cut into a portion of a cylinder of radius $2R$. In mounting, the crystal is deformed to radius R, thus stretching the outer layers of atoms and shrinking the inner, Fig. 5-5. The bent-crystal mounting eliminates some of the corrections that had to be applied when a plane diffracting crystal was used. Figure 5-6 shows one method of using a bent crystal. When the source is moved along an arc of radius R, the diffracted beam appears to come from a virtual source V.

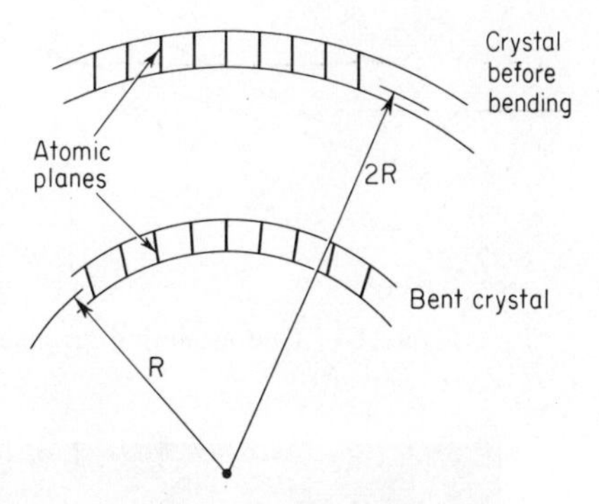

Figure 5-5. Deformation of the diffracting crystal in a bent-crystal spectrometer.

5.03 Electron Diffraction

It is common practice to use finely powdered crystals for X-ray diffraction studies instead of the carefully oriented single crystal required in the von Laue technique. Because of the random orientation of the small crystal grains, the individual spots seen in a von Laue photograph are spread out in concentric rings around the central beam, Fig. 5-7A. Interatomic spacings are calculated from the diameters of the diffraction rings.

An electron beam incident upon a thin foil produces a circular diffraction

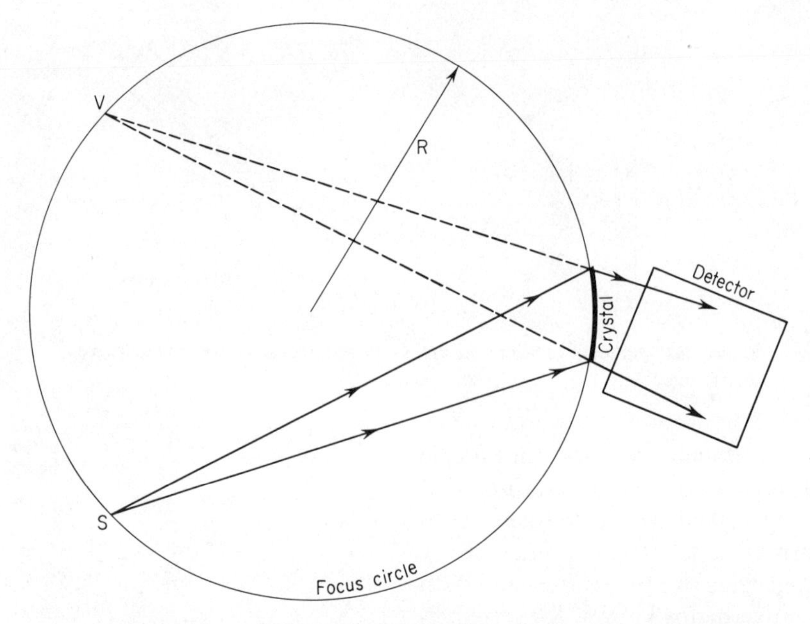

Figure 5-6. One mounting arrangement for a bent-crystal spectrometer.

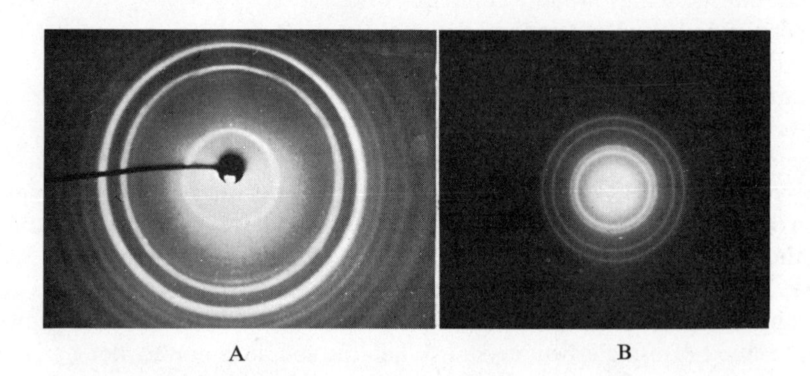

Figure 5-7. Crystal diffraction patterns. (A) X-ray diffraction by valeramide. (B) Diffraction of electrons by gold.

pattern with all of the characteristics of the powder X-ray pattern, Fig. 5-7B. Patterns of this type point inescapably to the wave nature of electrons. As with X rays, electron diffraction by crystals is readily demonstrated because the interatomic spacings are comparable to the wavelengths. Electron diffraction by slits as proposed in Sec. 5.01 occurs in principle but cannot be demonstrated because of the disparity of dimensions.

A series of measurements by Davisson and Germer, and by G. P.

Thomson, showed that electron wavelengths were accurately given by a relation proposed on theoretical grounds by de Broglie in 1925:

$$\lambda = \frac{h}{mv} \tag{5-2}$$

Electron diffraction was soon put to practical use in the electron microscope, which has a resolution limit far superior to that obtainable with optical microscopy. Diffraction effects limit the spatial resolution of any microscope to the separation of objects spaced at about the distance of one wavelength of the light being used. Thus a conventional microscope might resolve particles about 5×10^{-5} cm apart, which is the wavelength of green light. The ultraviolet microscope might improve the resolution by a factor of not more than 2. X-ray microscopy is little used because of the difficulty of obtaining monochromatic beams and because of the presence of incoherent scattering, Sec. 12.04.

Many electron microscopes operate at energies of 50–100 KeV. Resolutions of only a few angstrom units are obtained but examinations are limited to very thin sections because of the low penetration of the electrons. Energies of 1 MeV or more are now available, and correspondingly thicker sections can be examined.

Neutron diffraction has been demonstrated and has become a very powerful tool for the location of hydrogen atoms in crystal lattices. Diffraction has been observed for a number of particles such as protons, and helium and neon atoms. All demonstrate wave properties in accord with the predictions of the de Broglie wavelength relation.

5.04 Group and Phase Velocity

It is natural to inquire into the nature of the waves described by the de Broglie relation. They have been called "matter waves" but this tells nothing about the quantity whose amplitude would determine the energy content in a more conventional form of wave motion. They are certainly not electromagnetic waves, and with that negative statement we end the discussion of their nature.

Whatever their nature, these waves cannot be described by a relation of the type $A = A_0 \sin (kx - \omega t)$ or the more general $A = A_0 e^{i(kx-\omega t)}$. Both of these equations describe a wave that is moving along the x-axis. If the energy is to remain constant, A_0 must be constant and we have a wave of constant maximum amplitude extending without limit in either direction, Fig. 5-8A. Such a wave can scarcely describe a particle which is certainly localized to some extent, even though this localization may be a little indefinite.

Consider next a composite wave, made up of a bundle of constant-ampli-

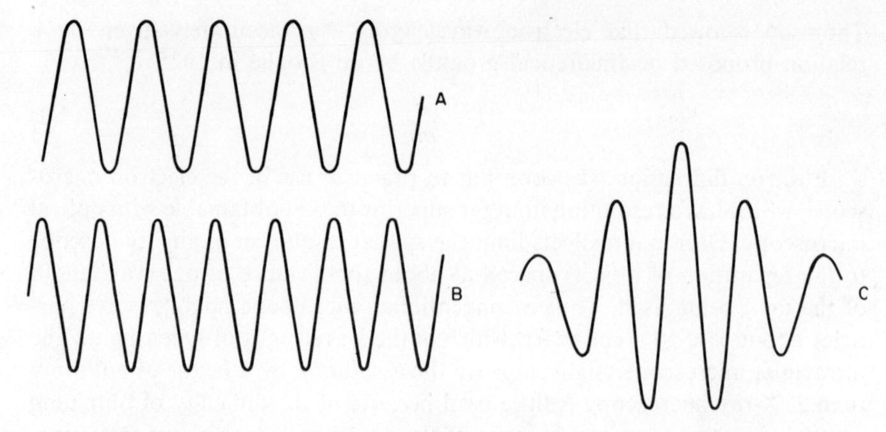

Figure 5-8. (A) A wave of constant amplitude extends indefinitely along the direction of propagation. A second wave of slightly different frequency, (B), will interfere with (A) to produce a space-restricted pattern, (C).

tude waves, each with a unique frequency slightly different from that of its fellows. Let these waves be propagated in a dispersive medium, which is to say that the velocity of propagation will be a function of the frequency. Figure 5-8C shows the situation at a particular instant of time for a bundle of two components, *A* and *B*. When the component amplitudes are added, a variable-amplitude or modulated wave will result. The position of maximum amplitude moves as the component waves progress, but with a quite different velocity.

The *group velocity v*, which is the velocity of the point of maximum amplitude, must be distinguished from the *phase velocity u*, which is the propagation velocity of one of the component waves. It can be shown that the group velocity is to be associated with the particle velocity in the classical sense. The two velocities are related by

$$v = \frac{c^2}{u} \tag{5-3}$$

and if $v < c$, it is necessary that $u > c$. The phase waves will have a velocity greater than the velocity of light but can carry no energy.

By combining several constant-amplitude waves, we have helped to localize the moving particle, but its exact location is still indefinite, and must remain so. The shape of the wave envelope, or the low-frequency modulation, depends upon the frequency range that we choose to incorporate into the wave bundle. If this frequency range is large, the successive amplitude maxima decrease rapidly and we have a good localization. On the other hand, if the frequency range is small, the amplitude decreases slowly and we have poor localization.

There are no criteria for selecting a frequency range and so we must consider various possibilities. It will be noted that as localization is improved by selecting a wide frequency range, information on the particle velocity becomes poorer because of the wider velocity spread in the component waves. Conversely, a poorer localization will be accompanied by a more precise knowledge of the velocity.

5.05 The Indeterminacy Principle

The discussion of the previous section led to a conclusion that was put into a more quantitative form by Heisenberg in 1927. According to classical mechanics, a simultaneous determination of position and momentum is sufficient to fix the state of a system, whose future behavior will then be governed by the force fields in which it moves. The indeterminacy principle states that it is impossible to simultaneously measure the two conjugate quantities with absolute precision. Any attempt at such a measurement will be subjected to the constraint of

$$\Delta p \, \Delta q \simeq \frac{h}{2\pi} = \hbar \tag{5-4}$$

where the Δ's are to be read as "the uncertainty in."

According to the Heisenberg relation, either variable may be known, in principle, with any desired precision, but any increase in the precision of one necessarily causes a loss of precision of the other. Note carefully that these uncertainties are the result of the basic nature of things, and not because of inadequate measuring equipment. The principle may also be written as

$$\Delta E \, \Delta t \simeq \hbar \tag{5-5}$$

where t is the time that a system will remain in an energy state E. Since according to the indeterminacy principle the initial state of a system cannot be specified exactly, it is obviously impossible to know its future behavior with certainty. As would be expected, this conclusion has far-reaching consequences.

Consider again the double-slit experiment, redrawn in Fig. 5-9, and let us seek to determine whether a particular

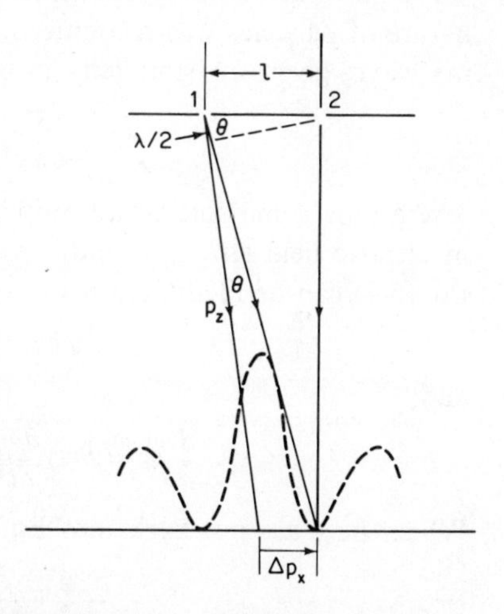

Figure 5-9. Geometrical relations in a double-slit experiment.

particle passes through slit 1 or 2. A particle moving toward the screen with momentum p_z will have some component of momentum p_x that will cause it to strike at one of the maxima in the diffraction pattern. Suppose now that a particle detector is placed below slit 1 to indicate the passage of a particle. We must consider the possibility that the act of detection will alter the particle momentum sufficiently to produce a change in the diffraction pattern. If the original pattern is to be preserved, the change in p_x, or Δp_x, must be less than the amount needed to shift the particle from a maximum to an adjoining minimum. From Fig. 5-9 this requires that $\Delta p_x/p_z \ll \theta$.

The requirements of constructive and destructive interference lead to $\theta = \lambda/2l$, and from Eq. (5-2) the wavelength–momentum relation is $p_z = h/\lambda$. Finally, the knowledge that a particle passes through slit 1 means that the uncertainty in the x-position, $\Delta x < l/2$. When these relations are combined and rearranged, we have

$$\Delta p_x \, \Delta x \ll \frac{h}{4} \tag{5-6}$$

The inequality expressed by (5-6) is a serious violation of the Heisenberg principle, and so it appears that we have asked a question to which there is no answer. Nature is so arranged that it is impossible to identify, with certainty, the slit through which a particle passes.

5.06 The Schrödinger Equation

Wave motions of many kinds had been treated in detail long before the wave nature of particles was recognized. The general equation describing a traveling wave, given for simplicity in one dimension, is

$$\frac{\partial^2 \psi}{\partial x^2} = \frac{1}{w^2} \frac{\partial^2 \psi}{\partial t^2} \tag{5-7}$$

where ψ is some sort of an amplitude such as displacement, overpressure, or electric field strength, and w is the velocity of propagation. Solution of Eq. (5-7) can be facilitated if the variables are separable. Then,

$$\left.\begin{aligned} &\psi(x, t) = \phi(x)\theta(t) \\ &\text{and} \\ &\frac{\partial^2 \psi}{\partial x^2} = \theta(t)\frac{d^2\phi}{dt^2} \qquad \frac{\partial^2 \psi}{\partial t^2} = \phi(x)\frac{d^2\theta}{dt^2} \end{aligned}\right\} \tag{5-8}$$

When these are put back into Eq. (5-7),

$$\frac{d^2\phi}{dx^2} \cdot \frac{1}{\phi(x)} = \frac{1}{w^2\theta(t)} \frac{d^2\theta}{dt^2} \tag{5-9}$$

If these separated variables are to be equal for all values of x and t, they must be equal to a constant known as the *separation constant*, conveniently taken as $-k^2$. This leads to two expressions:

$$\frac{d^2\phi}{dx^2} = -k^2\phi(x) \qquad \frac{d^2\theta}{dt^2} = -k^2w^2\theta(t) \tag{5-10}$$

These are two well-known equations with solutions in sin, cos, or exponential form. For the latter,

$$\phi(x) = e^{-ikx} \qquad \theta(t) = e^{-ikwt} \tag{5-11}$$

Now, $\phi(x)$ will be periodic when kx changes by 2π, and so there is a wavelength $\lambda = 2\pi/k$. Then,

$$\theta(t) = e^{-iw(2\pi/\lambda)t} = e^{-i2\pi\nu t} \tag{5-12}$$

where ν is the frequency of the wave motion. Differentiating and substituting into Eq. (5-7) gives

$$\frac{d^2\phi}{dt^2} + \frac{4\pi^2}{\lambda^2}\phi = 0 \tag{5-13}$$

Equation (5-13) is just another form of the classical wave equation. It is now time-independent, so it no longer refers to a traveling wave but rather to the spatial variation of a standing-wave pattern. Equation (5-13) will, in fact, give the maximum amplitudes of the standing-wave patterns as a function of the space variable x.

Quantum-mechnical concepts first appear when the wavelength in Eq. (5-13) is replaced by its wave-mechanical equivalent, h/p. Then Eq. (5-13) becomes

$$\frac{d^2\phi}{dx^2} + \frac{4\pi^2p^2}{h^2}\phi = 0 \tag{5-14}$$

If the system is *conservative*, total energy W will remain constant and equal to the sum of the potential and kinetic components. In the nonrelativistic domain,

$$W = \frac{p^2}{2m} + U \tag{5-15}$$

where U is the potential energy. Finally,

$$\frac{d^2\phi}{dx^2} + \frac{8\pi^2m(W-U)}{h^2}\phi = 0 \tag{5-16}$$

Equation (5-16) is the time-independent form of Schrödinger's equation in one dimension. In three dimensions the form is

$$\nabla^2\phi + \frac{8\pi^2m(W-U)}{h^2}\phi = 0 \tag{5-17}$$

where ∇^2 is the Laplacian differential operator

$$\frac{\partial^2}{\partial x^2} + \frac{\partial^2}{\partial y^2} + \frac{\partial^2}{\partial z^2}$$

The application of the Schrödinger equation to physical problems consists of putting in the proper values for U and then finding expressions for ϕ which satisfy the equation. In general there will be a series of solutions, known as *wave functions* or *eigenfunctions*, for various values of the total energy W. Solutions of Eq. (5-16) will be complex, with a real and an imaginary part, and with negative values. Such solutions cannot represent a real particle which must be real and positive at all times and places. However, the product of any value of ϕ with its complex conjugate ϕ^* will be real and the absolute value of the product will always be positive. The absolute value of the product, properly normalized so that the integral over all possible states is unity, is to be interpreted as the probability that the particle is located in a small volume $dx\,dy\,dz$ at position xyz.

This interpretation of the meaning of the wave functions is arbitrary, and its validity must be judged from its ability to deal properly with physical problems. In fact, the success of the wave-mechanical treatment has been so phenomenal that there can be little doubt of its validity.

5.07 The Linear Oscillator

Except for the very simplest systems, the solution of the Schrödinger equation requires the use of advanced and specialized mathematics. Detailed solutions are beyond the scope of this text, but some results will be given and some important conclusions pointed out.

Certain constraints must be placed on the wave functions if they are to represent a real, observable particle or system. Specifically, each wave function must be continuous, single-valued, finite at all points, and must vanish at infinity. As with any other differential equation, the general solution will be the sum of all possible special solutions. We anticipate, therefore, a series of wave functions, each probably valid for only one value of W. Thus we suspect a series of "quantized" wave functions and energy values even though each function itself must be continuous throughout space.

Consider again the linear harmonic oscillator treated by the old quantum theory in Sec. 4.13. If the restoring force is, as before, $F = -kx$, the potential energy will be $U = kx^2/2$ and Eq. (5-16) becomes

$$\frac{\hbar^2}{2m}\frac{d^2\phi_n}{dx^2} + \left(W_n - \frac{kx^2}{2}\right)\phi_n = 0 \tag{5-18}$$

where ϕ_n and W_n are written with subscripts to allow for a series of solutions. The detailed solution for even this simple system is too involved to present here. Only the essential results will be given.

The only acceptable solutions are those for which the energy of the system is given by

$$W = \left(n + \frac{1}{2}\right)\hbar\left(\frac{k}{m}\right)^{1/2} = \left(n + \frac{1}{2}\right)\hbar\omega \tag{5-19}$$

where n is a running integer 1,2,3, ... Each corresponding wave function consists of an exponential term of the form seen in the normal probability distribution, multiplied by one of the series known as the Hermite polynomials. In terms of a parameter $\alpha = \sqrt{km}/h$, the first few wave functions are

$$\left.\begin{aligned}
\phi_0 &= \frac{\alpha^{1/2}}{\pi^{1/4}}e^{-\alpha^2x^2/2} \\
\phi_1 &= \frac{\alpha^{1/2}}{2^{1/2}\pi^{1/4}}(2\alpha x)e^{-\alpha^2x^2/2} \\
\phi_2 &= \frac{\alpha^{1/2}}{8^{1/2}\pi^{1/4}}(4\alpha^2x^2 - 2)e^{-\alpha^2x^2/2} \\
\phi_3 &= \frac{\alpha^{1/2}}{48^{1/2}\pi^{1/4}}(8\alpha^3x^3 - 12\alpha x)e^{-\alpha^2x^2/2}
\end{aligned}\right\} \tag{5-20}$$

Let us first compare the energy states predicted by the wave mechanics with those of the quantum theory as given by Eq. (4-31). In the first place, the integer n, which was originally introduced rather artifically, now appears naturally as the term index in the polynomial solution of the Schrödinger equation. The second notable difference between Eqs. (5-19) and (4-31) is the replacement of n by $(n + \frac{1}{2})$. From the wave mechanics there is, then, an energy $h\nu/2$ even in the lowest or zero energy state. This result is in accord with the requirements of the indeterminacy principle, which forbids an exact value of energy under any condition. The spacings between the energy levels will be the same in the two theories, with each level predicted by the wave mechanics higher by $h\nu/2$.

Wave functions for the first three energy levels of the harmonic oscillator are plotted in Fig. 5-10, together with the wave mechanical probabilities $P_W = |\phi_n\phi_n^*|$ and the probability function P_c for the classical oscillator. The latter probability can be obtained from Eq. (4-26), replacing the amplitude a by the form required by the quantum restrictions. Then,

$$x = a\cos\omega t = \left(\frac{4n^2\hbar^2}{km}\right)^{1/4}\cos\omega t \tag{5-21}$$

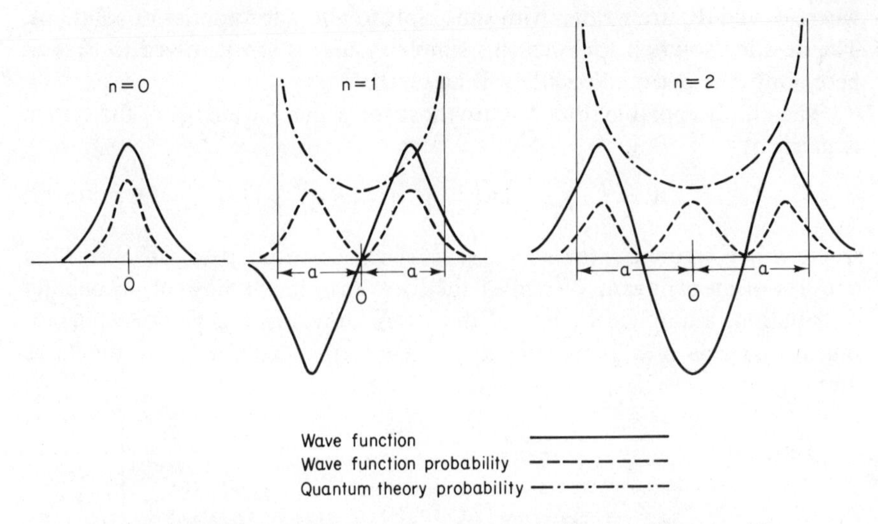

Figure 5-10. The first three wave functions for the linear oscillator, and the square of the functions, compared to the probabilities according to the classical quantization.

The probability of finding the particle in a small segment dx will be proportional to the time spent in the segment, which, in turn, will be inversely proportional to the particle velocity. Then,

$$P_c = \left|\frac{1}{dx/dt}\right| = \frac{1}{\omega}\left(\frac{km}{4n^2\hbar^2}\right)^{1/4} \sin \omega t \tag{5-22}$$

P_c will have its minimum value at the equilibrium point, where the velocity is the greatest, and will increase rapidly as x approaches its limits $\pm a$.

There is no classical counterpart for $n = 0$. The shape of the wave function is that of the normal probability exponential. Since the integral of P_W over all values of x must be normalized to unity, the peak of the P_W-curve will be less than unity, and so this curve will lie below the ϕ-plot. In spite of the fact that the system for $n = 0$ has zero energy, there is a small but finite probability that the particle will be found at a distance from the point of equilibrium.

When $n = 1$, the wave function becomes *odd*, or unsymmetrical about $x = 0$, because of the presence of a term in x. P_W will be positive, but surprisingly the probability that the particle will be at the equilibrium point is zero. Another surprising and important fact is that the particle is no longer limited by amplitude a. There is a small but finite probability that the particle will be found outside of a, an impossiblity under classical conditions.

An *even* or symmetrical wave function corresponds to $n = 2$. Again, there is a finite probability that the particle will be found outside of the classical amplitude. It is already evident at $n = 2$ that the wave-mechanical probability tends to approach the classical. This tendency increases with n but there will always be one distinct difference; wave mechanics always predicts some points of zero probability, where the particle is never to be found.

5.08 Parity

The alternating pattern of odd and even wave functions, seen for $n = 1$ and 2, continues for higher values of n. An even function, in which $\phi_n(x) = \phi_n(-x)$, is said to have an even or positive *parity*. An odd function, where $\phi_n(x) = -\phi_n(-x)$, has odd or negative parity.

Parity is a concept completely foreign to classical mechanics. It has become a most important parameter in specifying the state of a system. Parity has only the two possible values $+1$ and -1, in contrast to some of the other state parameters, such as n, above, which may take on a wide range of values.

Parity is sufficiently important to be included in the quantities that are thought to be conserved in nuclear reactions along with such basic parameters as energy and angular momentum. When a small violation of parity conservation was detected, Sec. 11.10, a detailed reappraisal of all conservation laws was initiated.

REFERENCES

Evans, R. D., *The Atomic Nucleus.* International Series in Pure and Applied Physics. McGraw-Hill, New York, 1955.

Fano, U. and L. Fano, *Basic Physics of Atoms and Molecules.* Wiley, New York, 1959.

Feynmann, R. P., R. B. Leighton, and M. Sands, *The Feynmann Lectures on Physics.* Addison-Wesley, Reading, Mass., 1965.

Schiff, L. I., *Quantum Mechanics.* International Series in Physics. McGraw-Hill, New York, 1955.

PROBLEMS

5-1. A sodium chloride crystal, molecular weight 58.45 and density 2.163 g cm^{-3}, is used to diffract the K_α line of copper, $\lambda = 1.54$ Å. Calculate the grating constant of the crystal and the angle at which the first-order diffraction maximum will be observed.

5-2. The grating constant for a crystal of CsCl is 2.06 Å. A first-order K_α maximum from an unknown element is observed at an angle Of 7°51′. What is the wavelength of the K_α emission? Identify the element in question.

5-3. A beam of monochromatic X rays, $\lambda = 0.213$ Å, is used in diffraction studies. What is the quantum energy of these photons in KeV? What voltage must be applied to an electron microscope to provide electrons of the same wavelength? What is the velocity of these electrons?

5-4. Electron microscopes designed for the study of thick specimens operate at a potential of 2 MV. What is the wavelength of these electrons? What is their mass? What is the quantum energy of photons of the same wavelength?

5-5. X rays have been produced in a linear accelerator by the impact of 2-GeV electrons on a target. Compare the wavelengths of the particles just before and just after the impact.

5-6. The maximum indeterminacy of a momentum cannot be greater than the momentum itself. Assume the limiting case and calculate the minimum uncertainty in position for the electron in Prob. 5-4.

5-7. Equation (5-5) permits a temporary violation of energy conservation provided that the violation is corrected within time Δt. How long can an energy violation of 10^{-3} eV be tolerated?

5-8. It is desired to repeat the double-slit experiment using a beam of 100-eV electrons instead of photons. What slit separation will be required if the limiting angle of observation for the first diffraction maximum is 2°?

5-9. Interference patterns were not seen with the sand experiment of Sec. 5.01 because the wavelengths of the particles were incompatible with the dimensions of the slit system. What is the wavelength of a 2 mg sand particle falling with a velocity of 85 cm sec^{-1}?

5-10. Calculate the momentum of the electrons in the linac of Prob. 5-5 and the radius of curvature when they are deflected by a magnetic field of 15,000 gauss. What is the momentum of the X-ray photon produced by the entire energy conversion of one of the electrons?

6

Atomic Structure

6.01 The Rutherford Scattering Experiment

The results of many early measurements agreed in showing that atoms have dimensions of the order of 10^{-8} cm, but no real progress was made in understanding the structure inside this distance until the Rutherford scattering experiments in 1911. Rutherford reasoned that charged particles, used as projectiles or probes against atomic targets, would be deflected or scattered by the atomic fields. The number of particles deflected, and their angular distribution, should serve to determine the strength and the spatial configuration of the field.

The only projectiles available to Rutherford were the alpha particles emitted by naturally radioactive nuclides such as radium. When these particles are directed against a heavy target, such as an atom of gold, the latter can be considered to be stationary throughout the encounter. Rutherford conceived of the atom as a small central nucleus containing a positive charge Z times the fundamental charge, and almost all of the atomic mass. A cloud of electrons was supposed to surround the nucleus, but their fields would have little effect on the path of the relatively massive alpha particle. Rutherford made the further assumption that the electric field produced by the nuclear charge was a coulomb or inverse-square law field.

Under these assumptions the force of repulsion between the two positive charges will cause the alpha particle to move along a hyperbolic path, Fig. 6-1, with the heavy nucleus at the outer focus of the hyperbola. In general, the heavy target nucleus will recoil, and the trajectory of the light particle will be hyperbolic only in a center-of-mass system whose origin is the moving center of mass of the two particles.

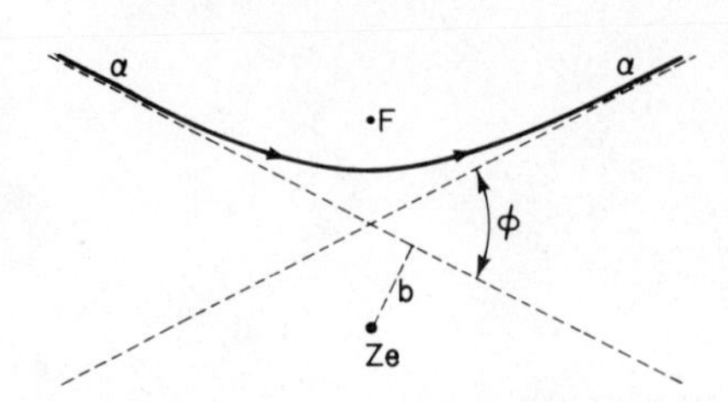

Figure 6-1. A concentrated positive charge Ze causes an approaching alpha particle to move along a hyperbolic path whose outer focus is located at the positive charge.

In the case under consideration, the scattering angle ϕ will be a function of the distance b, the *impact parameter*, which is the perpendicular distance from the scattering center to the original trajectory extended. The relation found by Rutherford is

$$b = \frac{Zze^2}{2T} \cot \frac{\phi}{2} \tag{6-1}$$

where Ze = nuclear charge
ze = charge on the approaching particle
T = kinetic energy of the particle

The angular distribution of the scattered particles is given by

$$n_\phi = n_0 N \left(\frac{Zze^2}{4T}\right)^2 \csc^4 \frac{\phi}{2} \tag{6-2}$$

where n_ϕ = particles per unit time and per unit angle scattered through angle ϕ
n_0 = number of incident particles per unit time
N = number of target atoms cm^{-2}

Both of these relations were developed on the assumption that the incoming particles were nonrelativistic.

Rutherford used the arrangement shown schematically in Fig. 6-2 to test the validity of his predictions. Collimated alpha particles were directed at a scattering foil made so thin that most of the alpha particles underwent only a single scatter on their way through. A zinc sulphide screen and microscope were arranged for viewing the scintillations at various scattering angles. A careful series of measurements confirmed Eq. (6-2) in almost every respect. The only discrepancy came from particles with small values of the impact parameter, which, as we now know, approach so close to the nucleus that they encounter a second, non-coulomb field of force.

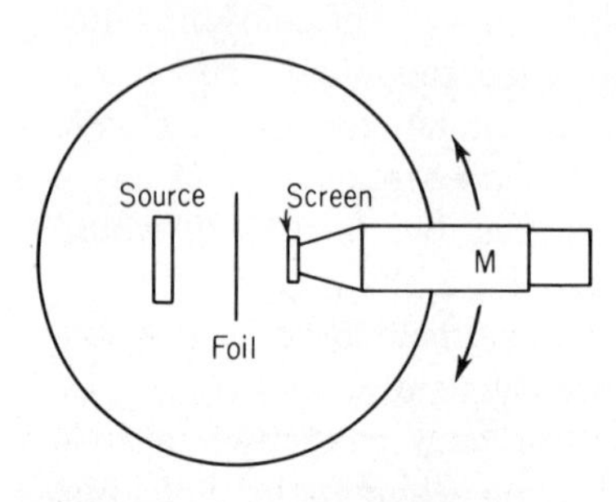

Figure 6-2. Source–microscope arrangement used by Rutherford to study the angular distribution of the scattered alpha particles.

Specifically, the measurements showed that the number of particles scattered through an angle ϕ will be

a. inversely proportional to v^4, where v is the initial velocity;

b. directly proportional to Z^2;
c. directly proportional to $\csc^4 (\phi/2)$;
d. directly proportional to foil thickness.

Values of Z were obtained for a series of scattering elements ranging from carbon to gold. Over a half-century later, the first positive identification of chemical elements on the surface of the moon was made by a Surveyor spacecraft, using the Rutherford alpha-particle scattering technique with modern pulse counters replacing the visual observations.

The distance of closest approach of a particle to a nucleus is simply calculated. When the impact parameter $b = 0$, the incoming particle will be heading directly toward the nucleus. It will continue to approach until all of its original kinetic energy has been expended in entering the potential field of the nucleus. This potential is Ze/d, so at the point of minimum separation,

$$\frac{mv^2}{2} = \frac{Zze^2}{d_{\min}} \tag{6-3}$$

Scattering studies have become the most important tool for determining the fields of force that exist inside atomic nuclei. In the general case, when projectile and target masses may be comparable, it is usually convenient to develop the mathematical relations of the collision in the *center-of-mass* coordinate system. Although this system has a moving origin, collision calculations are simplified by the fact that in it the total momentum is always zero. After the details of the collision have been worked out, they can be brought into the observational or laboratory system by the vector addition of the relative motion between the two systems.

6.02 Nuclear Constituents

The results of the Rutherford experiments suggested strongly that the nucleus of an atom of atomic number Z contains Z protons. This suggestion is obviously incomplete, because the proton masses will account for only about one-half of the atomic masses as determined from chemical atomic weights. To meet this difficulty it was assumed that the nucleus contains enough protons to account for the required atomic weight, combined in some unknown fashion with enough electrons to reduce the net charge to Z units.

We now have a mass of evidence against the existence of electrons inside the nucleus, Sec. 11.01. It is sufficient to state here that for any reasonable electron energy, the wavelength, calculated from Eq. (5-2), is far too great to be contained in a nucleus whose diameter is known to be of the order of 10^{-13} cm. We now know that the nucleus is made up of neutrons and protons, in combinations that satisfy both mass and charge requirements.

6.03 Atomic Spectra

For years before the Rutherford experiments, spectroscopists had been measuring the wavelengths of the visible and ultraviolet light emitted by atoms when they are excited by electric sparks or arcs. Each atomic emission spectrum consists of a series of spectral lines whose groupings seemed to show no regularities. Oscillations with harmonic overtones were known, but the spectral lines showed no evidence of any such relation, Fig. 6-3.

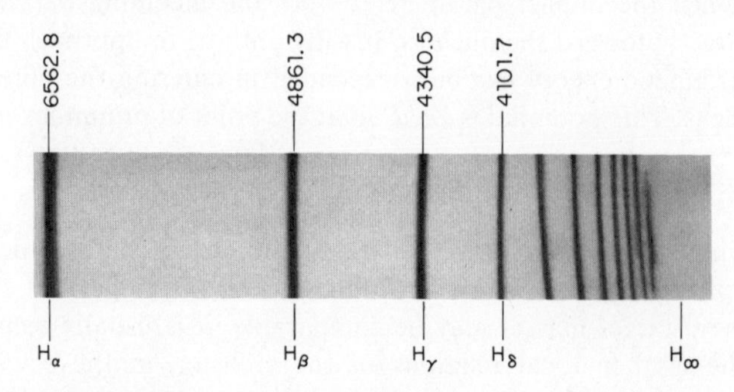

Figure 6-3. A portion of the emission spectrum of atomic hydrogen. The lines converge toward a short-wavelength limit, with no evidence of harmonic overtones. Wavelengths are in angstroms.

From gross observations of line appearance and persistence, four groups of lines—known as the sharp, principal, diffuse, and fundamental—had been identified. In the case of atomic hydrogen it was known that some of the emission lines formed a series whose wavelengths were given by

$$\frac{1}{\lambda} = R\left(\frac{1}{2^2} - \frac{1}{M^2}\right) \tag{6-4}$$

where M took on integral values 3, 4, 5, ...

The quantity $1/\lambda$ is the wave number $\tilde{\nu}$. R is known as the Rydberg constant and has a value of 109,677.6 cm^{-1} for hydrogen. Most of the lines described by Eq. (6-4) lie in the visible region of the spectrum and are known as the *Balmer series*.

In the ultraviolet region of the hydrogen spectrum there is a second group of lines called the *Lyman series*. The wavelengths of the components of this series are given by

$$\frac{1}{\lambda} = R\left(\frac{1}{1^2} - \frac{1}{M^2}\right) \tag{6-5}$$

where $M = 2, 3, 4, \ldots$. Other series of lines, all in the infrared, are recognized, and in each case the wavelengths are given by expressions having the form of Eqs. (6-4) and (6-5) with different values of the denominators inside the brackets.

The wavelengths of the lines in the various series have been measured with great precision. Measurements, and calculations from the empirical equations give a series of values which converge toward a *series limit*, obtained by setting $M = \infty$ in the equations. The series limits, expressed as wave numbers, form a series of *terms* and lists of these limits are known as *term tables*. According to the Ritz combination law, the emission of a spectral line is the result of a change from one term value to another, and the wave number emitted can be obtained by a simple subtraction of two term values.

Energy is lost by an atom when it emits a spectral line and the Ritz combination law strongly suggests that each series limit represents an atomic energy level. Transitions between these levels would then result in the radiation of the observed amounts of energy. It is important to note that this model calls for a number of discrete energy levels rather than a continuous distribution of energies.

6.04 The Bohr Atom

The Rutherford model of atomic structure did not explain the emission of spectral lines, nor, indeed, the stability of the proton–electron structure. In hydrogen, for example, a stationary electron would be promptly pulled into the nucleus by the attractive electric force. This could be prevented by assuming the electron to rotate about the nucleus. Such a rotation requires a central force which would be supplied by the electric attraction. According to classical electrodynamics any accelerating electric charge must radiate energy, and this requires that the rotating electron spiral in to the nucleus, radiating a continuum rather than discrete lines. Neither of these predictions is in accord with observation.

In 1913, the Danish scientist Niels Bohr proposed a solution to this problem by applying Planck's quantum hypothesis to an atomic system. As pointed out in Sec. 4.10, Planck assumed that electromagnetic energy is emitted in quanta of energy E given by the relation

$$E = h\nu$$

Bohr saw the relation between the Ritz combination principle and Planck's equation and proposed the relation

$$h\nu = E_1 - E_2 \tag{6-6}$$

where ν = frequency of the emitted electromagnetic radiation

E_1 = initial energy of the atom (prior to emission)
E_2 = final energy of the atom (after emission)

Since $\lambda\nu = c$, where c is the velocity of light, this equation may be written

$$\tilde{\nu} = \frac{1}{\lambda} = \frac{E_1}{hc} - \frac{E_2}{hc} \tag{6-7}$$

Equation (6-7) implies the existence of a set of *stationary states*, each having a definite energy value. Radiation is emitted from an atom only when it changes from one stationary state to another of lower energy. An atom may remain in a stationary state indefinitely, emitting no radiation.

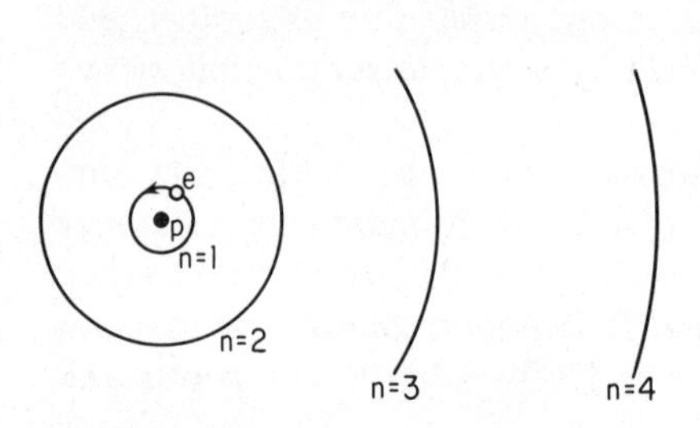

Figure 6-4. The Bohr concept of the hydrogen atom, with the electron quantized to move only in a series of specific orbits.

According to the Bohr model, the atom appears as in Fig. 6-4. The electron revolves about the nucleus, assumed to be stationary, in one of a set of discrete orbits which correspond to the stationary energy states of the structure. Bohr postulated that the only orbits that were allowed were those for which angular momentum was quantized in a fashion analogous to the quantization of linear momentum in the harmonic oscillator. The Bohr quantization rule was

$$mav = ma^2\omega = \frac{nh}{2\pi} \tag{6-8}$$

Here a is the radius of an allowed circular orbit and n is a running integer 1, 2, 3, ... Nonrelativistic velocities are assumed throughout.

The Bohr quantization rule is another example of the general quantum requirement

$$\oint p\, dq = nh$$

if q = angular momentum and p = angular velocity. Then,

$$\oint p\, dq = \int_0^{2\pi/\omega} ma^2\omega^2\, dt = nh \tag{6-9}$$

Integrating,

$$ma^2\omega = \frac{nh}{2\pi} = n\hbar \tag{6-10}$$

At first glance the Bohr rule appears to be completely arbitrary and without physical significance. However, a particle of momentum mv will have an associated wave system with a wavelength $\lambda_B = h/mv$. This wavelength is written here with a subscript to distinguish it from the electromagnetic wavelengths emitted by the radiating atom. Then, from Eq. (6-8),

$$n\lambda_B = 2\pi a \tag{6-11}$$

and we see that the quantization rule allows only those orbits whose total length is an integral number of the particle's wavelength. Allowable stationary states correspond, then, to stationary wave systems along the orbits.

The central force needed to keep the electron in a circular orbit must come from the electrostatic attraction between the charges. From elementary mechanics,

$$\frac{mv^2}{a} = \frac{e^2}{4\pi\epsilon_0 a^2} \tag{6-12}$$

The total energy of the electron will be the sum of its potential and its kinetic energies. The potential of the coulomb field surrounding the proton charge is $e/4\pi\epsilon_0 r$, and so the potential energy of an electron is $-e^2/4\pi\epsilon_0 a$ at distance a. The minus sign appears because we have chosen the zero-point of the energy scale at $r = \infty$ and because the attractive force will do work *on* the electron in bringing it in the radius a. The total energy is

$$E = \frac{mv^2}{2} - \frac{e^2}{4\pi\epsilon_0 a} = -\frac{e^2}{8\pi\epsilon_0 a} \tag{6-13}$$

Eliminating v and a from Eqs. (6-8), (6-12), and (6-13),

$$E = -\frac{me^4}{8\epsilon_0^2 n^2 h^2} \quad \text{joules} \tag{6-14}$$

The energies will be negative for all finite values of n because of our choice of the zero point of the energy scale. Energy values allowed by the theory will be proportional to $1/n^2$ through a constant which must have the dimensions of energy and which consists of terms whose values are all known.

From the Bohr condition, Eq. (6-6), the frequency of the light emitted in a transition will be given by

$$\nu = \frac{E_1 - E_2}{h} = \frac{me^4}{8\epsilon_0^2 h^3}\left(\frac{1}{n_2^2} - \frac{1}{n_1^2}\right) \text{sec}^{-1} \tag{6-15}$$

Equation (6-15) has the exact form of the empirical expression for the wave numbers of the Balmer series if $n_2 = 2$, $n_1 = 3, 4, 5, \ldots$, and if the Rydberg constant is taken as

$$R_\infty = \frac{me^4}{8\epsilon_0^2 h^2 c} = 1.09737 \times 10^7 \text{ m}^{-1} \tag{6-16}$$

The constant is written here as R_∞ to indicate that it refers to an infinitely heavy, fixed nucleus. In the actual case, rotation takes place around the common center of gravity of the two particles. An exact calculation for this situation yields a value of R_H slightly different from R_∞.

When the constant was evaluated for the case of hydrogen, the predictions of the Bohr theory were in excellent agreement with the observed wavelengths. Some discrepancies and failures led to extensions of the theory, and finally to its replacement by the quantum mechanics. In many respects, however, it continues to be a useful approximation to the more complete theory.

6.05 Excitation and Ionization

For spectroscopic applications the Bohr relation may be written in terms of wave numbers, and the constant evaluated in absolute units. In terms of energies, Eq. (6-15) becomes

$$E_1 - E_2 = \frac{me^4}{8\epsilon_0^2 h^2}\left(\frac{1}{n_2^2} - \frac{1}{n_1^2}\right) \tag{6-17}$$

and the constant evaluates to 13.60 eV in our more convenient energy unit.

Normally, the electron remains in the lowest energy level, or ground state, $n = 1$, with an energy content of -13.60 eV. If energy is supplied to the electron, usually through collisions, it may be raised to an excited state $n = 2, 3, 4, \ldots$. When the electron acquires 13.60 eV or more, $n = \infty$, the electron is freed from its nucleus, and an ion pair is formed. Thus 13.60 eV is the *ionization potential* of atomic hydrogen.

Excitation and ionization potentials are measured by bombarding a gas at low pressure with a controlled beam of electrons. In the case of hydrogen, ionization will occur when the bombarding electrons have an energy of 13.60 eV, and the entire line spectrum of hydrogen will be emitted as the electrons return to the ground state through the succession of lower energy levels.

As the energy of the bombarding electrons is decreased below the ionization potential, inelastic collisions will raise orbital electrons to those excited states that are compatible with the energy available. Just below the level of each excited state there will be a discontinuity in the current–voltage relations of the electron beam, as the transition becomes energetically impossible.

Excited states exist for a very short period of time before the orbital electron returns to the ground state, accompanied by the emission of the energy difference. Every electron need not, and in some cases cannot, stop at every energy level on its return to the ground state. Some transitions are *forbidden* by selection rules which cannot be discussed here. Transitions shown in Fig. 6-5 are drawn without reference to the probability of any transition.

Since the ionization potential of hydrogen is 13.6 volts, any hydrogen electron acquiring more than this energy will leave the atom to become, temporarily, a free electron. This fact can be used to explain the stability of the hydrogen atom against coalescence of electron and proton. According to the uncertainty principle, $\Delta x\,\Delta p \simeq h/2\pi$. If the electron attempted to enter the nucleus, Δx would have to be of the order of the nuclear diameter, 10^{-13} cm. Then Δp will be of the order of 10^{-14} g-cm sec^{-1}. The momentum must be at least as large as its uncertainty and the corresponding energy must be at least equal to $E = (\Delta p)^2/2m$. This leads to an energy of at least $10^{-28}/2 \times 9 \times 10^{-28} = 0.05$ erg $= 3 \times 10^{10}$ eV. This greatly exceeds the

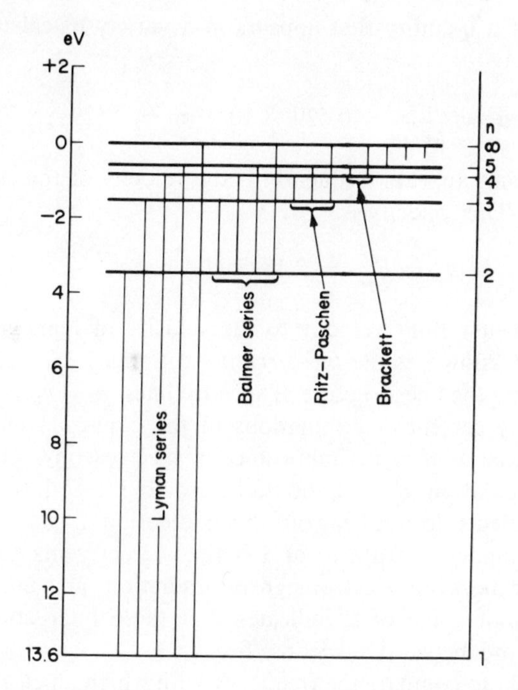

Figure 6-5. Energy levels and series relationships in atomic hydrogen.

ionization potential and it is evident that spontaneous ionization would occur rather than coalescence.

The uncertainty principle also provides a qualitative explanation for the preferred existence of hydrogen as H_2 molecules rather than as H atoms. At first glance it might appear that the force between two approaching H atoms would be repulsive because of the like charges of the orbital electrons. Actually as the two atoms approach and the two sets of allowable orbits overlap, the electrons can no longer distinguish between the two nuclei and can exchange positions. Because of this "to-and-fro" exchange, the uncertainty of position Δx increases. Correspondingly, the uncertainty of momentum (and, therefore, the energy) will decrease. Molecular hydrogen, therefore, has a lower energy content than two independent atoms and hence is the preferred state. The force binding the two atoms is called an *exchange force.*

6.06 Orbital Characteristics

Even though the Bohr model did not provide the ultimate description of the hydrogen atom, a good many of its parameters carry over into other domains. The radius of the ground-state orbit, a_H, known as the *Bohr radius,* is an example of a quantity that appears in a variety of calculations. This radius is

$$a_H = \frac{h^2\epsilon_0}{\pi e^2 m} = 0.529 \times 10^{-10}\text{ m} = 0.529\text{ Å} \tag{6-18}$$

Another factor that appears frequently is the velocity of the electron in the first Bohr orbit. This velocity is

$$v_H = \frac{e^2}{2h\epsilon_0} = 2.18 \times 10^6\text{ m sec}^{-1} \tag{6-19}$$

The ratio of the first Bohr velocity to the velocity of light gives a dimensionless constant known as the *fine-structure constant* because of its importance in explaining the fine structure of spectral lines: $\alpha = v_H/c = 0.00727 = 1/137.0388$. Many careful determinations of the constants entering into α have been made because of its importance in spectroscopy. There has been considerable speculation that $1/\alpha$ should be exactly integral, but increasingly precise measurements do not bear out this contention.

One of the important features of α is that it represents the strength of the interactions between electromagnetic radiation and moving electric charges. The small value of α indicates that these interactions are weak compared to those between nucleons, Sec. 7.03.

It is of interest to compare the frequency with which an electron traverses its orbits with the frequencies of the emitted radiations. For hydrogen,

$$v = \frac{e^2}{2nh\epsilon_0}, \qquad a = \frac{n^2h^2\epsilon_0}{\pi e^2 m},$$

and, since $f = v/2\pi a$,

$$f = \frac{me^4}{4n^3h^3\epsilon_0^2} \tag{6-20}$$

Bohr felt that there should be close ties between the predictions of the classical and the quantum theories, particularly at high quantum numbers where energy exchanges are small. This idea is expressed in the *correspondence principle*. We have already seen that the wave-function probabilities for the harmonic oscillator approach the classical values as the quantum numbers increase.

Equation (6-15) may be written in the form

$$\nu = \frac{me^4}{8\epsilon_0^2h^3}\left(\frac{(n_1 - n_2)(n_1 + n_2)}{n_1^2n_2^2}\right) \tag{6-21}$$

If both n_1 and n_2 are large, and if we let $n_2 = n_1 + 1$,

$$\nu \simeq \frac{me^4}{8\epsilon_0^2h^3}\left(\frac{2}{n_1^3}\right) = f \tag{6-22}$$

Thus at high quantum numbers the radiated frequencies will coincide with the orbital frequencies, and the radiations will be harmonic, corresponding to $n_2 - n_1 = 1, 2, 3, \ldots$.

6.07 Extensions of the Bohr Theory

The Bohr theory can be extended readily to atoms with nuclear charge Ze which have been ionized so as to leave only one orbital electron. The only change in the treatment involves replacing e in Eq. (6-12) by Ze. This leads to the general relation

$$E = -\frac{Z^2 me^4}{8n^2h^2\epsilon_0^2} \tag{6-23}$$

This relation applies to such systems as singly ionized helium He^+, or to doubly ionized lithium Li^{++}. When more than one electron is in the orbits, they will screen or reduce the effective nuclear charge in a complicated fashion, and the simple extension given in Eq. (6-23) will no longer apply.

Consider now the general case of a hydrogenlike system where the mass of the nucleus, M, is comparable to that of the electron. Such a system will rotate about the center of mass of the system of two particles, rather than about the center of the nucleus, as we assumed previously. Let a now represent the distance of the electron from the center of rotation, Fig. 6-6. This changes the distance between the two charges to $a(1 + m/M)$. Paralleling the Bohr treatment of the case of the infinitely heavy nucleus, we have

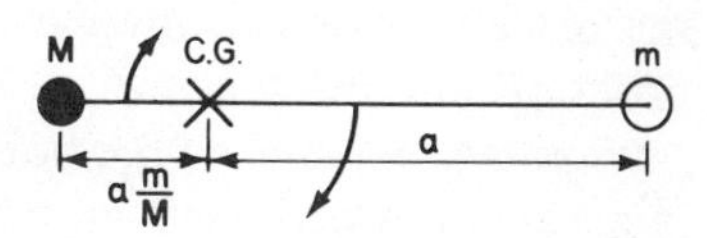

Figure 6-6. Two coupled masses will revolve about the common center of gravity, which may well be outside of either body.

$$\frac{mv^2}{a} = \frac{Ze^2}{4\pi\epsilon_0 a^2(1 + m/M)} \tag{6-24}$$

for the central-force requirement. The expression for energy becomes

$$E = -\frac{Ze^2}{8\pi\epsilon_0 a(1 + m/M)} \tag{6-25}$$

The total angular momentum of the system must be quantized. This leads to the requirement

$$mav\left(\frac{1 + m}{M}\right) = n\hbar \tag{6-26}$$

and the quantized energy expression becomes

$$E = -\frac{Z^2e^4 mM}{8\epsilon_0^2 n^2h^2(m + M)} \tag{6-27}$$

Equation (6-27) is similar to that obtained for the simpler case with the electron mass m replaced by the *reduced mass* $mM/(m + M)$. It is this factor which leads to the difference between the Rydberg constant for an infinite mass and that for the actual hydrogen atom. Slightly different values of R will obtain for He^+ and Li^{++}.

Two other cases are of practical importance. A hydrogenlike structure, *muonium*, is formed when a μ-meson is temporarily bound to a proton. Another hydrogenlike structure, *positronium*, is formed by the transient combination of a positron and a negative electron just prior to mutual annihilation. Each of these structures has an extremely short life but line spectra in accord with theoretical predictions have been observed from them.

6.08 Multiple Quantum Numbers

The Bohr model was remarkably successful but it left many spectral details unexplained. A logical extension of the theory was to consider elliptical orbits, since circles are special cases of ellipses, and Kepler had shown that the planets, under an inverse-square gravitational force, move in elliptical orbits around the sun.

Sommerfeld applied the Bohr principles to the case of an electron moving in an elliptical orbit around the nucleus at one focus. The electron velocity v, Fig. 6-7, can be resolved into two components, radial, v_r, and azimuthal, v_a. Each of these two components must be quantized separately, and there will be then two phase integrals $\int pdq$. Two new quantum numbers, k and r, are thus introduced.

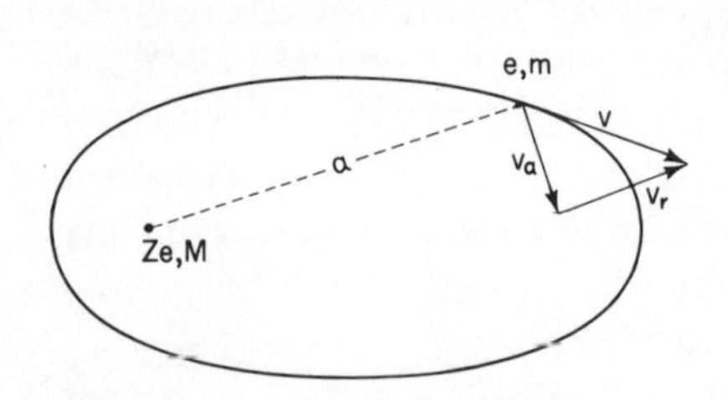

Figure 6-7. In an elliptical orbit, two quantum numbers are required to specify the motion.

An analysis of the elliptical orbits yielded only the original Bohr energy levels with n^2 replaced by $(k + r)^2$. Because of this interrelation, only two of the three quantum numbers are independent, and n and k proved to be the most useful pair. Total energy depends upon n, while k controls the angular momentum or the ellipticity of the orbits. Sommerfeld excluded the possibility $k = 0$ because this characterized a linear orbit, with no angular momentum, which requires the electron to pass directly through the nucleus.

Later work showed that still more quantum numbers were required to account for the observed spectra, and finally four parameters were specified. These are now denoted by

n = principal quantum number; this is the most important factor in specifying the total energy of the system; $n = 1, 2, 3, \ldots$

l = orbital quantum number; this quantizes the angular momentum of the system; $l = 1, 2, 3, \ldots (n - 1)$

m = magnetic orbital quantum number; this quantizes the spatial orientation of the orbital plane in the presence of an external field; thus only certain angles between field and orbit are allowed; $m = 0, \pm 1, \pm 2, \ldots \pm l$

s = spin quantum number; this takes account of the fact that the electron has an inherent angular momentum, as if it were spinning on its axis; s has only one magnitude, $\pm\frac{1}{2}$, but two directions, sometimes known as parallel and antiparallel.

Each electron, in each energy state, is characterized by this set of four numbers. For the ground state of hydrogen the electron has $n = 1$, which requires that $l = m = 0$. Two states are possible, for $s = \pm\frac{1}{2}$, and two forms of atomic hydrogen are known.

In elements beyond hydrogen, where two or more orbital electrons are present, the electrons will be grouped into energy levels or *shells* in accordance with the *Pauli principle.* In essence, the Pauli principle states that in any atomic structure no two electrons can have exactly the same set of quantum numbers. This turns out to be a requirement for particles with unsymmetrical wave functions—that is, for particles with negative parity.

If $n = 1$, l and m are necessarily zero and a maximum of two electrons is allowed. These two electrons form a *closed shell* known as the K shell. The Pauli principle requires that the electron spins be oppositely directed and so they will contribute nothing to the total angular momentum of the helium atom, of which they are a part. There are eight possible combinations for $n = 2$. Neon, $Z = 10$, has a closed K shell with two electrons and a closed L shell with eight. Again, there will be no spin contribution to the total angular momentum, because of the pairings of oppositely directed spins.

The concept of electron shells, each having a maximum number of available places for electrons, serves to explain the spectral similarities observed between elements similarly placed in the periodic table. Lithium ($Z = 3$) has a closed K shell with two electrons and the observed spectra are due primarily to energy transitions of the single electron in the L shell. Sodium ($Z = 11$) has two filled shells with two and eight electrons, respectively, leaving one electron for the M shell. This structure results in a series of energy levels analogous but not equal to those observed in lithium. Chemical properties such as valence and reactivity are also similar in elements having the same number of electrons in the outer shell.

Shell closures at $Z = 2, 2 + 8, 2 + 8 + 8$, and so on, are reflected in the lack of chemical reactivity of the noble gases. A plot of ionization potentials against atomic number, Fig. 6-8, shows the tight electron binding in the closed shells, the loose binding of the first electron added in each succeeding shell, and the gradual rise in binding strength as each shell fills.

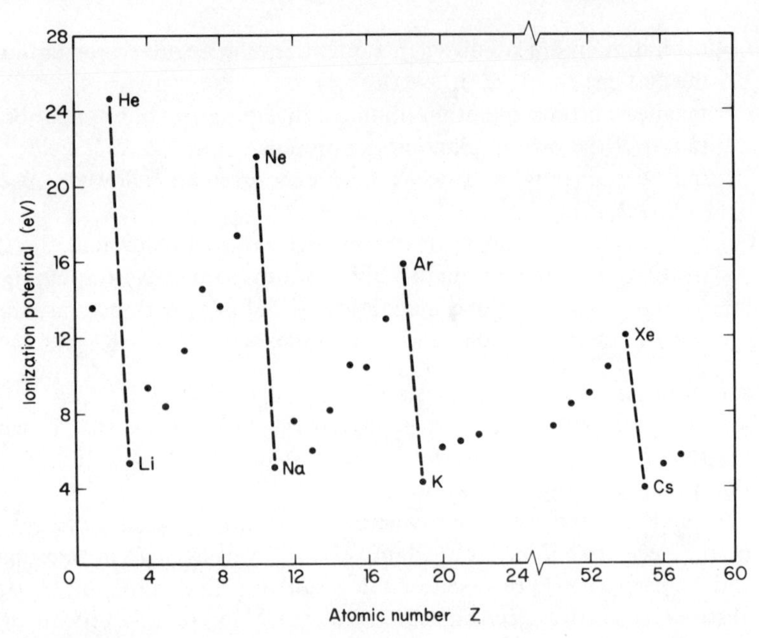

Figure 6-8. Ionization potential measurements show the tight electron binding at the closed shells and the looser binding when the shells are unfilled.

Calculations of energy levels for helium and heavier elements is complicated by the partial shielding or screening of the nuclear charge by inner electrons so that the electrons in outer shells do not move in the full nuclear field. Furthermore, electron–electron interactions make impossible simple calculations of the type given here. With proper conditions of excitation, helium gas can be ionized to He^+, which means that only one electron remains for the production of spectral lines. The He^+ spectrum would be expected to be hydrogenlike but displaced because of the nuclear charge $Z = 2$, and this is observed. Li^{++} also shows a spectrum analogous to that of hydrogen. Heavier elements have extremely complex spectra—iron for example, having more than 10,000 lines.

In the higher shells, when $n \geq 2$, it is convenient to divide the electron states into subshells in which each inhabitant has constant values of both n and l. Thus for $n = 2$, l can take on values of 0 and 1, and there will be two subshells containing two and six electrons, respectively. Electrons in the subshells are designated by the letters previously assigned from optical spectroscopy:

Orbital quantum number:	0	1	2	3	4	5	6	7	8
Letter assignment:	*s*	*p*	*d*	*f*	*g*	*h*	*i*	*k*	*l*

Lower-case letters refer to a single electron; capital letters represent a configuration of several.

Atomic electron configurations are given by specifying the principal quantum number, then the orbital quantum-number designation, and finally the number of electrons in that configuration as a superscript. In the case of neon, $Z = 10$, the specification is $1s^2 2s^2 2p^6$, which means that there are two electrons in the K shell, each with orbital quantum number 0. There are a total of eight electrons in the L shell, two with $l = 0$, six with $l = 1$. For magnesium, $Z = 12$, with a partially filled M shell the configuration is $1s^2 2s^2 2p^6 3s^2$. This configuration is also frequently written as 10Ne core $3s^2$. The shell-filling order tends to be $s, p, d, f, \ldots$, but this is not universally followed at the higher numbers.

6.09 Coupling

Quantum numbers l and s both refer to angular momentum and may be combined into a resultant quantum number j. In a complex electronic structure, interactions between electrons prevent a unique specification of individual values of n, l, j, and m. Total values, particularly of J and M, are of value in classifying spectra.

The interelectronic interactions, or *couplings*, have varied strengths or importance, and some may be neglected in a first approximation. In Russell–Saunders, or LS, coupling, for example, the magnetic interactions between spin and orbital angular momentum are assumed to be very weak, leaving only electrostatic interactions between electrons and coupling between individual spins. In Russell–Saunders coupling, total angular momentum L and total spin S are independent. This leads to

$$|L - S| \leq J \leq L + S$$

with integral steps between consecutive J values. If $L \geq S$, then J can take on $2S + 1$ values, and if $L < S$, then J takes on $2L + 1$ values. In either case, the value of $2S = 1$ is known as the *multiplicity* of the level. For an even number of electrons, S will be an integer and the multiplicity will be odd. An odd number of electrons will give an even multiplicity value.

In the absence of spin–orbit coupling, all energy values for a given L and S coincide, but with coupling they separate to produce singlets, doublets, triplets, and so on. Thus coupling gives rise to the fine structure in spectral lines.

The opposite to LS coupling is known as jj coupling. Here there is a strong spin–orbit interaction between l and s of the same electron. Individual l values and individual s values do not couple. Optical spectroscopy recognizes a progression of couplings ranging from pure LS in light elements through intermediate stages to nearly pure jj coupling in heavy elements with large values of orbital angular momentum.

6.10 Magnetic Properties of Hydrogen

The movement of an electric charge around a closed orbit is equivalent to an electric current, and this current will produce a magnetic field. The strength of any magnet is measured by a *dipole moment*, which is the product of the pole strength and the distance between the real or virtual poles. It can be shown that a current i flowing around the periphery of an area A is equivalent to a dipole of strength

$$\mu = iA \tag{6-28}$$

Electric current is equal to the amount of charge transported per second. In the case of an orbiting electron,

$$i = ef = \frac{e\omega}{2\pi}$$

where f is the frequency with which the electron traverses its orbit.

The dipole moment of an electron orbiting in a hydrogen atom will then be given by

$$\mu = \frac{n\hbar e}{2m} \qquad \text{joules tesla}^{-1} \tag{6-29}$$

For the Bohr ground-state orbit, $n = 1$ and the moment is

$$\mu_B = \frac{\hbar e}{2m} = 9.27 \times 10^{-24} \qquad \text{joule tesla}^{-1} \tag{6-30}$$

where μ_B is known as the *Bohr magneton,* used as the unit of magnetic-dipole strength in the atomic domain. Many magnetic measurements have been made in the CGS rather than in the MKSA system. In expressing the results of these measurements it is customary to use e in CGS electrostatic units. Equation (6-30) then becomes

$$\mu_B = \frac{\hbar e}{2mc} = 9.27 \times 10^{-21} \qquad \text{erg gauss}^{-1}$$

According to Eq. (6-29) the magnetic-dipole field of the hydrogen atom is quantized, with only integral multiples of the Bohr magneton allowed. We recall that $n\hbar$ is equal to the quantized angular momentum, so Eq. (6-29) can be written

$$\mu = \frac{e}{2m} \times \text{angular momentum} \tag{6-31}$$

These magnetic relations were developed from considerations of a point charge moving in a macroscopic orbit. There is no assurance that these considerations will apply to the atomic situation, particularly to that part of the angular momentum resulting from electron spin. To take care of possible differences it is customary to write Eq. (6-31) as

$$\mu = g\frac{e}{2m} \times \text{angular momentum} \tag{6-32}$$

The factor $e/2m$ is known as the *gyromagnetic ratio*, denoted by γ; g is a dimensionless multiplier, the *atomic g factor*.

Measurements show that for orbital motions the predictions of Eq. (6-31) are accurately fulfilled, which is to say that g(orbital) has a value of precisely 1. Spectral-line spacings show that the spin quantum number must be multiples of $\frac{1}{2}$, but the magnetic moment due to spin is found to be almost 1 Bohr magneton. Thus measurement requires that $g(\text{spin}) \simeq 2$. An involved theoretical treatment predicts that g(spin) should be given by $2(1 + \alpha/2\pi)$ where α is the fine-structure constant. The predicted value is experimentally confirmed.

6.11 Larmor Precession

Because of the dipole moment created by the orbiting electron, the angular-momentum vector (and consequently the plane of the orbit) can assume only quantized angles with respect to an external magnetic field H. These angles will be governed by a magnetic quantum number m. A magnetic dipole whose axis makes an angle with an external field will precess about the field vector, just as a spinning top will precess because of the torque exerted on it by the earth's gravitational field.

We shall adopt the convention that a rotation or torque will be represented by a vector directed in the direction that a right-handed screw would progress with the rotation. Consider now a spinning top, Fig. 6-9, whose angular-momentum vector p is in the X–Z plane. The gravitational field exerts a torque around the Y-axis, and this torque will be represented by vector $\mathbf{T}$ directed along Y. In a short interval of time dt, the torque will produce a change in angular momentum $d\mathbf{p}$ which will be in the same direction as the vector T. By the laws of rotational motion, $d\mathbf{p} = \mathbf{T}\,dt$. The new angular-momentum vector will have the original magnitude but it will be changed in direction, Fig. 6-9. From the figure we see that the effect of $\mathbf{T}$ is to cause $\mathbf{p}$ to precess about the Z-axis. From the figure,

$$d\phi = \frac{d\mathbf{p}}{\mathbf{p}\sin\theta}$$

where θ is the angle between the angular-momentum vector and the vector representing the external force (the minus Z-axis in the case of the gravitational force). Then,

$$d\phi = \frac{\mathbf{T}\,dt}{\mathbf{p}\sin\theta} \quad \text{or} \quad \frac{d\phi}{dt} = \frac{\mathbf{T}}{\mathbf{p}\sin\theta}$$

Since $d\phi/dt$ is the angular velocity of precession ω_p,

$$\mathbf{T} = \omega_p \mathbf{p} \sin\theta$$

In the case of the hydrogen atom, the torque exerted on a dipole of strength μ will be $\mathbf{H}\mu \sin\theta$, Fig. 6-10. Then,

$$\mathbf{H}\mu \sin\theta = \omega_p \mathbf{p} \sin\theta \tag{6-33}$$

and

$$\omega_p = \mathbf{H}\frac{\mu}{\mathbf{p}} = \mathbf{H}\frac{e}{2mc} \tag{6-34}$$

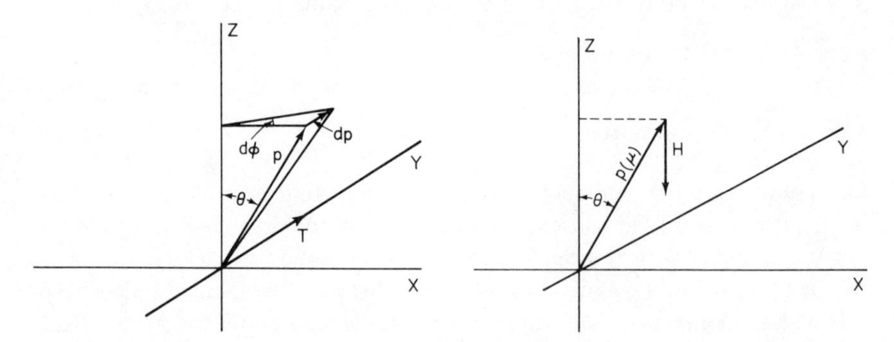

Figure 6-9. A spinning top with an initial angular momentum p will precess around the Z-axis under the influence of a torque T.

Figure 6-10. The magnetic dipole of a hydrogen atom will precess because of the torque exerted by the magnetic field H.

This is the expression of the Larmor precession of an atomic magnetic dipole in an external magnetic field. Note that the frequency of precession is independent of the quantum numbers. As external excitations cause values of m to change, energy will be absorbed from the exciting source, and this energy can be detected. For a constant external field H, however, the frequency of precession will remain constant.

Equation (6-34) was obtained on the assumption that the charges were free to respond to the field. In general, chemical bonding will modify the simple relationship by an amount that is characteristic of each bond. This effect can be used in magnetic-resonance absorption measurements to detect the presence of bonds whose characteristics have been cataloged.

REFERENCES

Evans, R. D., *The Atomic Nucleus.* International Series in Pure and Applied Physics. McGraw-Hill, New York, 1955.

Hughes, V. W., "Muonium." *Ann. Rev. Nuc. Sci.* **16,** 445, 1966.

Snipes, W., ed., *Electron Spin Resonance and the Effects of Radiation on Biological Systems.* National Academy of Science, Washington, D.C., 1966.

White, H. E., *Introduction to Atomic Spectra.* International Series in Physics. McGraw-Hill, New York, 1934.

PROBLEMS

6-1. A photon incident upon a hydrogen atom ejects an electron from the second excited state with a kinetic energy of 11.3 eV. What is the energy of the photon? What kinetic energy will this photon impart to an electron in the ground state of hydrogen?

6-2. Calculate the electrostatic force of attraction between a proton and an electron in the first Bohr orbit.

6-3. What spectral lines will be emitted from atomic hydrogen when it is excited with 11.2-eV electrons?

6-4. What is the loss in mass resulting from the formation of a hydrogen atom in its ground state from a free proton and electron?

6-5. Calculate the number of electrons allowed in the *M* shell by the Pauli principle, and show their configuration.

6-6. To what current, in amperes, is the electron in the ground state of hydrogen equivalent? What is its magnetic moment?

6-7. At what temperature will the mean thermal energy of hydrogen atoms be just sufficient to ionize the gas?

6-8. Calculate the wavelengths and the quantum energies of the first three lines of the "Lyman" series of positronium.

6-9. What is the distance between the proton and meson in an atom of muonium in the ground state?

6-10. The most energetic projectile that Rutherford had at his disposal was the 7.7-MeV alpha particle from ^{214}Po. What is the closest approach of this particle to a nucleus of aluminum?

6-11. Gravitational forces were neglected in deriving the Bohr relations for hydrogen. How much will gravitational attraction contribute to the energy of an electron in the first Bohr orbit?

6-12. Calculate the magnetic-resonant frequency of Lamor precession for an electron in a steady field of 1 gauss.

6-13. What will be the frequency of Larmor precession of a proton in a field of 1 gauss?

6-14. How much error was introduced into the evaluation of the ionization potential of atomic hydrogen by assuming the proton to be infinitely heavy? Locate the center of revolution of the actual structure.

6-15. Apply the law of conservation of momentum to the emission of the first line of the Lyman series and calculate the energy and velocity of the recoiling proton.

6-16. Some of the components of cosmic radiation appear to be heavy nuclei completely stripped of electrons. They should emit hydrogen like spectra when they acquire the first orbital electron. Calculate the energies of the first lines of the "Lyman," "Balmer," and "Paschen" series for ${}_{16}F^{25+}$.

6-17. The presence of two K electrons in ${}_{3}Li$ reduces the effective nuclear charge to $(Z - S)$, where S is a screening factor. Calculate the screening factor for ${}_{3}Li$ from the knowledge that the ionization potential is 5.39 eV.

7

Structure of the Atomic Nucleus

7.01 Nuclear Stability

Any atomic or nuclear configuration will be stable if it is unable to transform to another configuration without the addition of energy from the outside. Under normal environmental conditions, atomic structures are absolutely stable. Orbital electrons will remain indefinitely attached to their nucleus. Each orbital, undisturbed, will occupy the lowest possible energy state, except for those electrons which have been trapped in long-lived states of higher energy by some previous excitation. Orbital electrons may exchange, but it will be only that, with each of the exchanging systems maintaining its identity and energy content.

Many nuclei also appear to be stable, but others transform or decay spontaneously to structures of lower energy content. There are several kinds of nuclear instability. Heavy nuclei may decay with the emission of an alpha particle. Light nuclei, on the other hand, appear to be stable against alpha-particle emission, but may undergo beta-particle decay. A good many nuclear structures appear to be absolutely stable against any form of spontaneous decay, but it is not certain that there is a sharp dividing line between stability and instability. The detection of instability depends upon the sensitivity of our measuring techniques. Lanthanum-138, once thought to be a case of odd Z–odd N stability, has been found to undergo a long-lived radioactive decay. Future improvements in techniques may may lead to the discovery of very slow decay rates in structures now thought to be stable.

Figure 7-1 is a neutron–proton or N–Z plot of all of the relatively abundant stable nuclei. The most striking feature of this plot is the tendency for

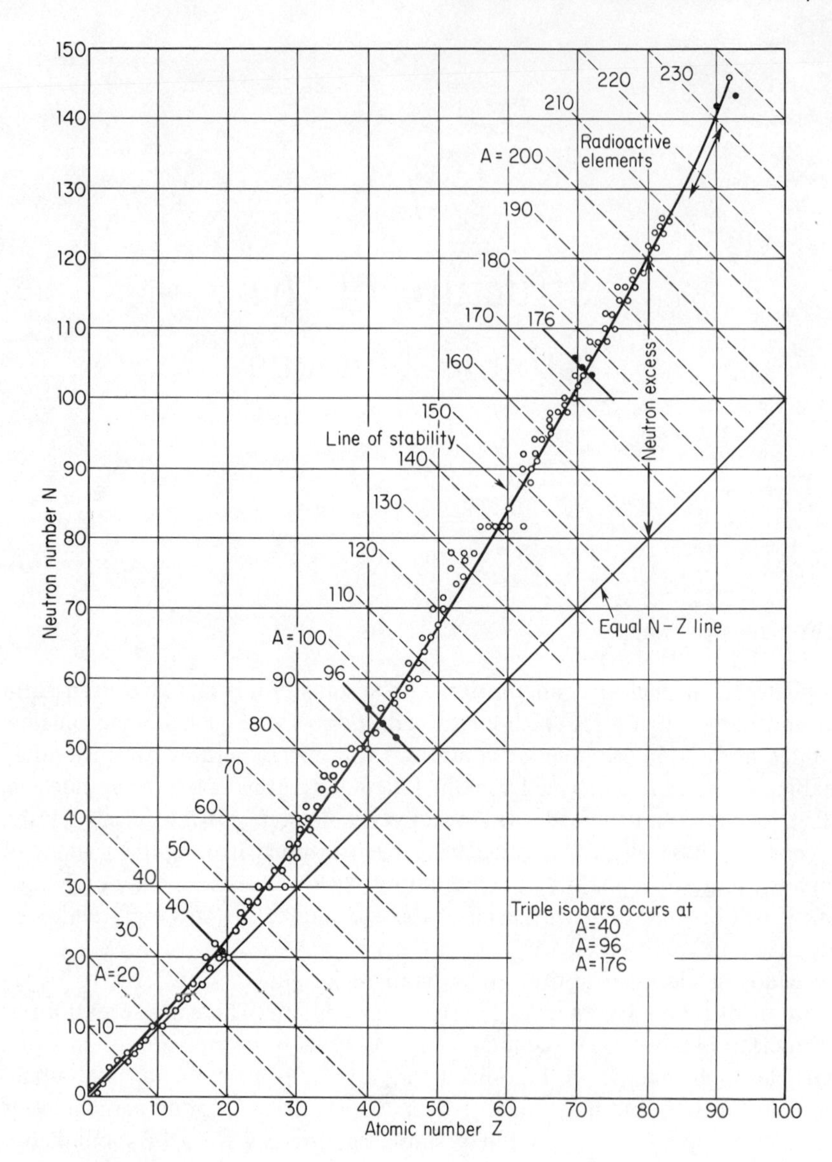

Figure 7-1. A neutron–proton plot of the most common stable nuclides. For simplicity, isotopes with a relative abundance of less than 20 per cent have been omitted.

the stable nuclei to fall along a narrow band. The solid line drawn through the points is known as the *line of stability*. Initially the slope of this line is 45°, because of the tendency for light nuclei to contain equal numbers of protons and neutrons. For heavier structures the N/Z ratio increases to about

1.5. Only one stable nucleus, ${}^3_2$He, is known in which $Z > N$. It is not evident from the figure, but there is at least one known stable structure for every value of A from 1 to 209 except for 5 and 8.

Nuclei with equal values of *mass number A* are *isobars*, which lie on perpendiculars to the $N = Z$ line in Fig. 7-1. The term isobar refers only to equal numbers of nucleons. The exact masses of isobars need not be identical, and in general they are not. Structures with equal atomic numbers Z are *isotopes*, those with equal neutron numbers N are *isotones*. In the N–Z plot these groups lie along vertical and horizontal lines, respectively. Less common, *isodiapheres* are nuclei with equal neutron excess $(N - Z)$.

The trilinear or A–$(N - Z)$ plot, Fig. 7-2, is useful since it shows simultaneously A–Z, N–Z, and A–N occupancy. In the trilinear plot:

Isotopes lie on a line with a slope of 45° upward
Isotones lie on a line with a slope of 45° downward
Isobars lie on a vertical line
Isodiapheres lie on a horizontal line

An inspection of Fig. 7-2 reveals the predominance of even Z–even N occupants among the stable structures. Thus there is only one occupant at $Z = 53$, 55, and 59, and only two at $Z = 57$. There is no stable isotope of promethium, $Z = 61$. The longest-lived radioactive isotope of this element

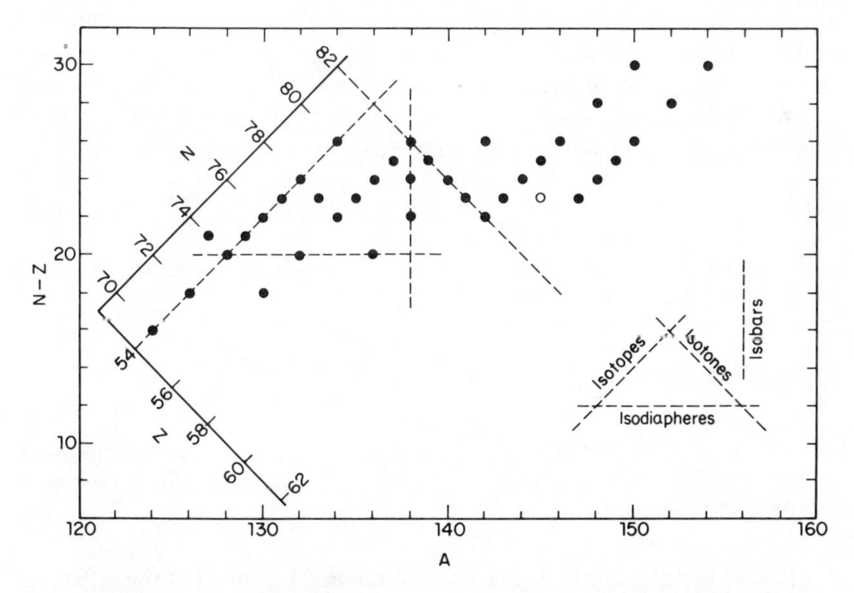

Figure 7-2. A trilinear plot of the stable nuclides in the region $Z = 76$, $A = 140$. Note the relative abundance of even Z–even N nuclides. The open circle at $Z = 61$, $N = 84$ represents a radioactive isotope of promethium, which has no stable isotope.

is shown in Fig. 7-2 by an open circle at $N = 84$. These population values are to be contrasted with the number of occupants of $Z = 54, 56, 58$, and 60.

A triple isobar occurs at $A = 138$, and a triple isodiaphere at $(N - Z) = 20$. The five isotones at $Z = 82$ reflect the unusual stability associated with this number.

7.02 Nuclear Mass and Binding Energy

It is a general principle that the most stable state of any system is that with the lowest energy content, which, from the Einstein mass–energy equivalence, is that with the least mass. As an example consider the nucleus of ^4_2He, composed of two protons and two neutrons, and known to be stable. According to the stability principle the mass of the assembled nucleus must be somewhat less than the sum of the masses of the disassembled constituent particles. A loss of mass, accompanied by the emission of energy, is to be expected during the assembly.

One measure of the loss of mass in an assembled nucleus is the *mass defect* Δ, defined as the difference between the atomic mass M and the mass number A. Thus

$$\Delta = M - A \tag{7-1}$$

where Δ will be positive for light nuclei and for those with $A > 215$, and negative for intermediate mass numbers. Because of this change in sign, Δ is sometimes known as the mass *excess* instead of *defect*. Another measure, the *packing fraction* f, is simply the mass defect per nucleon, multiplied by 10^4 to bring the small numbers to more convenient magnitudes:

$$f = \frac{M - A}{A} \times 10^4 \tag{7-2}$$

If no energy were gained or lost in the assembly of a nucleus, the mass of a Z–N structure would be

$$W_N = Zm_p + Nm_n \tag{7-3}$$

and the corresponding atomic mass would be

$$W = Zm_p + Nm_n + Zm_e \tag{7-4}$$

The total energy associated with either W_N or W is expressed in ergs when the mass values in grams are multiplied by c^2. It is usually more convenient to work with universal mass units and MeV, using the conversion factors in Eq. (4-7).

Another useful quantity is the *mass decrement* δ, defined as the difference between the calculated sum of the constituents W and the measured mass M:

$$\delta = W - M \tag{7-5}$$

Illustrative Example

Calculate the mass decrement of ${}^4_2$He. $Z = N = 2$, so

$$W = 2 \times 1.007276 + 2 \times 1.008665 + 2 \times 0.0005487 = 4.032979 \text{ u}$$

The measured atomic mass of ^{4}He is 4.002603 u, and hence

$$\delta = 4.032979 - 4.002603 = 0.030376 \text{ u}$$

The energy equivalent is $0.030376 \times 931.48 = 28.28$ MeV

The energy calculated in the illustrative example is the energy that would be radiated from the system during the synthesis of a nucleus of ^{4}He from two protons and two neutrons. The helium atom (or the helium nucleus, since it alone is involved in the energy relations considered here) is, therefore, stable against complete disruption by 28.28 MeV. In general, a nucleus is stable against complete disruption by the energy

$$E_B = 931.48(W - M) \tag{7-6}$$

where E_B is known as the *binding energy*. If the total binding energy is considered to be distributed uniformly over all of the constituent nucleons, we have the important quantity *binding energy per nucleon*, E_b, which is

$$E_b = \frac{931.48(W - M)}{A} \tag{7-7}$$

In the case of ^{4}He, $E_b = 28.28/4 = 7.07$ MeV. This is the energy required to extract a single nucleon from the ^{4}He nucleus.

The calculations of the preceding illustrative example can be done equally well in mass equivalents, or MeV, and it is sometimes advantageous to do so. Nuclear masses may be used instead of atomic masses.

Illustrative Example

Calculate the binding energy of the nucleus ^{4}He. Equivalent mass values will be found in the Appendix Table 9 and in Table 4-1.

$$\begin{aligned} W_N &= 2 \times 938.256 + 2 \times 939.550 = 3775.612 \text{ MeV} \\ M_N &= 3728.337 - 2 \times 0.511 \qquad\quad\; = \underline{3727.33} \\ &\qquad\qquad \text{Binding energy } E_B \qquad 28.28 \end{aligned}$$

Tables of nuclear data frequently list values of the mass defect instead of the actual mass or the mass equivalent. For ^{4}He, tabulated values list $\Delta = M - A = 2.4248$ MeV, whence

$$M_N = 4 \times 931.48 + 2.4248 - 2 \times 0.511 = 3727.33 \text{ MeV}$$

as above.

It is evident from the above definitions that nuclei having large values of E_b are most stable against the loss of a neutron or a proton, but this does not imply stability against all forms of nuclear decay. ^{8}Be, for example, has a greater binding energy per nucleon than ^{9}Be, but the former spontaneously splits up into two alpha particles, while the latter is stable.

Figure 7-3 shows the main features of the variation of E_b with A. Without considering here all possible types of instability, we see from the figure that

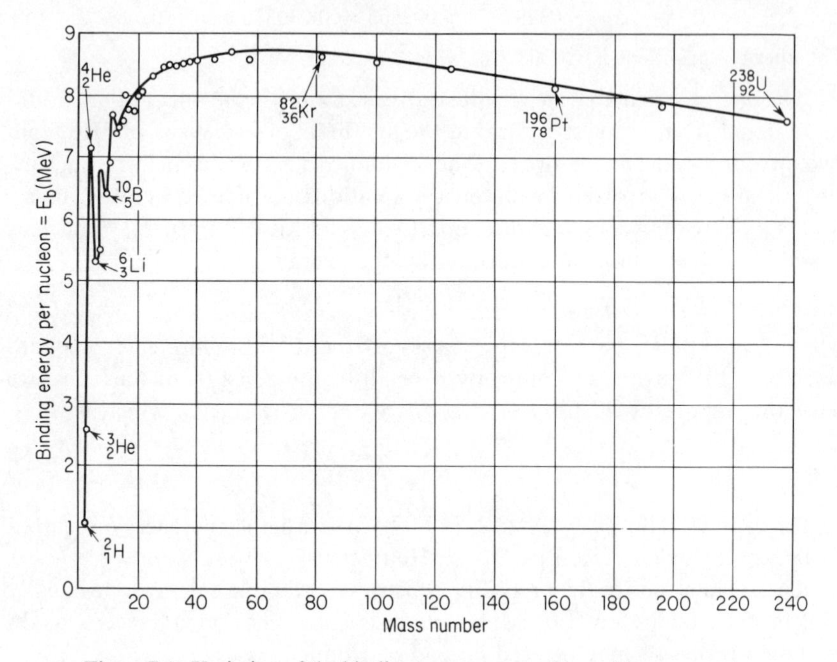

Figure 7-3. Variation of the binding energy per nucleon with mass number.

nuclei of intermediate mass number are more stable than those at either end of the mass-number range. Stability increases sharply with A for the lightest elements, becomes fairly constant at $A = 50$, and decreases slowly from a broad maximum.

For values of $A < 30$, the curve is not smooth but is marked by large irregularities. Thus ^{4}He, ^{12}C, and ^{16}O have higher values of E_b, and hence a greater stability, than do nuclei immediately adjacent to them. This behavior suggests that nuclei which are composed of exact multiples of the alpha-particle structure are more stable than other groupings. In the heavier nuclei this tendency is less apparent.

Other departures from the smooth binding-energy curve occur at Z or $N = 20$, 28, 50, 82, and 126, the so-called *magic numbers*. These energy-poor structures are presumably the result of the closure of some sort of nuclear shells analogous to the closure of orbital electron shells at $Z = 2$, 10, 18, ... Two sets of shell closures, N and Z, are observed in the nucleus.

This leads to a nucleus such as $^{208}_{82}Pb$, which is said to be *doubly magic* since $Z = 82$ and $N = 126$. As we shall see later, other nuclear properties also show discontinuities at the magic numbers.

7.03 Possible Nuclear Forces

The fact that nuclei are stable, with binding energies of several MeV, requires the existence of some sort of an attractive force between the nucleons. A nucleon entering an existing structure will be captured by this force, work will be done by the force, and energy will be radiated. The energy loss from the system will be reflected in a final nuclear mass that is less than the sum of the masses of the two separated components.

Gravitational forces are observed to be always attractive and must be considered as possibly responsible for the nuclear binding. Gravitational forces obey an inverse-square law:

$$F = G\frac{m_1 m_2}{r^2} \tag{7-8}$$

where G, the gravitational constant, has a value of 6.67×10^{-11} newton-m^2 Kg.$^{-2}$ With this force field the work done in attracting a mass m_1 to a distance r from a mass m_2 will be Gm_1m_2/r joules. For two protons separated by a distance of 5×10^{-15} m, the work amounts to -2.3×10^{-37} MeV. The minus sign indicates that work has been done by the system and lost from it.

In the nuclear domain the length 10^{-15} m forms a very convenient unit that is used frequently. Because of this it is known as 1 *fermi*, abbreviated F.

Electrical forces cannot account for the nucleon–nucleon attractions because the only nuclear charges are those associated with the protons. These like charges will produce mutual repulsions rather than the attractions that are required. As in the gravitational case, the electrical potential energy will be given by a $1/r$ law. A calculation for the case of two protons separated by 5 F leads to an energy of 0.28 MeV. This energy will be positive because it represents work done on the system, and hence gained by it.

Magnetic forces between nucleons also exist, the direction depending upon the relative directions of the interacting spins. The magnitudes of the magnetic energies are of the same order as the electrical energy just calculated.

The total energy of a nucleon bound in a nucleus can be estimated. Each nucleon must be confined in a volume whose dimensions are of the order of 5 F, the separation that we have used in the previous calculations. Then the de Broglie wavelength must be of the same order and hence

$$\lambda = \frac{h}{mv} = 5 \times 10^{-15} \text{ m}$$

$$T = \frac{p^2}{2m} = 5.3 \times 10^{-12} \text{ joule} = 33 \text{ MeV}$$

This estimate lacks the accuracy of the two previous calculations but it is sufficient to show that the nuclear force is the strongest of the three by at least one order of magnitude. Nuclear attractive forces are *strong interactions*, to be contrasted with the *intermediate interactions* of electric and magnetic origin, and the *weak interactions* of gravitation.

Gravitational forces fail to provide the required binding energy by a factor of about 10^{38}, and have an inverse-square dependence that is at variance with the facts. Electrical and magnetic forces are ruled out as the nuclear force because they are too small, are repulsive, and obey the inverse-square law. Obviously some new type of force must be involved in the nucleus.

There is one sharp distinction between atomic and nuclear forces. In the former case the orbital electrons move in the centrally directed field of a relatively massive nucleus. Inside the nucleus there is no unique central entity with which all other nucleons interact. Here, forces must act between individual nucleons rather than between them and a central body.

7.04 Characteristics of the Nuclear Force

Although the Japanese physicist Yukawa correctly predicted, in 1935, the origin of an attractive nuclear force, our knowledge of this force is still largely empirical and observational. Several unique characteristics serve to distinguish this nuclear force from the more familiar force fields obeying the inverse-square law. The characteristics may be summarized as follows:

1. Although the tangible effects indicate that the nuclear force is attractive, there must also be a repulsive component effective at extremely small distances. Without this repulsive component the individual nucleons would coalesce into a single particle.
2. Individual nuclear forces are considerably greater than the corresponding coulomb forces. If this were not true, nuclei would be disrupted by electrostatic repulsion. Because of the long range of the coulomb forces, however, they may predominate in the aggregate at high values of Z.
3. Nuclear forces are not strongly dependent upon the nature of the interacting nucleons. In other words, there is little difference between *n–n*, *n–p*, and *p–p* forces. This indicates an essentially nonelectrical origin.
4. Nuclear forces have a very short range, which is to say that they approach zero much more rapidly than the inverse square of the distance.
5. Little is known about the variations of the force at distances below 1 fermi, except that there must be a component of repulsion. At about 0.4 F this force becomes very large, forming an impenetrable *core*.

6. As a consequence of the short range, nuclear forces can act only on adjacent nucleons.
7. Nuclear forces exhibit saturation. Groups of four nucleons form unusually stable structures, in which the nuclear forces appear to be completely satisfied. The nonexistence of a five-particle nucleus such as ^{5}He or ^{5}Li indicates the saturation nature of the nuclear force.
8. Nucleons tend to form into closed shells with relatively weak interactions between shells. This is shown by the nonexistence of stable ^{8}Be, which would consist of two saturated ^{4}He shells.
9. Although groups of four nucleons appear to be saturated, there is some tendency for nucleon pairing.

Several important properties can be inferred from the characteristics of the nuclear force. Among these are the following:

a. Nuclei tend to be spherical in shape. This shape gives the most effective ratio of volume to surface and hence makes the most effective use of the short-range attractive forces. The lowest energy state will be attained when the smallest number of nucleons have unsaturated forces because of their position at the surface of the nucleus.
b. Nuclear material appears to be almost incompressible. This is shown by the fact that nuclear radii are given by

$$r = r_0 A^{1/3} \tag{7-9}$$

where r_0 has a value ranging from 1.1 to 1.6 F, depending upon the way in which the radius is defined. Thus nuclear material has an essentially constant density. Spectroscopic evidence from the isotope shift indicates that about the same nuclear volume is available to the neutrons and the protons.

c. Nuclear charge tends to spread nearly, but not exactly, uniformly throughout the nuclear volume. If charge is spread nonuniformly, an appreciable quadrupole moment, Fig. 7-4, or a pole of some other order, will be formed. These moments can be detected experimentally, because each will have a potential which varies in a quite different way than the inverse first power of distance, a characteristic of point or uniformly distributed charges. The results of the Rutherford scattering experiments agree well with the pre-

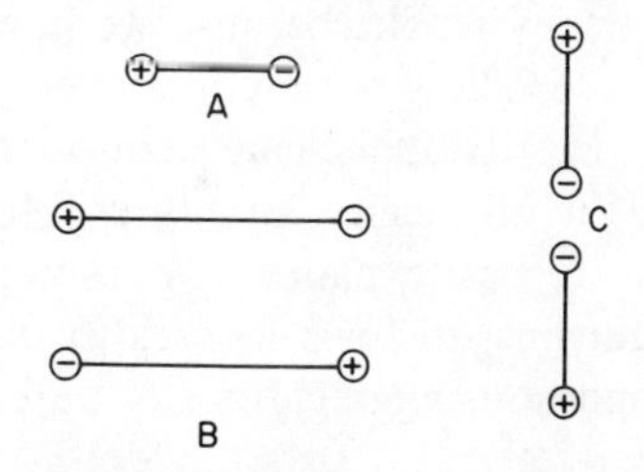

Figure 7-4. Charges or magnetic poles may exist as dipoles, (A), and these in turn may be arranged as quadrupoles, (B) or (C), or as poles of still higher order.

dictions of an inverse first-power potential, indicating that quadrupole moments, if they exist, are small. In fact, quadrupole moments have been detected and are responsible for some of the hyperfine structure seen in atomic and molecular spectra. The fields associated with the noncentral forces are *tensors*, rather than vectors or scalars, which are adequate to describe simpler quantities and distributions.

d. The relative charge-independence of the nuclear force can be seen by considering *mirror nuclei*, which are isobars with reversed proton and neutron numbers. The simplest mirror nuclei are $^3_1$H and $^3_2$He. In ^{3}H there are two *n–p* and one *n–n* interactions; in ^{3}He the *n–n* is replaced by a *p–p*.

The two binding energies, calculated with Eq. (7-6), are

$$E_B(^3\text{H}) = 8.48 \text{ MeV} \qquad E_B(^3\text{He}) = 7.72 \text{ MeV}$$

We know that in ^{3}He there is an energy due to coulomb repulsion that is not present in ^{3}H. This energy is readily calculated, using a separation distance of 2 F as suggested by various bits of evidence. This calculation leads to

$$E_C(^3\text{He}) = 0.72 \text{ MeV}$$

If the coulomb force did not exist in ^{3}He, we would expect

$$E_B(^3\text{He}) = 7.72 + 0.72 = 8.44 \text{ MeV}$$

This value is very close to the binding energy for ^{3}H and hence we infer that there is very little difference between the strengths of the *n–n* and the *p–p* interactions. Thus the nuclear force is essentially charge-independent.

Simple binding-energy considerations alone predict the existence of both the di-proton $^2_2$H, and the di-neutron, ^2_0n. The binding energy of the deuteron, $^2_1$H, is known to be about 2.2 MeV, and with charge independence this should equal the non-coulomb binding energy in either the di-proton or the di-neutron. No coulomb forces are present in the di-neutron, which should, then, be about as stable as the deuteron. The di-proton should also be energetically stable, because we have just seen that the *p–p* energy of repulsion is well below the binding energy of the deuteron.

More sophisticated considerations appear to rule out stability for each of the di-nucleon structures. However, the radiations emitted by some stars can be best explained by assuming that the source consists of closely packed neutrons, to form a so-called *neutron star*. It may well be that under some conditions aggregations of pure neutrons are possible.

7.05 Origin of the Nuclear Force

The short range and the saturation shown by the nuclear force suggest that it may arise from a situation analogous to that leading to molecular binding by *exchange forces*. In molecular hydrogen H_2, for example, two individual atoms are bound into a stable diatomic molecule by *homopolar bonds* formed by the exchange or sharing of electrons between the two nuclei. This sharing leads to a lower energy state, and hence to a binding energy, which does not exist if each electron remains permanently attached to one proton only. The molecular-exchange force shows saturation with only two particles, since the tendency to form triatomic hydrogen, H_3, is very much weaker that that leading to the diatomic form.

Yukawa proposed that the nuclear force came from the sharing of a then-undetected nuclear particle. He was able to predict that the exchange particle would have a mass intermediate between that of the electron and the proton.

The first particle of intermediate mass to be detected was the μ-meson ("meson" from the Greek for "middle") or *muon*, with a mass of 105.66 MeV or $206.6m_e$. Muons were observed as a very penetrating component of cosmic radiation and as such their properties could not be reconciled with those needed for the exchange particle. A particle capable of penetrating several hundred meters of water or earth must interact only very weakly with nucleons, whereas strong interactions are required of the exchange particle.

Later, π-mesons, or *pions* were detected and found to have the properties required by the Yukawa theory. Charged pions, $\pi^{\pm}$ with a mass of 139.58 MeV, and a neutral pion with 134.97 MeV have been identified.

The exchange reactions involved are:

$$\left.\begin{array}{ll} p \rightleftarrows n + \pi^+ & n \rightleftarrows p + \pi^- \\ p \rightleftarrows p + \pi^0 & n \rightleftarrows n + \pi^0 \end{array}\right\} \tag{7-10}$$

It will be noted that each one of the reactions in (7-10) violates the law of energy conservation. Thus a proton with a mass equivalence of 938.26 MeV becomes a neutron with 939.55 MeV and ejects a pion with 139.58 MeV! In fact, this lack of energy conservation lasts for only a very short time, because the exchange pion will quickly become attached to a neutron to form another proton, and then the energy needed in the first step will be recovered. A similar situation exists for each of the other reactions listed in (7-10).

The temporary failure of energy conservation can be used to obtain a theoretical estimate of the mass of the pion. According to the uncertainty principle, an uncertainty in energy ΔE can exist for a time Δt according to

$$\Delta E \, \Delta t \simeq \hbar \tag{7-11}$$

In the pion exchange, ΔE can be taken as the energy equivalent of the pion mass, considered to be unknown. Δt will be the time between the ejection of the pion by one nucleon and its capture by another, at which time the energy imbalance will be removed. The nuclear force is known to have a range of about 1.5×10^{-13} cm, and if the pion travels this distance at the velocity of light, then

$$\Delta t = \frac{1.5 \times 10^{-13}}{3 \times 10^{10}} = 0.5 \times 10^{-23} \text{ sec}$$

$$\Delta E = \frac{6.58 \times 10^{-22}}{0.5 \times 10^{-23}} = 132 \text{ MeV}$$

This value is surprisingly close to the measured value of the pion mass. There is no longer doubt that the π-mesons are the exchange particles required by the Yukawa theory. Further discussions of pions and other subnuclear particles will be deferred to a later section.

7.06 Nuclear Potential Barrier

Most of our information about nuclear structure has come from scattering experiments, in which high-speed particles are directed against target nuclei. Highly sophisticated versions of the original Rutherford experiments provide information on the relations existing between nucleon, nuclear force fields, and subnuclear particles.

For many studies, protons have been the probing particles of choice. They are readily produced, can be accelerated to almost any desired energy, and interact strongly with nucleons. Very-high-energy electrons have also proved useful in studies of nuclear structure.

The existence of the attractive nuclear force does not appear to alter the coulomb field due to the nuclear protons. The combined force field will, then, be simply the sum of the two components. A positively charged particle, such as a proton, moving toward a nucleus will first experience a retarding

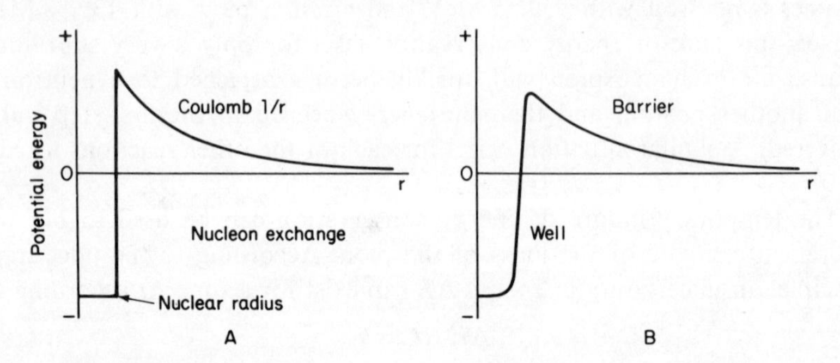

Figure 7-5. Idealized, (A), and more probable, (B), potential distribution around a nucleus.

force due to the relatively long-range coulomb field. As the proton moves in against this field, it loses kinetic and gains potential energy.

According to classical theory, the retarding field forms a *potential barrier* which must be surmounted if the positively charged particle is to enter the nucleus. If the particle originally had sufficient energy to carry it over the top of the potential barrier, it will come under the influence of the attractive nuclear force and will be pulled into the *potential well* of the nucleus. The distance at which the nuclear force just balances the coulomb repulsion may be taken as one measure of the nuclear radius. Figure 7-5A is an idealized diagram of the situation arising from the two forces. In the actual case the transition is undoubtedly less abrupt, Fig. 7-5B.

To a good approximation, the maximum height of the potential barrier can be calculated from the usual coulomb potential, Eq. (6-3), using a nuclear radius calculated from Eq. (7-9):

$$V_m = \frac{Zze^2}{r_0 A^{1/3}} \tag{7-12}$$

where Z and z are the atomic numbers of target and probe, respectively, and r_0 may be taken as 1.4 F. More precise values of V_m can be obtained by considering the radius of the probing particle as well as that of the target nucleus. Calculations of barrier heights for protons lead to values of several MeV, even in the light elements, with even higher values for heavier targets.

It has long been known from experiments that particles with energies well below the maximum height of the potential barrier are quite capable of penetrating it. This behavior, which is impossible according to classical mechanics, is readily explained by quantum mechanics. According to this, there is a finite probability that a particle of lower energy will penetrate or *tunnel* through the barrier instead of being forced to go over the top. As would be expected, the probability of penetration will depend upon the initial energy of the particle in relation to the height and thickness of the potential barrier. The probability of penetration increases as the particle energy approaches the height of the barrier, but does not become a certainty even for particles whose energy exceeds the barrier height. This behavior, like that of barrier tunneling, is completely unexplained by classical mechanics.

Quantum mechanics associates the probability of penetration with the uncertainty of position determination, which, in turn, depends upon the de Broglie wavelength of the incoming particle. According to the uncertainty principle, a 1-MeV proton will have at best an uncertainty in position of 4.5 F. This distance is of the order of nuclear dimensions, and if the incoming particle reaches a point on the potential barrier where the thickness is less than this, the particle might be located either inside or outside the nucleus. Similarly, a particle with an energy greater than the barrier height may not with certainty enter the nucleus.

7.07 The Nuclear Magneton

High-resolution optical spectroscopy disclosed a *hyperfine structure* in atomic spectra that could be explained in terms of neither the coulomb forces nor the magnetic properties of the orbital electrons. Pauli, in 1924, deduced correctly that this hyperfine structure was the result of a nuclear magnetic moment, arising from quantized levels of angular momentum in the nucleus. Considerations analogous to those which lead to the Bohr magneton, Eq. (6-30), serve to define a nuclear magneton:

$$\mu_N = \frac{\hbar e}{2M} \text{ joules tesla}^{-1} \tag{7-13}$$

The presence of M in the denominator means that the unit of nuclear magnetic strength is smaller than the Bohr magneton by a factor of about 1836. This is consistent with the measurements on hyperfine structure, which indicate that nuclear magnetic fields are relatively weak.

Corresponding to Eq. (6-32) for the atomic case we have

$$\mu = g\frac{e}{2M} \times \text{angular momentum} \tag{7-14}$$

By analogy, $e/2M$ is the *nuclear gyromagnetic ratio* γ and g is the *nuclear-g-factor*. Even before the concept of electron spin was utilized, Pauli had suggested that the angular momentum in Eq. (7-14) consisted of two parts, an orbital motion and a nucleon spin.

Neither the orbital nor the spin contributions to the nuclear magnetic moment is an integral number of nuclear magnetons, and thus the nuclear g factors have values differing substantially from unity. Three values are of particular interest:

$$\mu_p = 2.793 \qquad \mu_n = -1.913 \qquad \mu_d = 0.858$$

The magnetic-dipole moment of the proton is considerably larger than one nuclear magneton, but this is not very surprising in view of our incomplete knowledge of nuclear forces. More surprising is the fact that the neutron, with a net electric charge of precisely zero, has a magnetic moment $\frac{2}{3}$ as large as that of the proton. The negative sign of the neutron moment means that it has the direction expected from a negative charge moving in the spin direction. A comparison of the three values given shows that magnetic-dipole values are not additive. The measured value for the deuteron is distinctly smaller than the sum of the moments of its two constituent particles. Obviously much remains to be learned about the magnetic properties of the nucleus.

7.08 Nuclear Quantum Numbers

Although the nuclear magnetic moments are not integral multiples of the nuclear magneton, the general relations seem to parallel those of the orbital electrons. Because of the many apparent similarities between atomic and nuclear affairs, it seems logical to adapt the atomic scheme of quantum numbers to the nucleus with as few changes as possible.

We have, then, a principal quantum number n which can take on positive integral values 1, 2, 3, ... As in the atomic case, n is the sum of a radial quantum number v and an azimuthal or orbital quantum number l. In the absence of a strong central field, the principal quantum number does not play the dominant role in determining nuclear energy states.

The orbital quantum number l is a very important parameter in the nuclear domain. l can take on positive integral values 0, 1, 2, . . . , $(n - 1)$, corresponding to orbital angular momenta $\hbar\sqrt{l(l+1)}$. As before, l values are designated by the letter sequence $s, p, d, f, \ldots$, although now these do not have the original spectroscopic significance.

Spatial orientation of the orbital motion is quantized by the orbital magnetic quantum number m_l, which can take on values of 0, 1, 2, . . . , l.

Spin quantum number s has a magnitude of $\frac{1}{2}$, as in the case of electrons. Two spatial orientations of spin are possible in the presence of an external applied field, so the *magnetic-spin quantum number* m_s will have values of $\pm\frac{1}{2}$.

The total angular momentum of a nucleon will be the sum of the orbital and spin contributions, $j = l + s$. If $l = 0$, s can only equal $+\frac{1}{2}$ because a negative angular momentum appears to be without physical significance. For $l = 1$, $j = l \pm \frac{1}{2}$, or $\frac{1}{2}$ and $\frac{3}{2}$, and so on, which leads to the sequence of j values $\frac{1}{2}, \frac{3}{2}, \frac{5}{2}, \ldots$.

In quantizing spatial orientations both positive and negative quantum numbers are allowable, and so the *total magnetic quantum number* m can have a total of $(2j + 1)$ values $j, (j - 1), (j - 2), \ldots, (-j)$, each value being half-integral.

We have already noted that the proton and neutron may be simply two different states of a single common particle, the nucleon. This concept has led to the introduction of the *isobaric-spin* or isotopic-spin quantum number T. T_n for the neutron is arbitrarily given a value of $+\frac{1}{2}$, and T_p for the proton $= -\frac{1}{2}$. The total isotopic-spin quantum number for a Z–N nucleus will be

$T = \frac{1}{2}(N - Z)$. Isospin is a useful and an important quantum number that is conserved in nuclear transformations.

As we shall see in detail presently, the Pauli principle applies to nucleons. This principle may be stated: In a nucleus no two nucleons of the same type may have the same set of independent quantum numbers. The phrase "nucleons of the same type" serves to distinguish between neutrons and protons. Thus neither two neutrons nor two protons can have the same set of quantum numbers n, l, m_l, and m_s or their equivalents. However, one neutron and one proton may exist with identical sets of quantum numbers. In a sense, the difference in charge is a fifth quantization that assures that there are not two particles in identical states.

Nuclear quantum states can be designated by the same scheme that was used for designating electron states. Because of the decreased importance of the principal quantum number n in the nuclear case, it is preferable to replace it in the designation. Customarily, n is replaced by $\nu = (n - l)$. Letter designations are retained for l values, and j values are written as subscripts to the right. According to this scheme a nucleon with $n = 4$, $l = 2$, $s = -\frac{1}{2}$ would have $\nu = 2$, $j = \frac{3}{2}$, and would be written as $2d_{3/2}$.

7.09 Spin–Orbit Coupling

As in the case of quantized electron states, the quantum number for an aggregation of particles will be some sort of a sum of the quantum numbers for each constituent particle. As before, the quantum numbers for an aggregation will be denoted by capital letters instead of the lower-case letters used for individual designations.

There is no clear guide to the proper method of summing the l and s values of individual nucleons to obtain L, S, and finally I, the total nuclear angular-momentum quantum number. Two extreme summing methods can be distinguished, as in the atomic situation. Russell–Saunders or LS coupling assumes a weak interaction between the l and s momenta for each nucleon. Individual l's and s's are strongly coupled to form summed values L and S, representing total orbital and spin momenta, respectively. L and S then couple weakly to form I. Whatever the value of L, there will be $(2S + 1)$ values of spin and hence the same number of values of I. The factor $(2S + 1)$ is known as the *multiplicity* of the state.

At the opposite extreme, jj coupling assumes that each l and s couple strongly to form a series of j values. The j values are then summed to obtain I.

The two coupling schemes lead to identical values of total angular momentum and multiplicity. Different energy values are predicted by the two schemes, and it is thus possible to distinguish between them experimentally.

It would seem that *LS* coupling would be favored because the small magnetic forces in the nucleus would seem to preclude *jj* coupling. In fact, only *jj* coupling is observed in heavy nuclei, with some intermediate forms in light structures. The reasons for the absence of *LS* coupling are yet to be determined.

Nuclear angular momenta and dipole moments cannot be measured directly. These quantities can be determined from spectroscopic observations, primarily on the intensity patterns in rotational band spectra. Some important generalizations emerge from the experimental results.

In every even Z–even N nucleus the total nuclear angular momentum $I = 0$. This must mean that the lowest energy state is attained by the pairing of like nucleons, proton pairs and neutron pairs, in such a way that both orbital and spin-angular momenta cancel. In an odd Z–even N or an even Z–odd N assembly, angular momenta cancel in pairs, leaving the odd nucleon of either type to be responsible for the observed angular momentum, which will be $\frac{1}{2}, \frac{3}{2}, \frac{5}{2}, \ldots$. Only four stable odd Z–odd N nuclei are known and in each of these I is observed to have integral rather than half-integral values. Thus, angular-momentum cancellation does not extend to different nucleons. In these cases the lowest energy state occurs with angular-momenta adding. A comparison of I values with the natural abundances of the various Z–N combinations, Sec. 4.05, shows the stability achieved with low values of angular momentum.

7.10 Nuclear Parity

In the simple case of the linear oscillator we have noted that the eigenfunctions may be symmetrical or antisymmetrical with a change in sign of the spatial coordinate. The state of symmetry was denoted by the term *parity*, with a value of *even* or + for symmetry and *odd* or − for antisymmetry. In the general case with three spatial coordinates and one spin coordinate,

$$\psi(x, y, z, s) = \psi(-x, -y, -z, s) \quad \text{even or } +$$

$$\psi(x, y, z, s) = -\psi(-x, -y, -z, s) \quad \text{odd or } -$$

Note that the signs of the spatial coordinates only have been changed, the spin designation remaining constant.

Wave-function symmetries depend upon the values of the orbital angular-momentun quantum number l. It can be shown that the parity of a motion with an even value of l is even, while odd values of l give odd values of parity. The parity of a particle is the product of the parity of the angular-momentum state and the *intrinsic parity* of the particle. Table 7-1 lists the intrinsic parities for some of the commonly encountered particles.

TABLE 7-1
SOME INTRINSIC PARITIES

Electron	+	Pion	−
Proton	+	Muon	+
Neutron	+	Photon	+ or −
Neutrino	+		

The parity of the photon is particularly interesting, because it can have either value, depending upon the mode of generation. An electromagnetic wave can be generated in any one of several ways. Thus a photon may originate from an oscillating electric dipole, a process denoted as E_1; from an oscillating magnetic dipole, M_1; from an oscillating electric quadrupole, E_2; and so on to higher multipoles. Every photon has a nonzero value of angular momentum, described by a quantum number l as it is for other types of particles. For dipole radiation, $l = 1$; for quadrupoles, $l = 2$; and so on.

The parity of a photon will depend upon the value of l, and also upon whether its origin was electric or magnetic. Photon parities are given by

$$E \text{ parity} = (-1)^l \qquad M \text{ parity} = -(-1)^l$$

Some photon parity values are displayed in Table 7-2.

TABLE 7-2
SOME PHOTON PARITIES

Photon origin	l	*Photon parity*	*Nuclear P change*
Electric dipole E_1	1	−	Yes
Magnetic dipole M_1	1	+	No
Electric quadrupole E_2	2	+	No
Magnetic quadrupole M_2	2	−	Yes
Electric octupole E_3	3	−	Yes
Magnetic octupole M_3	3	+	No

The last column of Table 7-2 displays the yes–no notation commonly used to indicate parity changes in a nucleus undergoing some transformation.

A system of A particles such as a nucleus will have a parity equal to the sum of the parities of the individual components. That is,

$$\sum_1^A l = \text{even, parity even} \qquad \sum_1^A l = \text{odd, parity odd}$$

A system will have an odd parity only if it contains an odd number of odd-parity particles. Parity values for either a particle or an aggregation are frequently indicated by a right-hand superscript. Thus the $2d_{3/2}$ particle of Sec. 7.07 could be written $2d^+_{3/2}$.

The importance of parity lies in the fact that it is conserved in many types of nuclear transformations. For some time it was thought that all types of nuclear reactions conserved parity and hence were *P-invariant*. In 1957 experiments showed, following theoretical predictions by Lee and Yang, that parity is not conserved in weak interactions such as are involved in beta-particle decay or in some meson decays. The implications of this invariance failure, which is theoretically related to charge and time invariance (CPT invariance), is yet to be understood.

7.11 Quantum Statistics

The mathematical difficulties in the quantum-mechanical treatment of even a single particle preclude its application to many complex systems. Recourse must then be had to a statistical application of quantum-mechanical principles, just as statistical mechanics must be used to study many-body problems in the atomic or molecular domain.

One important difference distinguishes quantum statistics from the classical statistics of atoms and molecules. In the latter case each elementary particle is, in principle, distinguishable from every other. Only the size of the population under study prevents a detailed knowledge of each inhabitant. Quantum statistics, on the other hand, is applied to particles which are theoretically and practically indistinguishable. In the realm of nuclear particles our knowledge of position and momentum is so limited by the indeterminancy principle that the individual interacting particles can not be identified. For example, consider a scattering experiment in which a proton beam is being used to bombard a target containing protons. In such a situation it is impossible to tell whether the scattered proton was the incident particle or the target.

The inability to distinguish individuals has important consequences in quantum statistics. One of the important functions of statistical theory is to determine the distribution laws which govern the distribution of energy among the particles. The ability or inability to distinguish individuals makes an essential difference in the way in which populations are counted. When particles are distinguishable, statistical theory leads to the classical Maxwell–Boltzmann law of statistical mechanics.

Two cases arise when the basic particles are indistinguishable. It can be shown that an antisymmetric wave function cannot be constructed unless there is a change of sign in the solution when at least one pair of particles interchange completely (in x, y, z, and s). If two particles can exist with identical quantum states and numbers, they can be interchanged with no change of sign in the equations, and so, in this case, no antisymmetrical wave function can be constructed. Antisymmetric wave functions are known, and so these must require that not more than one particle of a given type be in any quantum state. This requirement will be recognized as the Pauli exclusion principle, which we have invoked on several occasions. Symmetrical wave functions, on the other hand, do not cause a sign reversal when particle pairs are interchanged, and in this case multiple occupancy is allowed in each quantum state.

The mathematical treatment that allows antisymmetrical wave functions is known as the Fermi–Dirac statistics; the case of symmetrical wave functions is handled by Bose–Einstein statistics. Now, antisymmetric wave functions are associated with half-integral values of spin, while particles with integral spins will have symmetric wave functions. The electron, proton, neutron, neutrino, and muon have half-integral spins, obey Fermi–Dirac statistics, and are known as *fermions.* Photons, pions, and the deuteron obey Bose–Einstein statistics and are called *bosons.* As we have seen, every nucleus with an odd number of nucleons will have a half-integral spin value and will be a fermion, while nuclei with even A values will be bosons.

The differences between the two types of particles can lead to important differences in physical properties. The nuclei of ^{4}He are bosons, and as such do not obey the Pauli exclusion principle. It is this fact that leads to some of the peculiar properties of liquid helium, properties that are not possible with a system of fermions.

7.12 Nuclear Models

In spite of years of intensive theoretical and experimental work, no completely satisfactory model of the atomic nucleus has been developed. Many models have been proposed, each capable of explaining some, but not all, of the nuclear characteristics. Even the most satisfactory of the proposed structures leaves something to be desired.

From the cyclic behavior of the lower portion of the nuclear-stability curve, it is tempting to think of a nucleus as an aggregation of alpha particles, with each group of four nucleons rather loosely bound to other similar groups. Alpha particles are ejected from the nuclei of some radioactive atoms and so it is apparent that these groups have at least a fleeting existence as an entity prior to emission. Such groups are, however, so transient, that one of the factors that enters into the probability of an alpha-particle emission is the probability of a pre-emission $2p$–$2n$ grouping. A considerable body of experimental evidence rules out the alpha-particle model, the strongest coming from scattering studies.

Each of two nuclear models, based on almost diametrically opposite postulates, has been quite successful in explaining, both qualitatively and quantitatively, a good many of the observed nuclear characteristics.

1. The *independent-particle model* assumes that the constituent particles interact only weakly with each other. Although there is no dominating central force field in the nucleus, each nucleon is assumed to move independently in a pseudo-central field produced by the other nucleons. The actual force field may indeed be nearly central in many cases, as shown by measurements of quadrupole electric moments.

Each nucleon moves independently in this field in a manner quite analogous to the movement of orbital electrons around a central nucleus. Each nucleon is characterized by quantized values of angular momentum, energy, and spin. Individual nucleon–nucleon interactions must be weak, a requirement that must be justified in the light of the known binding energies (2.2 MeV for the deuteron).

According to the independent-particle model, an incoming nucleon would interact with one, or at most a very few, of the nuclear occupants. The newcomer may be either scattered or captured and a strong interaction would be expected. However, with a quantized system in which the Pauli exclusion principle is operative, a normally strong interaction may be prevented from becoming manifest. If all possible quantum states are already occupied, there is simply no place for the incident particle to go. This situation is analogous to the atomic case where an incoming photon has a low probability of absorption if it has only enough energy to raise an orbital electron up to higher energy states which are fully occupied in accordance with the Pauli principle. In each case the Pauli principle acts to make a strong interaction appear weak.

Experimental evidence for discontinuities in many nuclear parameters at the magic numbers points to the existence of nuclear shells, each with a specific maximum number of occupants. As in the atomic case, the order of shell filling will depend upon the type of coupling that exists between the orbital and spin angular momenta. As in the atomic case, nuclear spin–orbit coupling may range from *LS* to *jj*, in which spin and orbit are strongly coupled in each nucleon, and the resultants are then summed to obtain the total angular momentum for the whole nucleus. There is no good theoretical reason to assume *jj* coupling in the nucleus, but this assumption leads to a prediction of shell occupancy that agrees with the concept of shell closures at the experimentally observed magic numbers.

In addition to correctly predicting the sequence of magic numbers, the shell model has been quite successful in dealing with nuclear angular momenta, parity determinations, and some of the nuclear magnetic moments. A good many nuclear properties cannot be explained by the independent-particle model, but it is clear that any complete nuclear model must include some sort of a shell structure.

2. The *strong-interaction*, or *liquid-drop, model* of the nucleus is unable to account for many of the properties properly predicted by the shell model.

On the other hand, the liquid-drop model does handle satisfactorily such properties as constant nuclear density, nuclear masses and binding energies, the energetics of alpha decay, beta decay, and fission, and the cross-sections for nuclear reactions. These properties do not respond to treatment by the independent-particle model, and it is evident that a complete nuclear model must contain a component with strong interactions.

According to the liquid-drop model, internucleon interactions are strong rather than weak. Nuclear quantum numbers apply in this case to the nucleus as a whole and not to a group of particles moving in a pseudo-central force field. A particle entering a liquid-drop nucleus will react promptly with the entire nucleus rather than with a single nucleon.

The two most successful nuclear models are based on such disparate assumptions that a merging into a single, complete theory seems very difficult or impossible. Much work in this direction needs to be done, and it well may be that the final, complete model will be quite different from anything currently visualized.

7.13 The Weizsäcker Semiempirical Mass Formula

No satisfactory complete theory for calculating nuclear masses and binding energies is available, but surprisingly good values can be obtained from a semiempirical formulation based on the liquid-drop model. A nuclear structure of A particles—Z protons and N neutrons—is related by analogy to a spherical liquid drop composed of many molecules. Weizsäcker first developed the formula by writing a series of terms, each with the analytic form known to be involved in the energy representation, but with undetermined coefficients. Best values of the coefficients were then determined from experimental data. With only five adjustable constants, the semiempirical relation is able to predict quite accurately nuclear masses for all nuclei except those at or near a magic number.

The Weizsäcker equation for the mass of a nucleus Z–N is

$$M = \underbrace{m_n(A - Z)}_{(a)} + \underbrace{(m_p)Z}_{(b)} - \underbrace{a_1 A}_{(c)} + \underbrace{a_2 A^{2/3}}_{(d)} + \underbrace{a_3 \frac{Z^2}{A^{1/3}}}_{(e)} + \underbrace{a_4 \frac{(A - 2Z)^2}{A}}_{(f)} + \underbrace{\delta}_{(g)} \quad \text{(term)} \tag{7-15}$$

The terms in Eq. (7-15) are interpreted as follows:

(a) is simply the neutron contribution to the mass.

(b) is similarly the mass contribution of the nuclear protons.

(c) is the mass equivalent of the energy lost from the system when the particles are brought together. In the liquid analog this term corresponds to the energy of cohesion that holds the molecules together. In the nucleus, (c) is the energy lost at assembly due to the work done by the nuclear force. This energy will be proportional to the nuclear volume, and consequently to A, and will be negative.

(d) arises because (c) has overestimated the energy lost through the nuclear force. Some of the particles, in either the liquid or the nuclear case, will be at the surface of the assembly. In this peripheral position they will not be subjected to the maximum forces of attraction that are exerted on more central particles completely surrounded by others. This "surface-tension" effect will be proportional to the surface area of the nucleus and hence to $A^{2/3}$, and will be positive.

(e) is the mass equivalent of the energy required by coulomb repulsion to bring Z charges together in the nuclear volume. With Z discrete charges, the energy is strictly $Z(Z-1)/A^{1/3}$ rather than the analytically simpler form given here. This term will be positive since it represents work done on the system.

(f) is another correction on term (c), to take into account the empirical fact that in the light nuclei the most stable structures have even Z–even N values with equal numbers of neutrons and protons, as ${}^{4}_{2}\mathrm{He}$, ${}^{12}_{6}\mathrm{C}$, ${}^{16}_{8}\mathrm{O}$, and ${}^{20}_{10}\mathrm{Ne}$. The effect becomes smaller with increasing A, and so a factor A is required in the denominator of (f).

(g) represents an odd–even effect, required because of the tendency of nucleons to pair off in forming states of lower energy, leaving any odd nucleon less tightly bound. The term δ has the form $1/A^{3/4}$, reflecting the decreasing importance of the pairing effect at the higher mass numbers. The pairing correction is

$$\left.\begin{array}{r}\text{subtracted for even } Z\text{–even } N\\ \text{added for odd } Z\text{–odd } N\end{array}\right\}\text{ even } A$$

$$\left.\begin{array}{r}\text{zero for even } Z\text{–odd } N\\ \text{zero for odd } Z\text{–even } N\end{array}\right\}\text{ odd } A$$

When Eq. (7-15) is simplified and the constants are inserted, we have for the *nuclear* mass in u:

$$M = \underbrace{1.008982(A-Z) + 1.007593(Z)}_{\mathcal{N}} - \underbrace{0.0151(A) + 0.015(A)^{2/3} + 0.000639(Z^2A^{-1/3}) + 0.0207(A-2Z)^2A^{-1} + \delta}_{\mathcal{B}}$$

$$\delta = 0.036A^{-3/4} \tag{7-16}$$

and in MeV:

$$M = \underbrace{939.55(A-Z) + 938.26(Z)}_{\mathcal{N}} - \underbrace{14.1(A) + 13.8(A^{2/3}) + 0.595(Z^2A^{-1/3}) + 19.3(A-2Z)^2A^{-1} + \delta}_{\mathcal{B}}$$

$$\delta = 33.5A^{-3/4} \tag{7-17}$$

In Eqs. (7-16) and (7-17), the $\mathcal{N}$ group of terms accounts for the total unbound masses of the constituent nucleons. The $\mathscr{B}$ group represents the energy *gained* by the nucleus in the assembly. $\mathscr{B}$ must, of course, be negative if the nucleus is to be stable against spontaneous decay. When the $\mathscr{B}$ group is written with reversed signs, it becomes a positive quantity which is just the binding energy E_B. In MeV units,

$$E_B = 14.1(A) - 13.8(A^{2/3}) - 0.595(Z^2A^{-1/3}) - 19.3(A - 2Z)^2A^{-1} + \delta \qquad (7\text{-}18)$$

The sign of the δ term must now be reversed from the schedule used in the mass calculation. Now,

$$\delta = \begin{cases} + \text{ for even } Z\text{–even } N \\ - \text{ for odd } Z\text{–odd } N \\ 0 \text{ for odd } A \end{cases}$$

Equations (7-16) and (7-17) are little used in direct calculations of nuclear masses because almost all of these values either have been measured with high precision or have been calculated from precision measurements of related parameters. We shall find the semiempirical mass equation most useful for predicting the behavior of various nuclear configurations against spontaneous or induced disintegrations.

REFERENCES

Aller, L. H., *Abundance of the Elements*. Interscience Publishers, New York, 1961.

Elton, L. R. B., *Introductory Nuclear Theory*. W. B. Saunders, Philadelphia, 1966.

Evans, R. D., *The Atomic Nucleus*. International Series in Pure and Applied Physics. McGraw-Hill, New York, 1955.

Lederer, C. M., J. M. Hollander, and I. Perlman, *Table of Isotopes*. 6th ed. Wiley, New York, 1967.

Lee, T. D., "Space Inversion, Time Reversal, and Particle–Antiparticle Conjugation." *Physics Today* **19,** 23, 1966.

Mattauch, J. H. E., W. Thiele, and A. H. Wapstra, "Atomic Mass Tables." *Nuclear Physics* **67,** 1, 1965.

Mayer, M. G. and J. H. D. Jensen, *Elementary Theory of Nuclear Shell Structure*. Wiley, New York, 1955.

Preston, M. A., *Physics of the Nucleus*. Addison-Wesley, Reading, Mass., 1962.

PROBLEMS

7-1. Using tabulated mass values, calculate the total binding energy and the binding energy per nucleon for ^{6}Li, ^{9}Be, ^{11}B, and ^{12}C.

7-2. Using tabulated mass values, calculate the mass defect and the packing fraction for ^{12}C, ^{31}P, ^{40}Ca, ^{63}Cu, and ^{209}Bi.

7-3. What is the density of nuclear material?

7-4. Assume a neutron star to consist of condensed nuclear material with a radius equal to that of the earth, 6.37 Km. What will be the potential energy of a neutron brought from infinity to the surface of the star?

7-5. How much energy will be released by the synthesis of 1 gram of ^{4}He from neutrons and protons? from two deuterons? from one triton and one proton?

7-6. Through what potential must an alpha particle be accelerated in order to give it enough energy to completely disrupt a ^{107}Ag nucleus?

7-7. A slow neutron enters a ^{14}N nucleus to form the compound nucleus ^{15}N. Assume the compound nucleus to have a specific heat of 1 cal g^{-1} and calculate the temperature rise that occurs before the binding energy is released.

7-8. Calculate the least uncertainty in position for a thermal neutron, $T = 0.025$ eV.

7-9. Calculate the maximum height of the potential barrier as seen by a proton as it approaches a nucleus of ^{6}Li; ^{56}Fe; and ^{197}Au. What will be the corresponding heights as seen by an alpha particle?

7-10. Show that the energy required to assemble an electric charge Q into a sphere of radius r is $3Q^2/5r$ ergs.

7-11. Assume that all of the mass of the electron is due to the energy required to assemble the charge and calculate the radius of the electron. Repeat the calculation for the proton.

7-12. Calculate the binding energies for the mirror nuclei $^{39}_{19}K$ and $^{39}_{20}Ca$ from the semiempirical binding-energy equation. What do you infer from the energies about the relative stabilities?

7-13. Use the semiempirical binding-energy equation to calculate the binding energies for the isobars ^{64}Ni, ^{64}Cu, and ^{64}Zn. Compare the relative stabilities.

7-14. Our sun has a mass of 1.98×10^{33} g and a radius of 7.4×10^{5} Km. Assume the mass to suddenly shrink until it attains nuclear density. What would be the radius of the new sun? How much energy would be available from 1 gram of solar material falling from the old radius to the new?

7-15. Assume that our universe contains 10^{81} nucleons. How large a sphere would be needed to contain them at the density of nuclear material?

8

The Ionizing Radiations

8.01 The Discovery of X Rays

During the 1890s, a large number of physicists were studying the conduction of electricity in gases at low pressure. Many of these workers were able to promptly confirm and extend the results reported by Roentgen in his monumental paper of December, 1895. Interest in the new "X ray" was immediate and intense. In a few months, the first cases of injury due to radiation overexposure were seen by physicians who neither were aware of the cause of the injury nor knew of any effective therapeutic measures.

The X ray was put to use in medicine long before its nature was understood. For many years, X rays were generated by an electrical discharge between cold-metal electrodes in a partially evacuated tube. Huge spark coils or frictional-electricity machines were used to generate a few tens of kilovolts to maintain the discharge.

Some residual gas was needed to sustain the current flow. As the tube was used, some of the gas was adsorbed on the walls and the discharge became hard to start. When a tube could be persuaded to operate under these conditions, it produced radiations that were somewhat more penetrating than before. Thus penetrating radiations became known as *hard*, and the less energetic ones as *soft*, terms still in use today.

The old gas-discharge tubes operated in a rather erratic fashion, and it became customary for the physician to check the radiation output before each patient exposure. No measuring instruments were then available; operation was checked most conveniently by holding the left hand close to the tube and examining the outlines of the bony structure with a zinc sulphide screen held in the right hand. This procedure might be repeated several times each day and as a consequence many malignancies originated in the overexposed left hands of the radiologists.

In 1913, W. D. Coolidge introduced the basic design features of the tubes in use today, Fig. 8-1. Coolidge tubes operate in a very high vacuum;

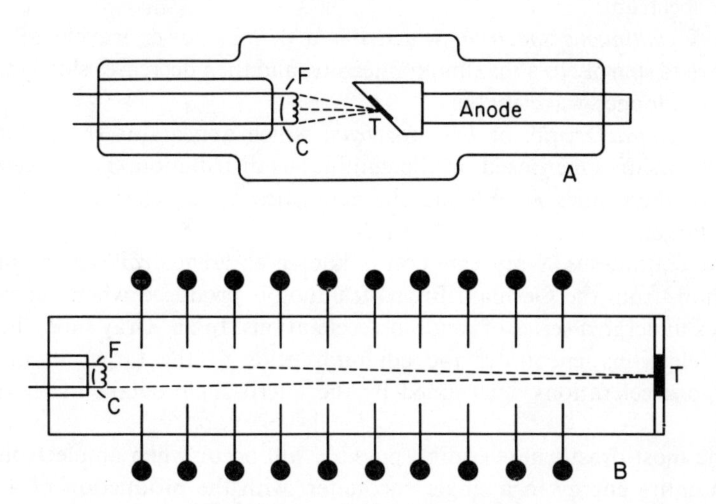

Figure 8-1. (A) A reflection target X-ray tube for voltages up to 500 KV. (B) A transmission target X-ray tube for voltages above 500 KV. The tube is sectionalized in order to distribute the potential evenly across the tube.

gas plays no role in the conduction of the current. Electrons are supplied by a hot cathode or filament, usually set in an electrode shaped to provide some focusing action for the electron beam as it is directed onto the anode or target. The anode–cathode voltage is so high that saturation current will flow—that is, every electron emitted by the filament will be accelerated to the target. Under these conditions the tube current, and consequently the quantity of radiation produced, is readily controlled by adjusting the filament temperature. X-ray quality can be independently varied by changing the accelerating voltage.

Recently, the cycle has been completed by the introduction of *field-emission tubes* in which the electrons are literally dragged out of a cold cathode by the intense electric field that exists near an electrode with a small radius of curvature, Eq. (2-1). In a field-emission tube, a high potential, derived from a charged condenser, is suddenly applied across a needlelike cathode and the anode. A current of many amperes will flow for about 10^{-6} sec, to produce a short pulse of high-intensity X rays.

8.02 Bremsstrahlung

When an X-ray beam is studied with a spectrometer, radiations covering a wide wavelength (or energy) range will be observed, even though a constant potential is applied to the tube. In general, two components can be identified in the spectrum:

1. A *continuous spectrum* which has a definite lower wavelength limit, rises sharply to a maximum intensity, and then decreases slowly toward the longer wavelengths.
2. A *characteristic*, or *line*, *spectrum* which appears as sharp peaks of emission superposed on the continuous distribution. The wavelengths of these lines are unique characteristics of the element used as the target.

The continuous X-ray emission is known as *bremsstrahlung*, or braking radiation, from the German. Bremsstrahlung is produced whenever moving charges undergo a series of random accelerations. In an X-ray tube, the high-speed electrons encounter the coulomb fields of the target atoms, and undergo accelerations determined by the microscopic details of the trajectories.

The most drastic acceleration possible will occur when an electron gives up its entire energy in a single encounter, with the production of a single photon. An electron accelerated through a potential V will have an energy Ve just as it enters the target. In general, the electron will be highly relativistic, although low-energy electrons are occasionally used to produce extremely soft X rays. For the limiting case,

$$Ve = h\nu_0 \tag{8-1}$$

where ν_0 is the maximum frequency that can be produced by potential V.

Encounters described by Eq. (8-1) are rare. The electron will more probably dissipate its energy over a series of interactions, to produce the continuous energy distribution that is characteristic of a bremsstrahlung spectrum. There is no theoretical lower limit to the photon energies produced by charge acceleration, but the probability of low-energy interactions is small, and the detection of the photons is difficult.

Kramers has derived an expression for the distribution of energy in a bremsstrahlung spectrum:

$$I_\lambda = \frac{KiZ}{\lambda^2}\left(\frac{1}{\lambda_0} - \frac{1}{\lambda}\right)d\lambda \tag{8-2}$$

where I_λ = energy emitted in wavelength interval from λ to $\lambda + d\lambda$
K = constant
i = electron current through the tube

Z = atomic number of the target
λ_0 = limiting wavelength from Eq. (8-1)

According to the Kramers relation the energy emitted will be proportional to the tube current, as would be expected. The appearance of Z in Eq. (8-2) reflects the effect of the large coulomb fields associated with the higher atomic numbers. Equation (8-2) predicts a wavelength of maximum energy emission, as can be seen by setting $dI_\lambda/d\lambda = 0$. This leads to

$$\lambda_M = \tfrac{3}{2}\lambda_0 \tag{8-3}$$

Figure 8-2 shows a comparison of the predictions of the Kramers theory with experiment.

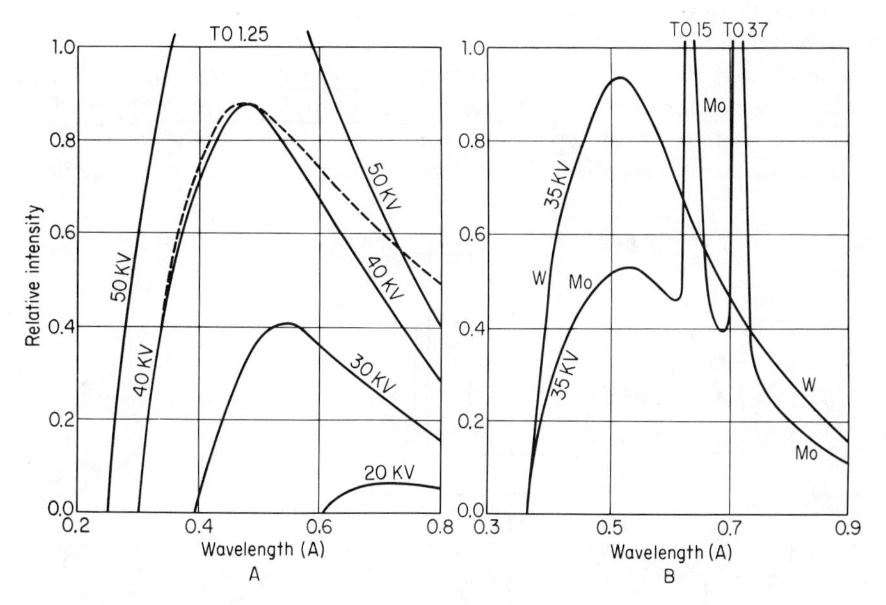

Figure 8-2. (A) X-ray spectra from a tungsten target. The dotted line is the spectrum as calculated from the Kramers relation for 40 KV. (B) X-ray spectra produced at 35 KV from a tungsten and a molybdenum target.

Kramers' analysis leads to a useful expression for the efficiency of total bremsstrahlung production:

$$\text{Eff} = \frac{\text{X-ray power output}}{\text{beam power input}} = VZ \times 10^{-6} \tag{8-4}$$

where V is the accelerating potential in KV. This expression cannot be exact, since it predicts an efficiency that increases with voltage without limit. It is, however, reasonably correct for potentials up to a few MV.

8.03 High-voltage Waveforms

The Kramers expression was derived on the assumption that a constant, or DC, voltage was applied to the X-ray tube. This is frequently far from the case. When spark coils were first replaced by transformers as high-voltage sources, the AC output of the transformers was applied directly to the X-ray tube. Electrons were accelerated only during the half-cycle when the target was positive with respect to the cathode. During this half-cycle the tube voltage varied from zero to some peak value denoted by KVP. Such an arrangement obviously produces a much greater proportion of low-energy photons than are predicted by the Kramers constant-potential relation. In general, these soft radiations are undesirable.

Next, half-wave or full-wave rectifiers were added, either with or without a condenser, Fig. 8-3. If the capacitance is sufficiently large, it will give up charge and will maintain the circuit voltage nearly constant during the low-voltage portions of the cycle, and will absorb charge at the peaks of the cycle. Such an arrangement is said to be a constant potential or CP circuit.

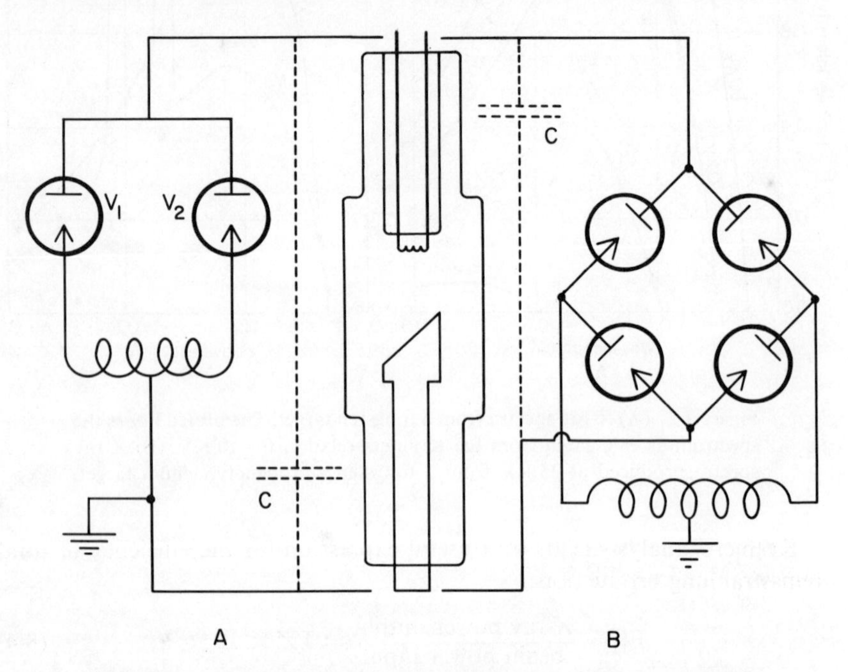

Figure 8-3. (A) A two-valve full-wave rectifier used either with or without a condenser C. (B) With a bridge rectifier the full voltage of the transformer is applied to the X-ray tube.

8.04 Characteristic X-ray Spectra

With spectrometers of moderate resolution, characteristic X-ray spectra are much simpler than the optical spectra of the same elements. Strong lines appear in groups or *series* whose wavelengths shift regularly with the atomic number of the emitting element. A given series of lines is essentially similar in all elements, except for the wavelength shift. The group of lines with the shortest wavelength is known as the K series, followed in succession by the L, M, N, series, and so on. Individual lines are identified as K_α, K_β, L_α, and so on.

Moseley, in an outstanding investigation showed, in 1913, that the wave number for any particular series line is given by

$$\tilde{\nu} = \frac{1}{\lambda} = R(Z - S)^2\left(\frac{1}{n_1^2} - \frac{1}{n_2^2}\right) \tag{8-5}$$

where Z is the atomic number of the target and where n_1 and n_2 are integers characteristic of the particular line. Except for the factor $(Z - S)$, Eq. (8-5) is identical in form to Eq. (6-4). Obviously, characteristic X-ray emission involves orbital-electron transitions closely analogous to those of optical spectroscopy. The factor Z is required to account for the increased coulomb field associated with Z nuclear charges. S is a *screening factor*, which is needed because orbital electrons in outer shells will be partially shielded from the nuclear charge by the negative charges in inner shells.

Moseley showed that the K_α wave numbers for a large number of elements were determined by putting $S = 1$, $n_1 = 1$, and $n_2 = 2$ in Eq. (8-5). Wave numbers for L_α lines came from $S = 7.4$, $n_1 = 2$, and $n_2 = 3$. Thus the K_α line is emitted when an electron in the L shell falls into a K vacancy. An L_α is produced by a transition from the M shell to an L vacancy. L electrons are less tightly bound than are those in the K shell; hence L-series lines occur at longer wavelengths than those of the K series.

Later work has shown that each of the lines reported by Moseley has a fine structure, being made up of several lines closely spaced in energy. Thus we have the $K_{\alpha 1}$, $K_{\alpha 2}$, $L_{\alpha 1}$, and so on. Each of these lines obeys a relation of the form of Eq. (8-5) with a characteristic value of S.

Shell occupancy is strictly controlled by the Pauli principle; so the first requirement for characteristic X-ray emission is the creation of a vacancy in an inner shell. This is readily accomplished by an energetic electron accelerated by a high potential in an X-ray tube. As this electron enters a target atom, it may eject an orbital electron as, say, an occupant of the K shell. In general, there will be no vacancy in outer shells and so the incoming electron must impart sufficient energy to the K occupant to remove it completely from the atom. This will require somewhat more energy than will be emitted at any one of the subsequent transitions. If the incoming electron lacks the energy required to eject a *K* occupant, there can be no *K* vacancy and hence no *K*-line emissions, Fig. 8-2B.

A *K*-shell vacancy is most readily filled from the *L* shell with the emission of K_α; if an *M* electron is involved, K_β will be emitted. Vacancies thus created in outer shells will be successively filled, each transition producing a characteristic spectral line, until the outermost or valence electron is finally restored. Figure 8-4 shows the electron shell structure of silver, $Z = 47$. The

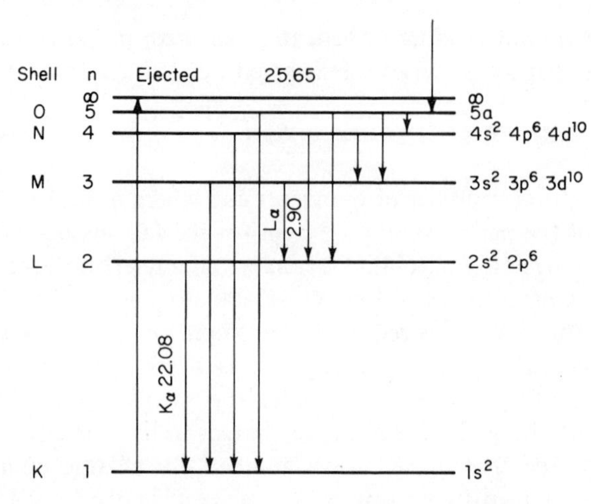

Figure 8-4. The electronic shell structure of silver, $Z = 47$. Each of the higher values of *n* will have subshells in accordance with the electron configurations shown, leading to fine structure in the X-ray line emissions. Energies are in KeV.

K, *L*, and *M* shells are filled and the 18 *N* shell electrons fill the *s*, *p*, and *d* subshells. In silver the energy levels are such that the last electron perfers *O*-shell occupancy to starting a new *N* subshell. Figure 8-4 is an energy-level diagram for silver showing the transitions and the associated X-ray emission lines.

8.05 The Discovery of Radioactivity

Within a year after Roentgen's discovery of X rays, Becquerel observed that uranium salts emitted radiations capable of exposing a photographic plate. The importance of this discovery was emphasized a few years later when Marie and Pierre Curie succeeded in isolating, from tons of uranium ore, small quantities of two substances far more active than uranium. This discovery of polonium and radium led the way to the isolation and identification of a large number of naturally radioactive nuclides.

It was soon observed that three types of radiations were emitted by radioactive elements. Some of the properties of these radiations differ only in degree. The most clearcut distinction comes from their behavior in a magnetic field. One type, called alpha particles, was deflected only slightly by a magnetic field in the direction expected of positive charges, Fig. 8-5. The large radius of curvature suggested a small value of e/m and consequently a relatively large mass.

Positive identification of the alpha particles was made by Rutherford. He enclosed a sample of radon gas in a glass tube whose walls were thin enough to permit the passage of the alpha particles, Fig. 8-6. The thin-walled

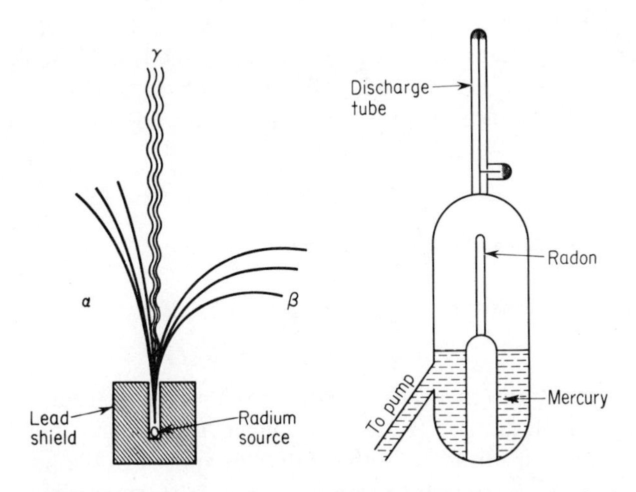

Figure 8-5. The response of the three types of radiation to a magnetic field.

Figure 8-6. The Rutherford apparatus for determining the nature of alpha particles.

tube was then sealed into a glass system equipped with electrodes to form a discharge tube. After several days, there was sufficient gas in the space, originally evacuated, to support an electrical discharge. The characteristic emission lines of helium were observed, furnishing positive evidence that the alpha particles were He^{++} ions.

Beta particles emitted by radioactive nuclides were deflected by magnetic fields as if they were high-speed electrons. Identity of charge and mass with electrons was quickly established. Many years later, Goldhaber showed that the particles are identical even in sophisticated wave-mechanical properties.

If beta particles differ from electrons in any property such as parity or spin, the Pauli exclusion principle would allow them to be captured in atomic energy states, even though these states had a full complement of electrons. Subsequent excitation should then lead to the emission of characteristic X rays differing from those normally observed. In fact, a completely normal X-ray spectrum is found, and the identity of beta particles and electrons is established.

The emissions that were undeflected by a magnetic field were identified as electromagnetic radiation, and were called gamma rays. Careful time-sequence measurements show that in most cases gamma-ray emission follows the emission of an alpha or a beta particle. In these transitions, gamma-ray emission is the mechanism by which a nucleus loses energy in going from an excited state to the ground state.

Except for the mode of generation, an X-ray photon and a gamma-ray photon of the same energy are identical in all respects. One difference in energy designation must be kept in mind. For example, every photon in a 1-MeV gamma-ray beam will have an energy of exactly 1 MeV. A beam of 1-MeV X rays, on the other hand, will contain photons with energies that range from 1 MeV down to zero, with an average energy well below the maximum.

8.06 Radioactive Decay by Alpha-particle Emission

An alpha-particle transition can be represented by the type reaction

$$ {}^{A}_{Z}X \longrightarrow {}^{A-4}_{Z-2}Y + {}^{4}_{2}\alpha + \gamma + Q \tag{8-6} $$

where X represents the chemical symbol for the parent element, and Y that for the daughter. Both mass or nucleon number A and atomic number Z are conserved in the transition. Note that Eq. (8-6) applies to the nucleus only, with no reference to the orbital electrons. When an alpha particle is emitted, the orbital structure must promptly lose two electrons, to agree with the lower atomic number of the daughter element, but this is not shown in the type reaction.

If the alpha-particle emission leaves the daughter nucleus in an excited state, the drop to the ground state will be accomplished by the emission of one or more gamma rays. In many transitions only a fraction of the alpha-particle emissions leave the daughter in an excited state; the rest lead to the ground state directly, after the emission of a more energetic alpha. The factor γ in Eq. (8-6) is symbolic only. In some transitions there will be no gamma emission, in others several photons may follow the particle emission. In general, the photon emission will follow the particle in 10^{-6} sec or less. Occasionally, the excited state will be *metastable*, and gamma emission may be delayed for hours or even days.

The Q term in Eq. (8-6) represents the total energy released in the transition. If the exact masses of all of the components are known, a calculation of Q will give the total kinetic energy of the alpha particle and the recoiling nucleus. Conversely, if the kinetic energy of the alpha particle is known, the recoil energy of the daughter nucleus can be calculated from the requirements of momentum conservation, and these values may then be used to determine the mass of one of the constituents. In this type of calculation, care must be taken to use consistent values of either nuclear or atomic masses.

Before energy calculations can be made it is necessary to know the *decay scheme* of the transition. A decay scheme is a detailed description of the steps by which a parent nucleus arrives at the ground state of the daughter. A typical alpha-particle decay scheme is shown in Fig. 8-7.

Positive-particle emission is conventionally depicted by an arrow drawn down and to the left, as in either Fig. 8-7A or 8-7B. Energy levels in the

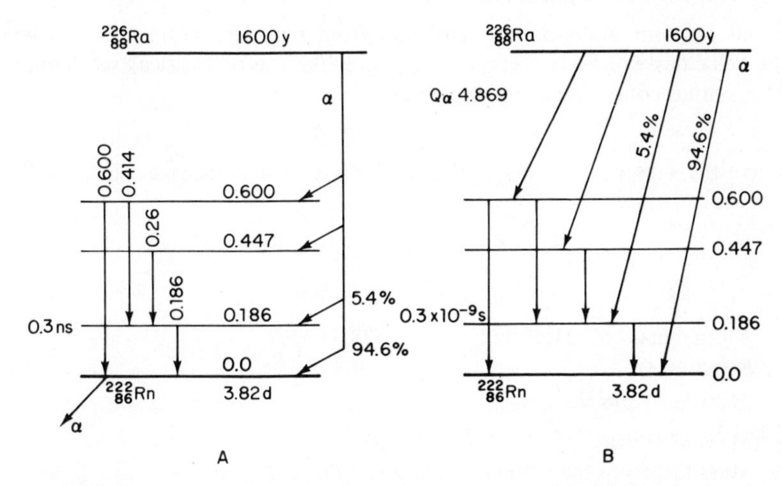

Figure 8-7. Two ways of depicting the decay scheme of ^{226}Ra. In both (A) and (B) positive-charge emission is represented by an arrow down and to the left.

daughter nucleus are shown schematically, with only a rough attempt to depict relative values. Each level will be labeled with its energy in MeV above the daughter ground state. Gamma-ray energies are sometimes given, or they may be obtained by a subtraction of the appropriate energy.

Relative percent occurrences are given when they are known and form an important part of the decay scheme. Thus in Fig. 8-7 the alpha transitions to the ground state and to the 0.186-MeV excited state account for essentially 100 per cent of the decays. The other two transitions are known but occur very infrequently. From the decay scheme we see that the most prominent gamma ray is the 0.186-MeV emission that follows 5.4 per cent of the alpha emissions with a delay of about 0.3 nsec. The intense gamma-ray activity associated with old sealed radium sources arises from the decay of the radon and subsequent daughter products.

The complete decay scheme will show other data relating to the transition which have been omitted from Fig. 8-7 for clarity. The missing features will be discussed later, in connection with the details of particle emission. Our present information is sufficient for making mass–energy calculations.

Illustrative Example

Calculate the atomic mass of ^{222}Rn from the following data:

Atomic mass of ^{226}Ra:	210,537.72 MeV
Atomic mass of ^{4}He:	3,728.34
Most energetic alpha from ^{226}Ra:	4.782

In all cases of alpha-particle emission from naturally radioactive nuclei, the velocities will be low enough to permit the use of classical mechanics. Momentum conservation requires that

$$m_\alpha v_\alpha = m_r v_r$$

where the subscript r refers to the recoiling daughter nucleus. Then, in the present example,

$$T_r = 4.782 \times \frac{4}{226} = 0.085 \text{ MeV}$$

$$Q = 4.782 + 0.085 = 4.867$$

Atomic mass of ^{4}He:	3,728.34 MeV
Reaction Q:	4.87
Mass lost from the system:	3,733.21
Mass of ^{226}Ra:	210,537.72
Mass lost from the system:	3,733.21
Mass of ^{222}Rn:	206,804.51 MeV

$$206{,}804.51 \times 1.074 \times 10^{-3} = 222.1080 \text{ u}$$

Note that by using the atomic mass of neutral helium instead of the mass of the alpha particle He^{++}, we accounted for the mass of the two orbital electrons released in the decay. The final value is the atomic mass of neutral ^{222}Rn.

8.07 Negative-beta-particle Emission

Although free electrons do not exist inside the nucleus, a beta particle originates there and must pass the nuclear potential barrier to escape. A negatron, or negative-beta-particle, transition is equivalent to the nuclear conversion

$$^1_0n \longrightarrow {}^1_1p + e^- \tag{8-7}$$

This leads to the type equation

$$^{A}_{Z}X \longrightarrow {}_{Z+1}^{A}Y + \beta^- + \nu + \gamma + Q \tag{8-8}$$

As in the alpha-particle case, mass number and atomic number are conserved. The loss of a negative charge from the nucleus by Eq. (8-7) increases the atomic number by one unit, but leaves the mass number unchanged, since one nucleon has merely changed its identity. The γ term allows for the possibility that the daughter nucleus will be formed in an excited state. As before, this term is symbolic, since some transitions go directly to the ground state of the daughter, while others may require several photon emissions to reach the same energy level.

Alpha-particle emissions are essentially monoenergetic, with at most a few discrete energies as some of the transitions go to excited states. Beta particles, on the other hand, show a continuous energy distribution ranging from a characteristic maximum value down to zero. This energy distribution seemed to pose a serious threat to energy conservation, since both parent and daughter were in definite energy states, yet seemed to be connected by a variable energy difference.

This dilemma was solved by Pauli, who postulated the simultaneous emission of a second particle, the neutrino. Charge and mass conservation required that the neutrino be neutral and have a mass that at most would be a small fraction of the electron mass. The neutrino was required to have energy, momentum, and spin, yet its properties were such that it would interact very, very weakly with any matter through which it passed. These properties are just those that would make detection of the particle difficult, if not impossible. However, the particle has been detected, as a particle with zero rest mass, emitted with the velocity of light.

The constant amount of energy available at each beta transition is shared between a neutrino and a beta particle in accordance with calculable probabilities. Tabular values of beta-particle energies list the maximum energy of the transition, unless otherwise stated. Local energy deposition by beta particles, as measured by direct absorption in a calorimeter, is determined by the average value of the energy spectrum, rather than by the maximum.

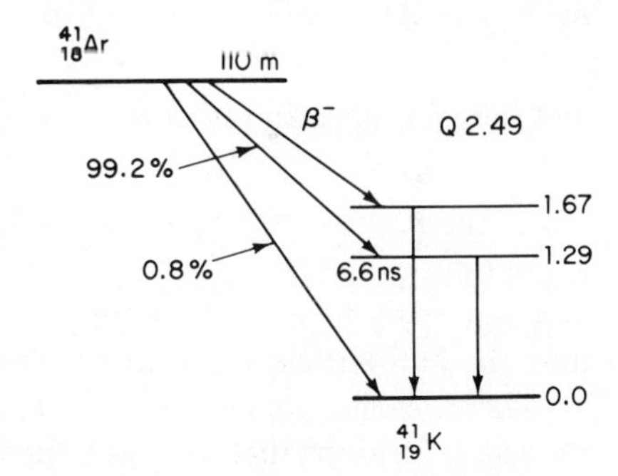

Figure 8-8. Negatron emission is indicated in the decay scheme by an arrow down and to the right.

The ν term in Eq. (8-8) represents the neutrino. Strictly, the particle that accompanies a negative beta particle is an *antineutrino* $\bar{\nu}$, but the distinction does not concern us here and it will not be made.

No account of the neutrino is taken in the energy diagram, Fig. 8-8.

Negative-particle emissions are depicted by arrows directed down and to the right, with vertical arrows representing gamma-ray emission.

In energy calculations relating to beta transitions, the kinetic energy of the recoiling daughter nucleus is usually neglected since it is negligible compared to the energy of the ejected particle. No account need be taken of the mass of the ejected negatron, since this is exactly compensated by the electron that must enter the orbitals to satisfy the increased atomic number.

Illustrative Example

Calculate the atomic mass of ^{41}Ar from the following data:

Atomic mass of ^{41}K:	40.97847 u	
Energy of most common beta:	1.20	MeV
Gamma-ray energy:	1.29	MeV

Atomic mass of ^{41}K:		40.97847 u
Beta-particle energy:	$1.20 \times 1.074 \times 10^{-3} =$	0.00129
Gamma-ray energy:	$1.29 \times 1.074 \times 10^{-3} =$	0.00139
	Atomic mass of parent ^{41}Ar	40.98115 u

This type of calculation permits the determination of many atomic masses not amenable to direct measurement.

8.08 Positron Emission

Negatron emission is an example of the radioactive decay of a nuclide with a neutron excess. This species of nuclide will lie to the left of the stability line, Sec. 7.01 and will tend toward stability by the conversion of a neutron to a proton according to Eq. (8-7).

Neutron-deficient nuclei are also known and these may emit a positive beta particle as they go to a lower-energy, more stable configuration. The nuclear conversion is equivalent to

$$^{1}_{1}p \longrightarrow {}^{1}_{0}n + e^{+} \tag{8-9}$$

and the type equation is now

$$^{A}_{Z}X \longrightarrow {}_{Z-1}^{\;\;A}Y + \beta^{+} + \nu + \gamma + Q \tag{8-10}$$

All of the terms in Eq. (8-10) have the same meanings as in previous usage. A continuous energy spectrum is seen in positron transitions, and again an accompanying neutrino is required. In this case it is truly a neutrino rather than the antiparticle associated with negative beta particles.

There is one important difference between negative and positive beta-particle transitions that does not appear explicitly either in the type equation or in the energy diagram. Since positron emission *decreases* the atomic number by one unit, the orbitals must lose one electron as soon as the nucleus ejects the beta particle. Thus the atomic mass of the daughter *must* be at

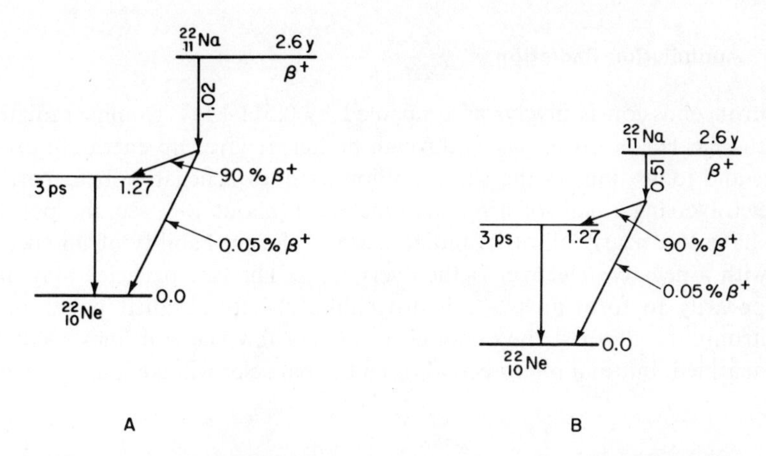

Figure 8-9. (A) Atomic decay scheme of positron emitter ^{22}Na. (B) Nuclear decay scheme of ^{22}Na showing a loss of only one electron mass.

least two electron masses lighter than the parent. There is no comparable requirement in negative-beta-particle emission.

The decay schemes previously presented refer to atomic energy levels. It is now instructive to compare atomic and nuclear energy levels in a positive-beta-particle transition. Figure 8-9A represents the atomic decay scheme of ^{22}Na. The first step in the decay is the loss of 1.02 MeV of mass, corresponding to the two electrons that have left the atom. The slant arrows to the left indicate positive-particle emission, with 90 per cent of the transitions going to the 1.27 MeV state of ^{22}Ne and a very small number going directly to the ground state. Electron capture, to be described later, accounts for the missing 10 per cent.

The nuclear energy diagram, Fig. 8-9B, shows first a loss of 0.511 MeV as the beta particle leaves the nucleus. From this point on, the nuclear diagram is identical with the atomic diagram, although the absolute values of the nuclear levels are lower by the energy of 10 orbital electrons.

Illustrative Example

Calculate the atomic mass of ^{22}Na from the following data:

Atomic mass of ^{22}Ne:	21.99982 u	
Energy of most common beta:	0.54	MeV
Gamma-ray energy:	1.27	MeV

$$
\begin{array}{llr}
\text{Atomic mass of daughter:} & & 21.99982\ \text{u} \\
\text{Mass of two electrons:} & 1.02 & 0.00110 \\
\text{Beta-particle energy:} & 0.54 \times 1.074 \times 10^{-3} = & 0.00058 \\
\text{Gamma-ray energy:} & 1.27 & 0.00136 \\
\hline
\textit{Atomic mass of } ^{22}\text{Na } \textit{parent:} & & 22.00286\ \text{u}
\end{array}
$$

8.09 Annihilation Radiation

Positron emission is always accompanied by 0.511-MeV gamma radiation. As the ejected positron passes through matter, it gives up energy to excitations and ionizations in the same fashion as does a negative beta particle. After traversing a path of a few millimeters in about 10^{-6} sec, the positron will have lost nearly all of its initial energy. The probability of an encounter with a negative electron is then very high. The two particles may unite temporarily to form an e^+e^-, hydrogenlike structure called positronium. Positronium may exist long enough to emit a few spectral lines, and thus be identified, but in a microsecond or so the particles will coalesce according to

$$e^+ + e^- \longrightarrow 2h\nu \tag{8-11}$$

The initial linear momentum of the system was nearly zero and hence the *annihilation radiation* must consist of two oppositely directed photons, each with an energy of 0.511 MeV. Two counters, arranged to intercept these photons, and connected to respond only to coincident pulses, can be used to locate a source that is emitting positrons.

Equation (8-11) is one example of the prompt annihilation of matter–antimatter pairs upon collision. Our earth, and indeed our neighbors in the solar system, appear to be made up of particles in the matter system. Any antimatter particle in this system suffers swift annihilation as soon as its velocity is low enough to permit an interaction of the type of Eq. (8-11). This is the reason that some tables give the life of the positron as about 10^{-6} sec. Inherently, the positron is as stable as its negative counterpart, and in an antimatter environment the latter particle would appear to have the fleeting existence.

8.10 Electron Capture

Some neutron-deficient nuclei must decay by a $p \longrightarrow n$ conversion, but have daughter products whose mass is greater than the maximum acceptable for positron emission. These nuclei can only decay by the capture of one of the orbital electrons. The negative charge thus acquired by the nucleus reduces the atomic number by one unit and yields the same daughter that would have been produced by positron emission, had this been energetically possible.

An electron-capture transition has no specific energy requirement, so this mode of decay is in competition with positron emission. It is this competing mode that accounts for the missing 10 per cent of the ^{22}Na transitions in Fig. 8-9.

The wave functions of the orbital electrons locate the *K*-shell occupants closest to the nucleus, and in some cases these electrons may spend part of their time inside it. *K*-electron capture is, therefore, more probable than *L*, *M*, ... capture, although some of these are known. Electron-capture transitions were formerly known as *K* captures, designated by *K* rather than the more general EC.

Electron capture is represented by the type reaction

$$^{A}_{Z}X + e^- \longrightarrow {}_{Z-1}^{A}Y + \nu + \gamma + Q \tag{8-12}$$

Although the daughter product in Eq. (8-12) is the same as that resulting from a positron decay, the neutrino emissions are quite different. In the positron case the energy of the transition will be divided between the positron and the neutrino, with the latter carrying off roughly two-thirds of the available energy. In an EC transition there will be a relatively simple *line spectrum* of neutrino energies, corresponding to the few exicted states of the daughter. Thus in Fig. 8-10 there are two neutrino energies, corresponding to decays to the ground state and to the 0.320-MeV state of ^{51}V.

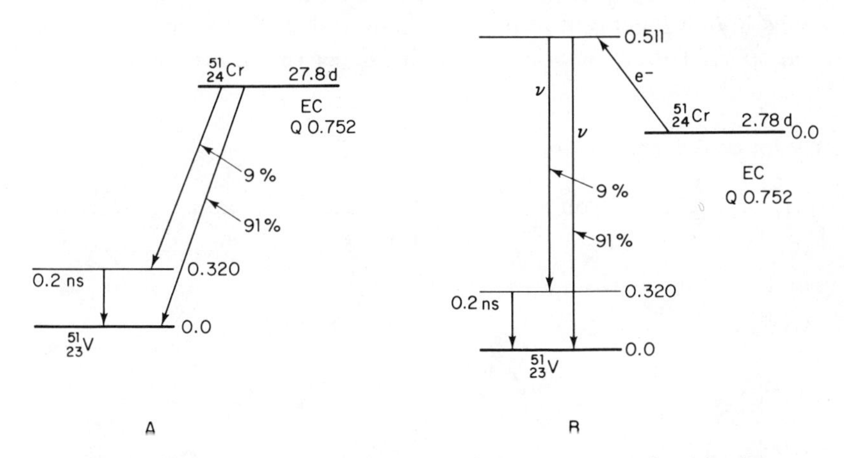

Figure 8-10. (A) Atomic decay scheme of electron capture in ^{51}Cr. (B) In the nuclear diagram the first step depicts the acquisition of one electron mass from the orbitals.

Neutrino emissions are shown more explicitly in the nuclear energy diagram, Fig. 8-10B. In the first phase of the decay, the nucleus gains 0.511 MeV as it captures an orbital electron. The subsequent neutrino emissions, since they involve no charge transfer, are shown by vertical lines.

The nuclear energy diagram suggests that a daughter nucleus might be formed by an EC transition with a mass even greater than that of the parent. Figure 8-11 shows the energy relations in the EC decay of ^{7}Be. Ten per cent

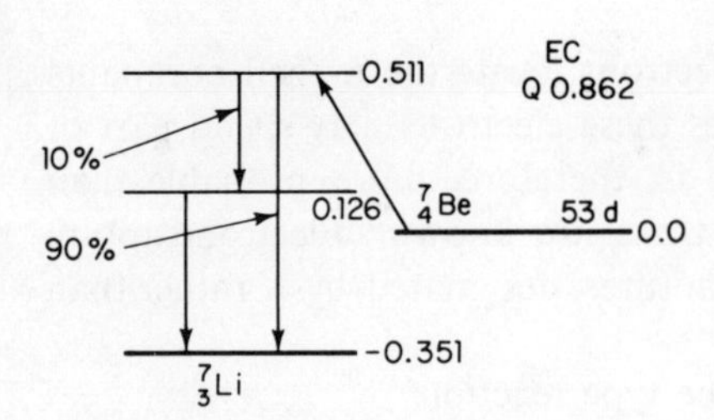

Figure 8-11. In the nuclear diagram the excited state of the ^{7}Li daughter is seen to have a mass greater than that of the parent.

of the decays go to an excited state that is 0.126 MeV heavier than the parent nucleus. The subsequent emission of an 0.477-MeV gamma ray drops the nuclear mass to 0.351 MeV below that of the parent. Note that in Fig. 8-11 all energies are referred to the parent instead of the daughter, as is customary.

Following electron capture there will be an orbital vacancy, usually in the *K* shell, and this will promptly fill from outer orbits. These transitions will produce the characteristic X-ray spectrum of the daughter element. If all of the neutrino transitions go to the ground state, the characteristic X rays and a weak bremsstrahlung radiation furnish the only means of detecting an EC transition. In general, these radiations are soft and detection is difficult. It is customary, for example, to assay ^{51}Cr, Fig. 8-10, through the 0.320-MeV gamma ray, even though it is emitted in only 9 per cent of the transitions, rather than to make use of any of the soft radiations associated with the 91 per cent transition.

8.11 Internal Bremsstrahlung

When an orbital electron is captured by a nucleus, there is an acceleration of the negative charge, and consequently bremsstrahlung will be produced.

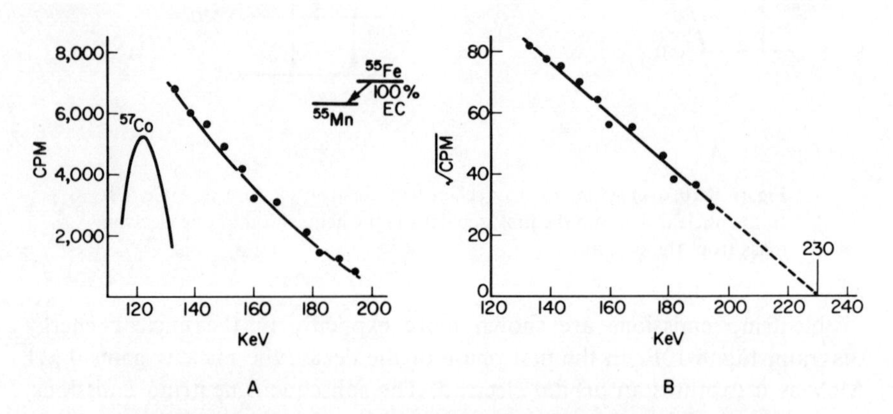

Figure 8-12. (A) Bremsstrahlung count rate from the EC transition in ^{55}Fe. (B) The square root plot extrapolated linearly shows a transition energy of 230 KV.

We would expect the accelerations to vary over a wide range, in accordance with the microscopic details of each individual capture. Correspondingly, we would expect a continuous bremsstrahlung spectrum, with a maximum energy equal to the energy of the transition. The photons thus produced are known as *internal bremsstrahlung*.

The energy distribution of internal bremsstrahlung is complicated at the lower energies, but near the upper end the square root of the probability (and hence the count rate of a sample) is nearly proportional to the energy. Internal bremsstrahlung can be best demonstrated in a transition that is uncomplicated by other photon emissions. Electron capture in ^{55}Fe is a good example, Fig. 8-12. As the energy diagram indicates, ^{55}Fe decays by electron capture, with 100 per cent of the transitions going to the ground state of ^{55}Mn. The square-root plot shown in the illustration extrapolates to an energy of 0.230 MeV, in good agreement with the accepted transition energy of 0.232 MeV. The 0.122-MeV gamma ray from ^{57}Co, also plotted, served to calibrate the energy scale.

8.12 Internal Transition

Any excited state in the decay schemes previously discussed had a transitory existence, as can be seen from the lifetimes shown. Some excited states may exist for long periods of time and in these cases the entire decay may be thought of as two separate events.

Figure 8-13 illustrates the decay of ^{99}Mo, with about 1 per cent of the beta transitions omitted for clarity. Gamma emission from the 1.10-MeV state is prompt, and so all beta transitions lead immediately to the 0.143-MeV state of ^{99}Tc. This state has a half-time for decay of 6 hours and is designated as a *metastable state*, ^{99m}Tc. Gamma radiation from this state leads by *internal transition* (IT) to the ground state, ^{99}Tc. This type of transition may be thought of as a case of nuclear isomerism, since there is no change except in the energy content of the nucleus.

The particular nuclide shown is used extensively in diagnostic medicine. A ^{99}Mo source with its 67-hr half-life serves as a generator of ^{99m}Tc, which is extracted periodically.

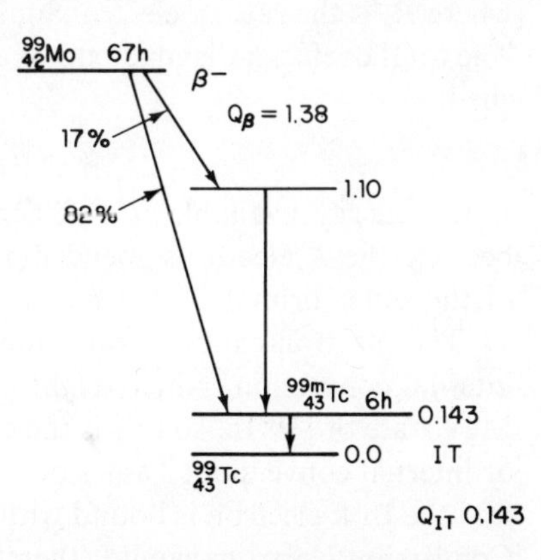

Figure 8-13. The 0.143 excited state of ^{99m}Tc decays by internal transition to the ground state.

8.13 Internal Conversion

According to quantum mechanics, some of the orbital electrons spend an appreciable time near, or actually inside, the nucleus. This fact permits electron capture and also allows the nucleus to exert strong forces on the orbitals.

When a nucleus is in an excited state, it tends to relieve the excitation by the emission of a gamma ray. In some cases, however, the gamma ray is not emitted, and the energy involved is transferred to one of the orbital electrons which will be ejected from the atom. Some of the available energy is required to supply the energy with which the electron is bound to its nucleus; the remainder goes into kinetic energy. Energetically, the process appears to be an internal photoelectric effect, with the energies divided as required by Eq. (4-21). In fact, the gamma photon is never emitted. The energy is transferred directly to the orbital electron and the process is known as *internal conversion.*

All energy states involved in internal conversion have definite values and the electron emission is monoenergetic. There will be a different electron binding energy associated with each shell, and hence a series of electron energies are to be expected, but the spectra will show lines rather than a continuum. No neutrinos are involved in internal conversion. As in electron capture, the process is completed by the emission of the characteristic X-ray spectrum.

In general, internal conversion competes with gamma emission. This competition leads to the definition of an internal conversion coefficient α, as

$$\alpha = \frac{R_e}{R_\gamma} \tag{8-13}$$

where R_e is the rate of electron emission and R_γ the rate of gamma emission. The total coefficient is made up of a sum of coefficients, one for each electron shell:

$$\alpha = \alpha_K + \alpha_L + \dots \tag{8-14}$$

If the energy available is sufficient, K-shell conversion is most probable, because the K electrons spend more time close to the nucleus than do any of the outer orbitals.

The electron-capture transition in ^{113}Sn provides a simple example of internal conversion. Ninety-eight per cent of the transitions go to the 0.393-MeV state of ^{113m}In, so this is the energy available for either gamma emission or internal conversion. Tables of critical X-ray absorption wavelengths show that the In K electron is bound with 27.93 KeV, the L electron with 4.25 KeV. Conversion electrons would, therefore, be expected with 365 and 389 KeV, respectively. The beta spectrum of the $^{113}\text{Sn} \longrightarrow {}^{113}\text{In}$ decay is shown in Fig. 8-14. Careful measurements show the total internal-conversion coefficient $e/\mathcal{A} = 0.43$ with $K/L = 4.0$.

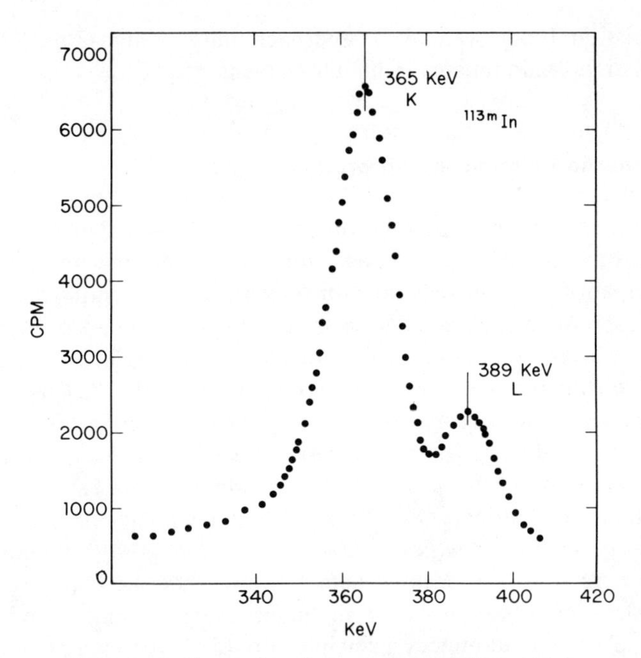

Figure 8-14. Conversion electron spectrum from ^{113m}In.

Internal conversion is not restricted to EC transitions, but may compete with any photon emission. Figure 11-2 shows the *K*- and *L*-conversion lines of ^{198}Hg, superposed upon the continuous beta-particle spectrum from the primary decay of ^{198}Au.

Another type of internal conversion can follow electron-capture transitions. In these transitions the orbitals are left with one vacancy, usually in the *K* shell. This vacancy may be filled from the *L* shell, and so on, with the emission of the characteristic *X*-ray line spectrum. Alternatively, the energy available for each of the X-ray photons may be transferred directly to one or more of the outer electrons, causing their ejection from the atom. Electrons so ejected are called *Auger electrons.*

As in gamma-ray conversion, the kinetic energy of the Auger electrons will be the residue after the original energy has supplied the orbital binding energy. In general, the Auger energies are very low, and their detection is difficult. They form a line spectrum rather than a continuum, and no neutrinos are involved. The process will be completed by the filling of the vacancies created by the loss of the Augers. The photon energies here will be very low, usually in the ultraviolet or even in the visible spectrum.

As in gamma-ray conversion, Auger emission competes with photon emission. *Fluorescent yield* is defined as the ratio of the number of photon emissions to the number of shell vacancies available for that particular photon emission. Fluorescent yields approach unity for high-Z nuclides, while Auger emission predominates in lighter elements.

8.14 Resonance Emission and Absorption

Some of the earliest spectroscopic observations showed that atoms absorb most effectively precisely those wavelengths which they emit upon excitation. This principle of resonance absorption has served to identify many of the elements responsible for the solar spectrum, through the absorption by the outer, cooler layers of lines emitted from deeper sources. The same principle of resonance absorption applies, with modifications, to the nucleus. A nucleus will absorb most effectively certain precise energies, either from incident photons or charged particles, and will emit other precise energies. This process is known as *resonance absorption* or *resonance fluorescence*, depending upon whether one considers the primary or the secondary photon.

The energy spread, or *natural width*, Γ, of a gamma-ray emission is very small. We note from the decay schemes shown that the lifetimes of most excited states are to be measured in nanoseconds or picoseconds. Then, according to the indeterminacy principle, Eq. (5-5), the energy uncertainty in a state of lifetime Δt will be

$$\Gamma = \Delta E = \frac{\hbar}{\Delta t} = \frac{6.58 \times 10^{-22}}{10^{-12}} = 6.58 \times 10^{-10} \text{ MeV} \tag{8-15}$$

This extremely narrow energy spread is not actually realized, since the thermal motions of a group of emitting nuclei will broaden the emission through the Doppler effect to perhaps 0.1 eV.

The energy available to the photon in the usual gamma emission is somewhat less than that required for absorption because some energy is imparted to the recoiling emitting nucleus. Consider a nucleus in an excited state, customarily denoted by *, moving prior to a photon emission. Momentum conservation requires

$$p^* = p_\gamma + p \tag{8-16}$$

The corresponding energy relations are

$$(E^* - E) + \frac{(p^*)^2}{2m} = E_\gamma + \frac{p^2}{2m} \tag{8-17}$$

Combining, we have, to a good approximation,

$$E_\gamma = (E^* - E) + \frac{pp_\gamma}{m} - \frac{p_\gamma^2}{2m} \tag{8-18}$$

The term in the parentheses is the energy available from the nuclear transition. The second term may be positive or negative, depending upon the relative directions of the momentum vectors p and p_γ. This is a Doppler-effect term which leads, because of the velocities of thermal agitation, to a broadening of the almost monoenergetic transition energy. The last term, always negative in an emission, represents the energy lost to the gamma ray because of the energy imparted to the recoiling nucleus. Figure 8-15A shows the Doppler broadening and the degraded gamma-ray energy in relation to the transition energy.

Similar considerations apply to gamma-ray absorption. Momentum conservation requires that some energy of recoil be imparted to the absorbing nucleus, and again there will be broadening because of thermal agitation. The latter will be represented by a term identical to the corresponding term in Eq. (8-18). In absorption, however, the recoil-energy term will be positive, reflecting the fact that the photon must have more than the energy of transition if it is to undergo resonance absorption. Energy relations in absorption are shown in Fig. 8-15B.

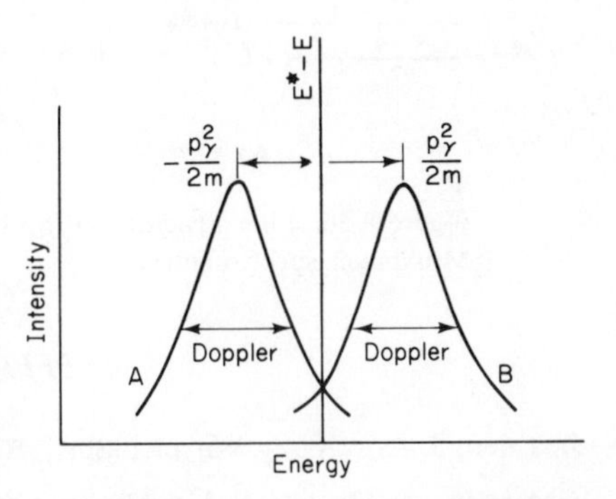

Figure 8-15. (A) The energy distribution of a gamma-ray emission compared to the transition energy E* − E. (B) The energy distribution required for energy absorption in the same nucleus.

Gamma-ray transitions involve recoil energies that are large compared to the energies of Doppler broadening. With only a small overlap of the emission and absorption energies there is only a small probability that a nuclear species will absorb the gamma rays that it emits. A gamma-ray energy will equal the energy of transition only if p has a magnitude and a direction such that the two momentum terms in Eq. 8-18 cancel. Resonance absorption is strong in optical spectra, since here the recoil energies are small compared to the Doppler broadening.

Recoil energies can be reduced by incorporating the emitter and absorber in rigid crystal lattices. Momentum conservation will still be fulfilled according to Eq. (8-16), but the energy will now be transferred to and from a massive, rigid crystal, and will be relatively small. In 1958, Mössbauer showed that exact resonance between emission and absorption can be achieved by moving the radiation source relative to the absorber. This movement creates a directed Doppler effect that cancels the reduced recoil effect, and greatly enhances resonance absorption.

The Mössbauer effect has proved to be a very powerful tool for studying the fine structure of nuclear energy levels. Figure 8-16 shows the energy diagram of ^{57}Fe as ordinarily given and as revealed by studies utilizing the Mössbauer effect.

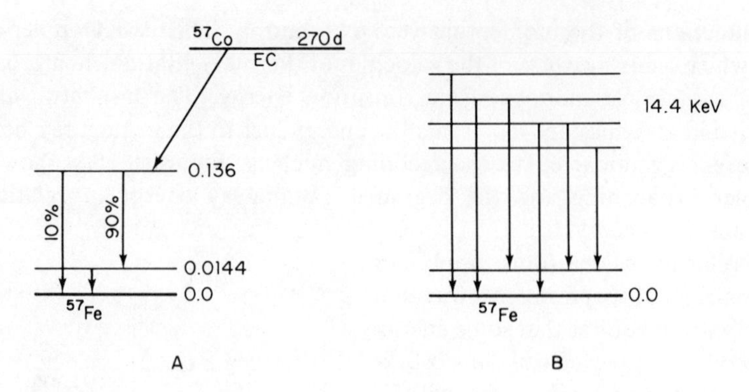

Figure 8-16. Fine structure in the 14.4 KeV state of ^{57}Fe as revealed by Mössbauer spectrometry.

REFERENCES

Bearden, J. A., "X-ray Wavelengths." *Rev. Modern Phys.* **39,** 78, 1967.

Compton, A. H. and S. K. Allison, *X-rays in Theory and Experiment.* 2nd ed. Van Nostrand, New York, 1935.

Kramers, H. A., "On the Theory of X-ray Absorption and the Continuous X-ray Spectrum." *Phil. Mag.* **46,** 836, 1923.

Lederer, C. M., J. M. Hollander, and I. Perlman, *Table of Isotopes.* 6th ed. Wiley, New York, 1968.

Mössbauer, R. L., "Recoilless Nuclear Resonance Absorption." *Ann. Rev. Nuc. Sci.* **12,** 123, 1962.

Roentgen, W. C., Communication to the Physikalisch Medicinscher Gesellschaft Wurzburg, Dec. 28, 1895. Translated by A. Stanton in *Science III* **227,** 726, 1896.

Rutherford, E., J. Chadwick, and C. D. Ellis, *Radiations from Radioactive Substances.* Macmillan, New York, 1930.

PROBLEMS

8-1. Derive Eq. (8-3) from the differential form of Kramers' equation.

8-2. Express the Kramers relation in terms of ν and sketch the form of the analytical relation between I_ν and ν. Relate this to the form of the relation in terms of λ.

8-3. Show that the Kramers relation leads to an expression for the efficiency of X-ray production of the form of Eq. (8-4).

8-4. An X-ray tube operates at 200 KVCP with a current of 25 mA on a tungsten target. At what rate must heat be removed from the target? At what rate must heat be removed from a betatron target when the machine is operating at 10 MeV with an average current of 0.05 mA to the target?

8-5. Calculate enough points to plot a representative curve of the spectral distribution produced from a tantalum target operating at 63.0 KV. Compare this distribution with that emitted from a tungsten target operating at the same voltage. How will the distributions be changed if the tubes are operated at 63.0 KVP half-wave rectified instead of at 63.0 KVCP?

8-6. The nucleus $^{74}_{33}$As is produced in an excited state which decays to the ground state with the emission of an 0.283-MeV photon. Negatron emissions of 0.725 and 1.360 MeV are observed, leading to stable ^{74}Se with an atomic mass of 68,857.161 MeV. Draw and label the decay scheme. What is the mass in u of the excited state of the parent atom? Is any other mode of decay open to the parent nucleus?

8-7. Calculate the mass defect and the mass decrement of the ground state of the parent atom in Prob. 8-6. What is the binding energy of the stable daughter?

8-8. The mass excess of ^{226}Ra is $+23.69$ MeV. Calculate the atomic mass of this structure and the mass excess of ^{222}Rn.

8-9. Starting with atomic mass values, calculate the maximum beta-particle energy emitted in the decay of ^{22}Na.

8-10. A radioactive nuclide $^{A}_{Z}X$ is observed to emit 1.27-MeV beta particles (93%), 0.79-MeV gamma rays (7%), and 0.511-MeV gamma rays (200%). Draw and label the decay scheme, identify all of the radiations emitted, and calculate the Q of the transition.

8-11. What radiations, in addition to the 0.320-MeV gamma ray, are available for the assay of ^{51}Cr? What are the energies of the neutrinos emitted in the transition?

8-12. Direct observations of many X-ray emissions and absorption edges are difficult because of the low energies involved. Calculate the energy of the L-absorption edge of zirconium from the knowledge that the K-absorption edge is located at 0.6888 Å and that the K-emission line is at 0.7860 Å.

8-13. Radioactive cerium is produced as $^{139m}_{58}$Ce, which decays by gamma emission (0.74 MeV, 55s). This product decays to $^{139}_{57}$La (140d) in a transition accompanied by an 0.166-MeV gamma ray (100%). Draw the decay scheme and label all known components. The mass defects involved are $\Delta(^{139}\text{Ce}) = 87.16$, $\Delta(^{139}\text{La}) = 87.43$ MeV.

8-14. An X-ray diffraction study requires the use of the fluorescent radiation from the K_α line of iron. What target material must be used in an X-ray tube whose K emissions will be used to excite the fluorescence? What voltage must be applied to the exciting tube?

8-15. An atom of ^{210}Po is bound in a crystal-lattice structure with bond strengths totaling 1.08 eV. Compare this to the recoil energy imparted to the daughter nucleus by the emission of the characteristic 5.3-MeV alpha particle.

8-16. A ^{14}C atom is incorporated into an organic molecule as a radioactive tracer. The bond strength is estimated from infrared spectra to be 0.97 eV. Compare this energy with the energy of recoil at the emission of the most energetic beta particle of 0.156 MeV.

8-17. The 0.014-MeV gamma ray that follows electron capture in ^{57}Co is used in a study of the Mössbauer effect. Assume that the recoil momentum is effectively distributed between 1000 neighboring atoms in the crystal lattice, and calculate the relative source–absorber velocity required to obtain resonance absorption.

8-18. Assume that all of the recoil momentum at the emission of an 0.320-MeV gamma ray by ^{51}Cr is imparted to the daughter nucleus. What relative source–absorber velocity will be required to obtain resonance absorption? Repeat the calculations for the emission of the first line of the Lyman series by an atom of hydrogen.

9

Natural Radioactivity

9.01 Uranium Ore Deposits

The intense research activity that followed the pioneer discoveries of Becquerel and Pierre and Marie Curie led to the discovery of a large number of radioactive species. Each of these *naturally radioactive* nuclides decayed with its own characteristic emission pattern and could be identified by its particular chemical behavior. Rutherford and Soddy were the first to suggest that radioactive atoms disintegrate into lighter structures as they emit the various radiations. This suggestion received powerful support with the discovery that the alpha particle was just an ionized atom of the element helium.

A good many of the newly discovered elements were found in various fractions of uranium-bearing ores, and this, the heaviest naturally occurring element, was soon suspected to be the parent substance. Thorium also was identified as a somewhat less important radioactive parent, with many decay products.

Prior to the Becquerel discovery, uranium had been mined on a limited scale for use as a coloring agent in the glass industry. Most of the material used in this way came from the Joachimstal mines in Czechoslovakia. The Curies used material from these mines in their original work that led to the discovery of radium and polonium.

The similarity between the penetrating gamma rays from radioactive elements and the radiations from Roentgen's gas discharge tube was quickly recognized. Medical uses of the X ray developed rapidly and increased demands for radium followed close behind. The search for uranium ore, desired at this point for its radium content, was stimulated and many new deposits were discovered. Commercially profitable deposits were located in Katanga Province in the Congo, and near Great Bear Lake, Canada.

Uranium occurs chiefly as pitchblende, a brown-black ore containing the metal as U_3O_8. In the Colorado Plateau region of the United States the most common form is the yellow carnotite. The average uranium concentration in the earth's crust is about 4×10^{-6} grams per gram of rock. Seawater concentrations range between 0.3 and 2.3×10^{-6} g liter^{-1}.

Intensive prospecting for uranium followed the discovery of methods for releasing nuclear energy from it, and many new deposits were discovered. As with other minerals, the richest deposits will be exploited first. The extent to which low-grade ores will be utilized depends upon future demands for uranium as a source of energy and upon the economics of procurement and purification.

9.02 Radioactive Decay

Early studies of radioactive materials showed that the activity of each species decreased at its own characteristic rate. Each rate of decrease was a *constant fraction* of the activity present. Because of the constancy of this fractional decrease, it was convenient to define the *half-life* of any radioactive nuclide as the time required for any amount of it (with an exception noted below) to decay to one-half of its original activity. Thus, if the activity of a sample was originally 100 units, it would decay to 50 units in 1 half-life. At the end of 2 half-lives, one-half of the 50 would have decayed, 25 would be left; and so on.

Assume that at some observation time t we have N atoms of some radioactive species. Then the law of constant fractional decay requires that over a short interval of time, dt, the number of atoms decaying, dN, will be

$$dN = -\lambda N\, dt \tag{9-1}$$

where λ, the constant of proportionality, is called the *decay* or *disintegration constant*. The minus sign is required because dN represents a decrease in N rather than an increase. An integration of Eq. (9-1) gives the relation between N and t:

$$N = N_0 e^{-\lambda t} \tag{9-2}$$

where N_0 is the number of atoms present at $t = 0$. Alternative forms of Eq. (9-2) are obtained by taking logarithms of both sides:

$$\ln\left(\frac{N_0}{N}\right) = \lambda t \tag{9-2a}$$

$$\log_{10}\left(\frac{N_0}{N}\right) = 0.434\lambda t \tag{9-2b}$$

According to classical mechanics, a charged particle in a potential well is faced with the same problem as a particle attempting to enter from the outside. Given sufficient energy, escape will be instantaneous; without sufficient energy, escape will be forever impossible. Such behavior is at complete variance with the observed decay process. The penetration of the potential barrier permitted by quantum mechanics leads to escape probabilities and to a decay law of the form given.

Since the decay process is based on probabilities, N_0 must be large enough to allow for statistical averaging. For example, Eq. (9-2) can scarcely apply to a single nucleus, which might decay almost at once or only after an extremely long time. Constant fractional decay also requires that each decay be an independent event, without any effect on the decay probabilities of other nuclei. For a given nucleus, the probability of decay must be independent of nuclear age. Measurements show that when the condition on N_0 is fulfilled, Eq. (9-2) is obeyed over an enormous range of decay rates. Half-lives of 10^{-6} sec and of 10^{21} years are known, and future techniques may extend both of these limits.

An important relation between decay constant λ and half-life T is obtained by putting $N = N_0/2$ in Eq. (9-2a). Then,

$$\ln\left(\frac{N_0}{N_0/2}\right) = \ln 2 = \lambda T$$

and

$$\lambda T = 0.693 \tag{9-3}$$

Units of λ and T may be any that are convenient to the problem at hand. T is usually given in either years, days, hours, minutes, or seconds, abbreviated as y, d, h, m, or s. Each value of λ will be the corresponding reciprocal y^{-1}, d^{-1}, h^{-1}, m^{-1}, s^{-1}, respectively.

Average lifetime, $\bar{T}$, is another term that is sometimes useful in describing radioactive decay. From Eq. (9-1), the number of nuclei decaying during the interval from t to $t + dt$ is $\lambda N\, dt$ and, since each of these nuclei had a lifetime t, the total lifetime associated with this interval is $\lambda Nt\, dt$. An integration of this factor over all values of t gives the total lifetime of all N_0 nuclei, and a division by N_0 gives the mean lifetime $\bar{T}$. The integral is

$$\bar{T} = \frac{1}{N_0}\int_0^\infty \lambda Nt\, dt = \int_0^\infty \lambda t e^{-\lambda t}\, dt = \frac{1}{\lambda} \tag{9-4}$$

$$\bar{T} = \frac{T}{0.693} = 1.44T \tag{9-5}$$

Half-life has a unique unchangeable value for each radioactive species, and geological evidence indicates that the presently observed values have obtained for a geologically long time. No change in the decay rates of particle emission has been observed over extreme variations of conditions such as temperature, pressure, chemical state, or physical environment. Very small variations can be produced in transitions involving orbital electrons. These changes are so small that they can be measured in only a few cases, and then only by using the most refined techniques.

9.03 The Units of Radioactivity

Equation (9-2) shows the time course of the number of radioactive nuclei of a species characterized by a decay constant λ. Since at every instant the rate of decay is λN, this rate, known as the *activity* A, will also follow an exponential decay law. That is,

$$A = \lambda N = \lambda N_0 e^{-\lambda t} \tag{9-6}$$

Originally, the unit of activity, the curie, was defined as equal to the number of disintegrations per second occurring in 1 gram of ^{226}Ra. Although radium was a relatively rare element, it could be prepared in adequate quantities in a reasonably pure state. Radium-226 has a long half-life, so prepared standards will change slowly, and corrections for decay can be applied. The half-life is too long for direct determination, so it is necessary to calculate it from the measured disintegration rate of a carefully purified weighed sample.

Illustrative Example

One series of measurements gave a value of 3.69×10^7 sec^{-1} as the most probable disintegration rate of 1.00 mg of ^{226}Ra.

In the 1 milligram there are

$$\frac{6.02 \times 10^{23} \times 0.001}{226.05} = 2.66 \times 10^{18} \text{ nuclei}$$

$$\lambda N = \frac{dN}{dt} = 3.69 \times 10^7 \times 3.15 \times 10^7 = 1.16 \times 10^{15} \text{ y}^{-1}$$

$$\lambda = \frac{1.16 \times 10^{15}}{2.66 \times 10^{18}} = 4.36 \times 10^{-4} \text{ y}^{-1}$$

$$T = \frac{0.693}{4.36 \times 10^{-4}} = 1590 \text{ y}$$

As originally defined, the curie was a changing standard, since it depended upon experimentally determined quantities. In 1950, the curie (Ci) was redefined as exactly 3.7×10^{10} disintegrations per second (dps). This definition leads to some uncertainty in the activity of 1 gram of radium-226, but this is less serious than the use of an uncertain standard for the curie. The curie is a relatively large unit; for most purposes the millicurie, mCi, and the microcurie, μCi, are more applicable.

The curie is defined in terms of the number of decaying nuclei and not in the number of emissions. Details of the decay scheme must be known in order to calculate a source strength from an emission measurement. For example, the rate of disintegration of ^{226}Ra, Fig. 8-7, will be the sum of all of the alpha-particle rates. If ^{226}Ra is assayed by a gamma-ray measurement, the *percent occurrence* of each gamma, and the response of the measuring system to it, must be known. A nucleus may have two modes of decay, each with its own characteristic decay constant. Thus ^{213}Bi can decay by either alpha or beta emission, Fig. 9-1, and the disintegration rate of the parent

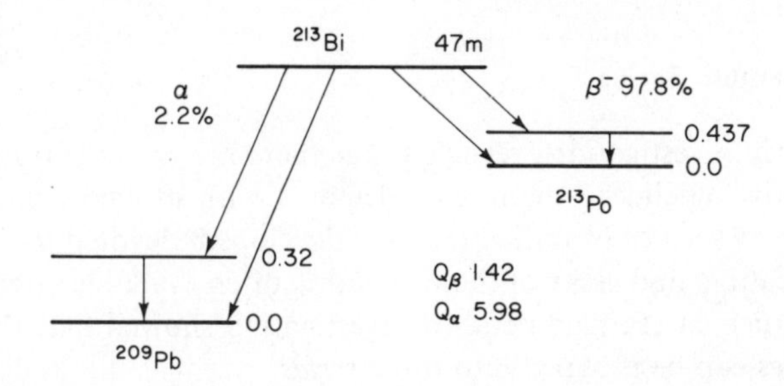

Figure 9-1. The decay scheme of ^{213}Bi which has two competing modes of decay.

will be the sum of the rates of the two modes. In the case of ^{213}Bi, measurements show that

$$\begin{aligned} \lambda_\alpha &= 0.00033\ \text{m}^{-1} \\ \lambda_\beta &= 0.01443 \qquad\qquad T = \frac{0.693}{0.01476} = 47\ \text{m} \\ \lambda &= 0.01476 \end{aligned}$$

Each radioactive species has an *intrinsic specific activity* (ISA) which is the activity of a unit mass of the pure material. When the radioactive isotope is diluted with a stable isotope of the same species, the corresponding calculation gives the *specific activity*.

Illustrative Example

Radioactive carbon ^{14}C, with a half-life of 5730 y, is found in nature mixed with stable ^{12}C and ^{13}C with an average activity of 16 dpm (disintegrations per minute), per gram of carbon. Then,

$$16\ \text{dpm} = \frac{16}{60 \times 3.7 \times 10^4} = 7.2 \times 10^{-6}\ \mu\text{Ci}$$

Specific activity $= 7.2 \times 10^{-6}\ \mu\text{Ci g}^{-1}$ of carbon

The intrinsic specific activity can be obtained by rearranging Eq. (9-1) and evaluating it for a mass of 1 gram:

$$N_1 = \frac{6.02 \times 10^{23}}{14} = 4.3 \times 10^{22}\ \text{g}^{-1}$$

$$\text{intrinsic specific activity} = \frac{dN_1}{dt} = \lambda N_1 = 0.693 \frac{N_1}{T}$$

$$\text{ISA} = \frac{0.693 \times 4.3 \times 10^{22}}{5.73 \times 10^3 \times 3.15 \times 10^7} = 1.65 \times 10^{10}\ \text{dps g}^{-1}$$

$$= \frac{1.65 \times 10^{10}}{3.7 \times 10^7} = 448\ \text{mCi g}^{-1}$$

9.04 The Uranium Series

Extensive early investigations revealed that there are many naturally occurring radioactive nuclides among the elements with atomic numbers from 81 to 92. A new science of radiochemistry developed, devoted to the separation, identification, and assay of these nuclides, often available in only minute quantities. Much careful and detailed investigation showed that these radioactive nuclides can be grouped into three *series*.

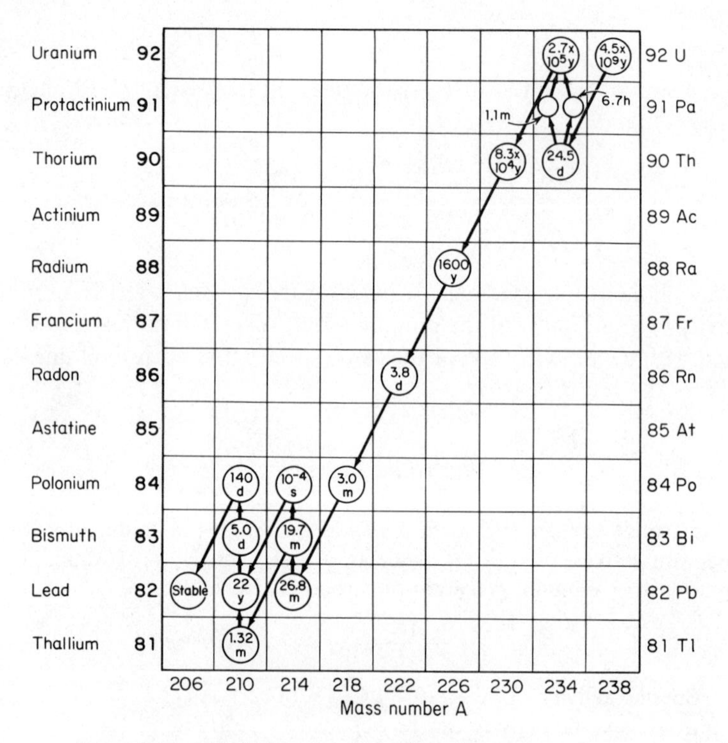

Figure 9-2. A Z–A plot of the transitions in the uranium series.

TABLE 9-1
THE URANIUM SERIES

Element	Symbol	Half-life	Energy (MeV)			Percent γ occurrence
			α	β	γ	
92 uranium	^{238}U	4.5×10^9 y	4.2	—	0.048	23
90 thorium	^{234}Th	24.1 d	—	0.19	0.09	4
91 protoactinium	^{234}Pa	1.17 m	—	2.29	1.0	0.6
92 uranium	^{234}U	2.5×10^5 y	4.8	—	0.05	28
90 thorium	^{230}Th	8.0×10^4 y	4.8	—	0.068	24
88 radium	^{226}Ra	1602 y	4.8	—	0.186	4
86 radon	^{222}Rn	3.82 d	5.49	—	0.5	0.07
84 polonium	^{218}Po	3.05 m	6.00	—	—	—
82 lead	^{214}Pb	26.8 m	—	0.65	0.24	4
83 bismuth	^{214}Bi	19.7 m	5.5	1.5	0.61	47
84 polonium	^{214}Po	160 μs	7.7	—	0.8	0.014
82 lead	^{210}Pb	21 y	—	0.016	0.046	81
83 bismuth	^{210}Bi	5.0 d	—	1.16	—	—
84 polonium	^{210}Po	138 d	5.30	—	0.80	0.001
82 lead	^{206}Pb			[*stable*]		

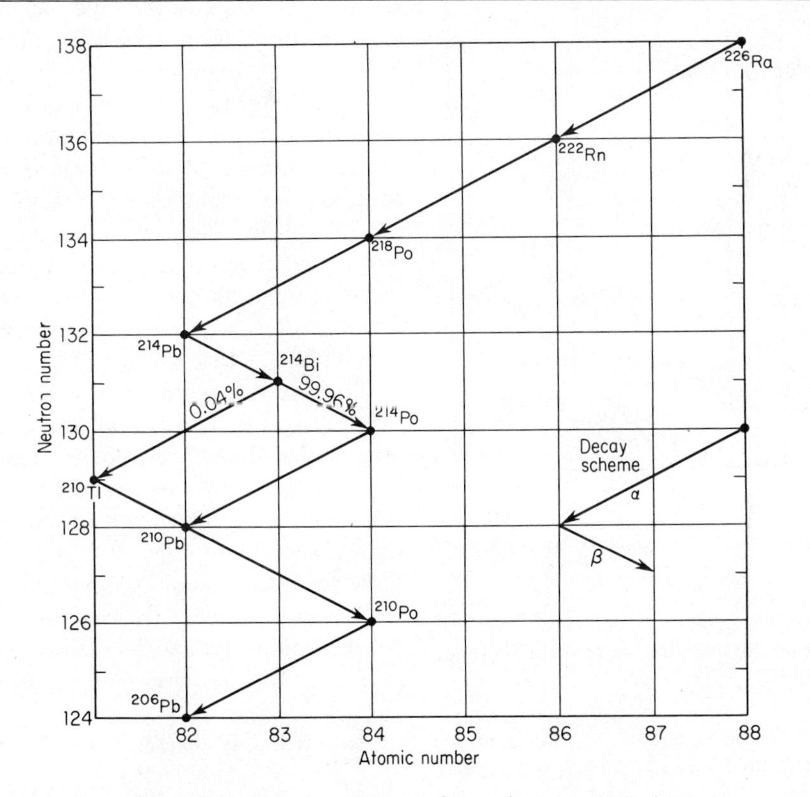

Figure 9-3. A portion of the uranium series on an *N–Z* plot.

The uranium series is the most important of the three, primarily because of the decay characteristics of two series members ^{226}Ra and ^{222}Rn. Radium in particular has been used extensively in medicine and in industry, and even today has not been completely supplanted by artificially produced radiation sources. Table 9-1 lists the members of the uranium series together with some of the more important decay characteristics. Almost every decay is far more complicated than is indicated in Table 9-1. Details of these and all other known radioactive transformations will be found in special compilations.*

Before isotope identifications were completed and series relationships understood, series members had been given generic names and symbols. These names and symbols do not, in many cases, properly identify the element, and so they should not be used.

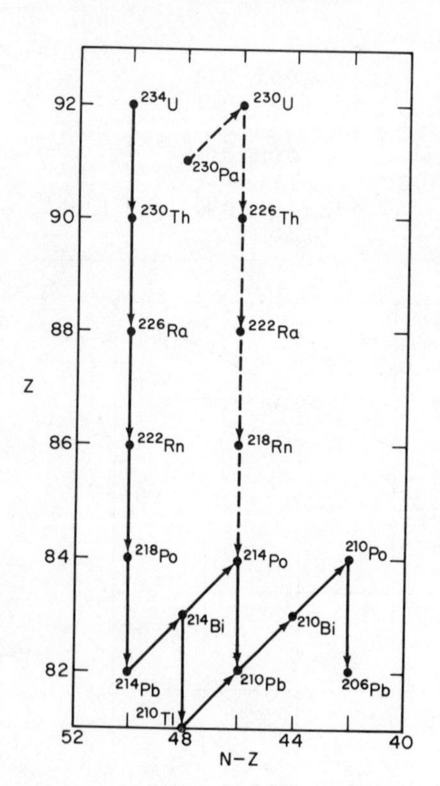

Figure 9-4. The uranium and protoactinium series on a Z–$(N-Z)$ plot. Alpha emission is shown by a vertical, beta by an upward slanting arrow. The protoactinium series joins the main series at ^{214}Po.

Figure 9-2 shows the decay characteristics of the members of the uranium series on a Z–A plot. The head of the series, ^{238}U, has a half-life that is long even on a geological time scale. This is a necessary requirement for an element existing today, if it was laid down at the time of formation of the earth's crust. Radon-222, the only gaseous member of the series, is a noble gas of the He, Ne, Ar, . . . sequence with closed orbital-electron shells. Like the other naturally occurring series, the uranium series ends in a stable isotope of lead, in this case ^{206}Pb.

For some purposes the N–Z plot of Fig 9-3 is more useful than the Z–A presentation. In an N–Z plot an arrow down and to the left indicates an alpha emission, a line down and to the right a beta. Both Figs. 9-2 and 9-3 show the phenomenon of *branching* at ^{214}Bi. About 99.96 per cent of the disintegrations go by beta emission, so for ^{214}Bi the *branching ratio* is

* C. M. Lederer, J. M. Hollander, and I. Perlman, *Tables of Isotopes*, John Wiley & Sons, Inc., New York, 1968.

99.96/0.04 = 2500. The emission characteristics of the two daughters are such that the series reunites at ^{210}Pb.

The uranium decay chain is known as the $(4n + 2)$ series because each member has a mass number of $(4n + 2)$, where n is an integer. In 1948, Studier and Hyde reported the discovery of a new $(4n + 2)$ subseries. They produced the head of the series, ^{230}Pa, by bombarding thorium with high-energy charged particles. After one beta decay to ^{230}U, the daughter products undergo four successive alpha decays, Fig. 9-4, to ^{214}Po, where the subchain joins the main series. This *protoactinium series* does not exist in nature because the longest half-life in it is only 20.8 days. To emphasize the main-series and subseries relationships, the decay chains in Fig. 9-4 are shown on a Z–$(N - Z)$ plot.

9.05 The Thorium, Actinium, and Neptunium Series

Two other series of naturally occurring radioactive elements have been known for many years. These are the $(4n)$ or *thorium series* and the $(4n + 3)$ or *actinium series*. For a good many years, $(4n + 1)$ was known as the missing series because none of its members had been identified. Today, all members

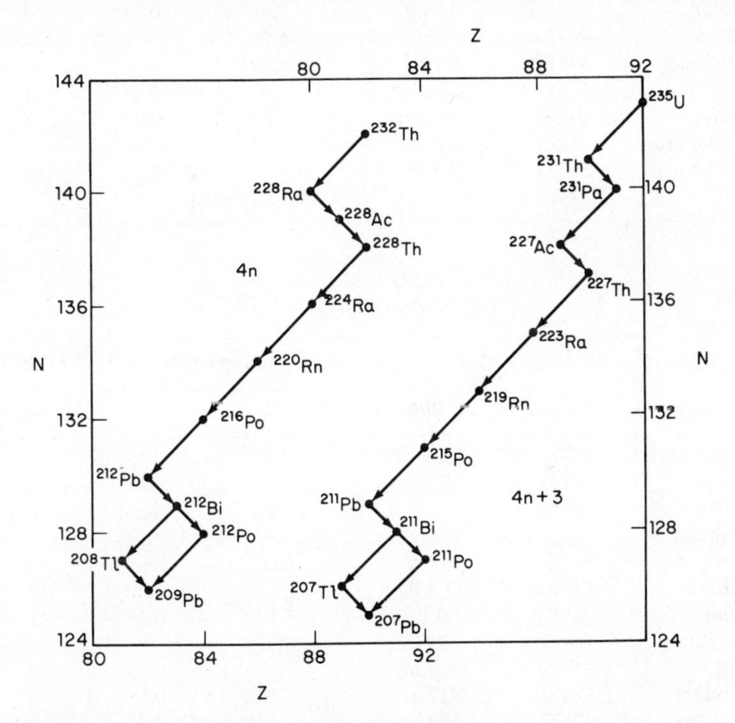

Figure 9-5. The thorium $(4n)$ and actinium $(4n + 3)$ series on N–Z plots. The upper Z scale applies to the actinium series only.

of this series are known and the series relationships are firmly established, although the $(4n + 1)$ series does not exist in nature.

a. *The thorium series*, $4n$. This series originates in nature with ^{232}Th, an alpha emitter with a half-life of 1.4×10^{10} y. As in the case of the other series, Table 9-2 lists only the most prominent features of the decays, most of which are complicated, particularly as far as the gamma emissions are concerned. The series relationships are shown in an N–Z plot, Fig. 9-5.

TABLE 9-2
THE THORIUM SERIES

			Energy (MeV)			*Percent*
Element	*Symbol*	*Half-life*	α	β	γ	γ *occurrence*
90 thorium	^{232}Th	1.40×10^{10} y	4.00	—	0.06	23
88 radium	^{228}Ra	6.7 y	—	0.054	—	—
89 actinium	^{228}Ac	6.13 h	—	1.11	0.90	30
90 thorium	^{228}Th	1.91 y	5.43	—	0.08	28
88 radium	^{224}Ra	3.64 d	5.68	—	0.24	5
86 radon	^{220}Rn	55 s	6.29	—	—	—
84 polonium	^{216}Po	0.16 s	6.78	—	—	—
82 lead	^{212}Pb	10.6 h	—	0.36	0.238	81
83 bismuth	^{212}Bi	60.6 m	6.05	2.20	0.04	17
84 polonium	^{212}Po	300 ns	8.78	—	—	—
81 thallium	^{208}Tl	3.1 m	—	1.79	2.62	100
82 lead	^{208}Pb			[*stable*]		

TABLE 9-3
THE ACTINIUM SERIES

			Energy (MeV)			*Percent*
Element	*Symbol*	*Half-life*	α	β	γ	γ *occurrence*
92 uranium	^{235}U	7.1×10^{8} y	4.38	—	0.185	12
90 thorium	^{231}Th	25.5 h	—	0.30	0.25	90
91 protoactinium	^{231}Pa	3.2×10^{4} y	5.06	—	many	
89 actinium	^{227}Ac	21.6 y	4.95	0.046	many	
90 thorium	^{227}Th	18.2 d	6.04	—	many	
88 radium	^{223}Ra	11.4 d	5.86	—	many	
86 radon	^{219}Rn	4.0 s	6.82	—	0.27	11
84 polonium	^{215}Po	1.78 ms	7.38	—	—	—
82 lead	^{211}Pb	36.1 m	—	1.36	0.83	20
83 bismuth	^{211}Bi	2.15 m	6.62	0.59	0.35	?
81 thallium	^{207}Tl	4.79 m	—	1.44	0.90	0.16
82 lead	^{207}Pb			[*stable*]		

A good many similarities will be noted between the thorium and the uranium series. Each has one gaseous member, and branching occurs at ^{212}Bi in a manner analogous to that at ^{214}Bi in the uranium series.

b. *The actinium series*, $(4n + 3)$. The plot of these series relationships along with the thorium series in Fig. 9-5 emphasizes the close parallelism between the two. Here again the series contains a gaseous member with a relatively short half-life, and again there is branching at a bismuth isotope, ^{211}Bi.

c. *The neptunium series*, $(4n + 1)$. This series was observed only after its precursors, ^{241}Pu and ^{241}Am, were artifically produced. A study of Fig. 9-6 and Table 9-4 will show that none of the half-lives involved in this series are long enough to insure measurable quantities today, if the series head existed at the time the elements were laid down in the earth's crust.

The $(4n + 1)$ series is unique in that it contains no gaseous member. Both astatine (85) and francium (87), long missing from the periodic table, are members of this series. Again, branching occurs at

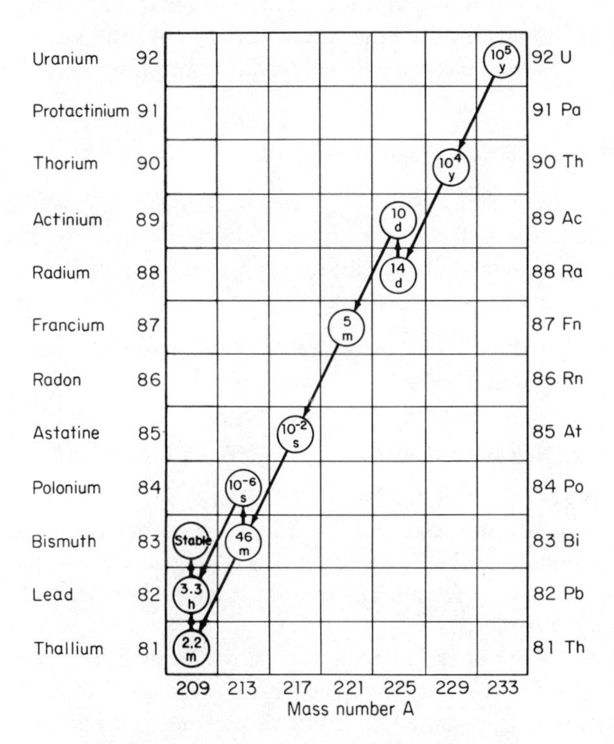

Figure 9-6. A *Z–A* plot of the neptunium series.

TABLE 9-4
THE NEPTUNIUM SERIES

Element	*Symbol*	*Half-life*	*Energy (MeV)* α	β	γ	*Percent* γ *occurrence*
93 neptunium	^{237}Np	2.1×10^6 y	4.8	—	0.09	50
91 protoactinium	^{233}Pa	27 d	—	0.26	0.06	0.20
92 uranium	^{233}U	1.6×10^5 y	4.8	—	0.04	14
90 thorium	^{229}Th	7.3×10^3 y	4.8	—	0.11	6
88 radium	^{225}Ra	14.8 d	—	0.32	0.04	33
89 actinium	^{225}Ac	10 d	5.8	—	0.03	28
87 francium	^{221}Fr	4.8 m	6.3	—	0.22	14
85 astatine	^{217}At	0.03 s	7.0	—	0.22	15
83 bismuth	^{213}Bi	47 m	6.0	1.39	0.44	?
83 polonium	^{213}Po	4.2 μs	8.4	—	—	—
82 lead	^{209}Pb	3.3 h	—	0.63	—	—
83 bismuth	^{209}Bi		[*stable*]			

a bismuth isotope, in this case ^{213}Bi. The series terminates with an isotope of bismuth instead of at lead, as is the case for the other three series. It should be noted that the termination of the neptunium series is at ^{209}Bi, which has a magic neutron number 126.

9.06 Similarities in Decay Sequences

We have already noted that there are some striking similarities between the decay characteristics of the four radioactive series. Still further similarities can be seen by examining the decay sequences, starting with the long-lived heads of the decay chains:

$$^{238}\text{U}\text{-}\alpha\text{-}\beta\text{-}\beta\text{-}\alpha\text{-}\alpha\text{-}\alpha\text{-}\alpha\text{-}\alpha\text{-}\beta\text{-}\beta\text{-}\alpha\text{-}\beta\text{-}\beta\text{-}\alpha\text{-}{}^{206}\text{Pb} \qquad \Delta Z = 10 \qquad \Delta A = 32$$

$$^{232}\text{Th}\text{-}\alpha\text{-}\beta\text{-}\beta\text{-}\alpha\text{-}\alpha\text{-}\alpha\text{-}\alpha\text{-}\beta\text{-}\beta\text{-}\alpha\text{-} \quad {}^{208}\text{Pb} \qquad \Delta Z = 8 \qquad \Delta A = 24$$

$$^{235}\text{U}\text{-}\alpha\text{-}\beta\text{-}\alpha\text{-}\beta\text{-}\alpha\text{-}\alpha\text{-}\alpha\text{-}\alpha\text{-}\beta\text{-}\alpha\text{-}\beta\text{-} \quad {}^{207}\text{Pb} \qquad \Delta Z = 10 \qquad \Delta A = 28$$

$$^{237}\text{Np}\text{-}\alpha\text{-}\beta\text{-}\alpha\text{-}\alpha\text{-}\beta\text{-}\alpha\text{-}\alpha\text{-}\alpha\text{-}\beta\text{-}\alpha\text{-}\beta\text{-} \quad {}^{209}\text{Bi} \qquad \Delta Z = 10 \qquad \Delta A = 28$$

Each decay series is initiated by an alpha decay followed by one or more beta emissions, which in turn are followed by a sequence of 3–5 alpha emissions near the middle of the decay chain. Each long alpha sequence is terminated by one or two beta decays, and there is at least one more alpha emission before stability is reached.

Some of the decay characteristics can be made plausible from a consideration of the N–Z plot of stable nuclei, Fig. 7-1. On an N–Z plot, an alpha-particle emission is represented by a line down and to the left at 45°, while beta emission is depicted by a line down and to the right, Fig. 9-3.

An alpha-particle emission by a nucleus that is located near the top and to the left of the stability line will lead to a daughter that is farther away from the line of stability. This is necessarily so because at large values of A and Z the stability line has a slope considerably greater than 45°. Beta emission, typically that from ^{234}Th and ^{234}Pa in the uranium series, brings the daughter nuclei back toward the stability line. Alpha-particle emission may still be possible, as at ^{234}U, and again the daughter moves away from the line of stability. During the sequence of five alpha-particle emissions from ^{234}U to ^{214}Pb, the points depicting the nuclei move steadily away from stability. This departure is reflected in the half-life sequence: 2.5×10^5 y, 8×10^4 y, 1.6×10^3 y, 3.8 d, 3.05 m. At ^{214}Pb, two beta-particle emissions enter to bring the nuclei back toward the stability line. Somewhat similar sequences are observed in the other series.

9.07 Serial Transformations

We have seen that the decay law for a single radioactive species, leading to a stable daughter product, is a simple negative exponential. In the radioactive series all products except the final, stable daughter will also be radioactive. It is necessary, then, to examine the relations that exist when a nuclide is being produced and is also decaying.

Consider a parent radioactive nuclide that contained N_{10} nuclei at $t = 0$, N_1 nuclei at some later time t, and that is characterized by a decay constant λ_1. N_2 represents the number of daughter nuclei at time t, and λ_2 is the decay constant of the daughter. At any arbitrary time t there will be both production and decay of the daughter, and the net gain in N_2 in a short interval of time will be

$$dN_2 = (\lambda_1 N_1 - \lambda_2 N_2)\, dt$$

From the simple decay law for the parent,

$$dN_2 = (N_{10}\lambda_1 e^{-\lambda_1 t} - \lambda_2 N_2)\, dt \tag{9-7}$$

This can be readily integrated, and if we let $N_2 = 0$ at $t = 0$,

$$N_2 = N_{10}\frac{\lambda_1}{\lambda_2 - \lambda_1}(e^{-\lambda_1 t} - e^{-\lambda_2 t}) \tag{9-8}$$

Equation (9-8) is perfectly general provided only that the initial conditions on N_2 and on the size of the sample are met.

N_2 will also be zero at $t = \infty$, since at that time all radioactive nuclei of both species will have decayed. Equation (9-8) should, then, have a maximum value for some finite time t_m. Differentiating Eq. (9-8) with respect to time and setting the derivative equal to zero gives

$$t_m = \frac{\ln(\lambda_2/\lambda_1)}{\lambda_2 - \lambda_1} \tag{9-9}$$

There will be a positive, finite value of t_m for all values of λ_1 and λ_2 except when $\lambda_1 = 0$ or $\lambda_1 = \lambda_2$. The first exception is very nearly realized when the parent has a half-life that is very long compared to that of the daughter. This leads to $t_m \longrightarrow \infty$, as will be seen presently.

The case of $\lambda_1 = \lambda_2$ does not arise practically, but the value of t_m can be approximated for $\lambda_2 \longrightarrow \lambda_1$. If we let $\lambda_2 = k\lambda_1$, Eq. (9-9) becomes

$$t_m = \frac{\ln k}{\lambda_1(k-1)} \tag{9-10}$$

This can be evaluated by the method of indeterminate forms as $k \longrightarrow 1$:

$$t_m(\lambda_2 \longrightarrow \lambda_1) = \lim_{k \to 1} \frac{1/k}{\lambda_1} = \frac{1}{\lambda_1} \tag{9-11}$$

Equation (9-11) shows that as the two decay constants approach equality, the time of maximum amount of the daughter species approaches $1/\lambda_1$, which is just the mean lifetime, $1.44T$.

Three special cases of Eq. (9-8) are of interest:

a. *Extremely long-lived parent*: $\lambda_1 \ll \lambda_2$. This is the case when the parent has such a long half-life that it decays at an essentially constant rate during the time under consideration. Then, in Eq. (9-8) we may set $e^{-\lambda_1 t} = 1$ and $\lambda_2 - \lambda_1 = \lambda_2$, which leads to the approximation

$$N_2 = N_{10}\frac{\lambda_1}{\lambda_2}(1 - e^{-\lambda_2 t}) \tag{9-12}$$

According to Eq. (9-12), N_2 approaches an equilibrium value:

$$N_2\,(\text{eq}) = N_{10}\frac{\lambda_1}{\lambda_2} \tag{9-13}$$

which it only exactly reaches in an infinite time.

The most important practical application of this approximation is to the buildup of ^{222}Rn from an initially pure sample of ^{226}Ra, Fig. 9-7.

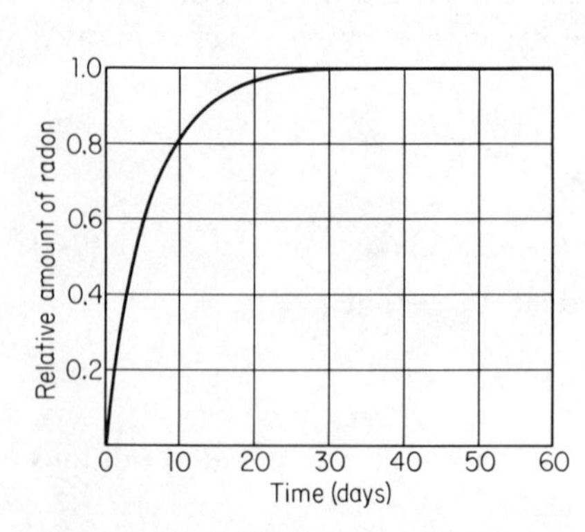

Figure 9-7. The build-up of 3.8d ^{222}Rn from the extremely long-lived parent ^{226}Ra.

b. *Relatively long-lived parent*: $\lambda_1 < \lambda_2$. For small values of t, N_2 must be calculated from the general relation, Eq. (9-8). After a sufficiently long time, $e^{-\lambda_2 t}$ will be negligible compared to $e^{-\lambda_1 t}$ and then Eq. (9-8) reduces to

$$N_2 = N_{10}\frac{\lambda_1}{\lambda_2 - \lambda_1}e^{-\lambda_1 t} = N_1\frac{\lambda_1}{\lambda_2 - \lambda_1} \tag{9-14}$$

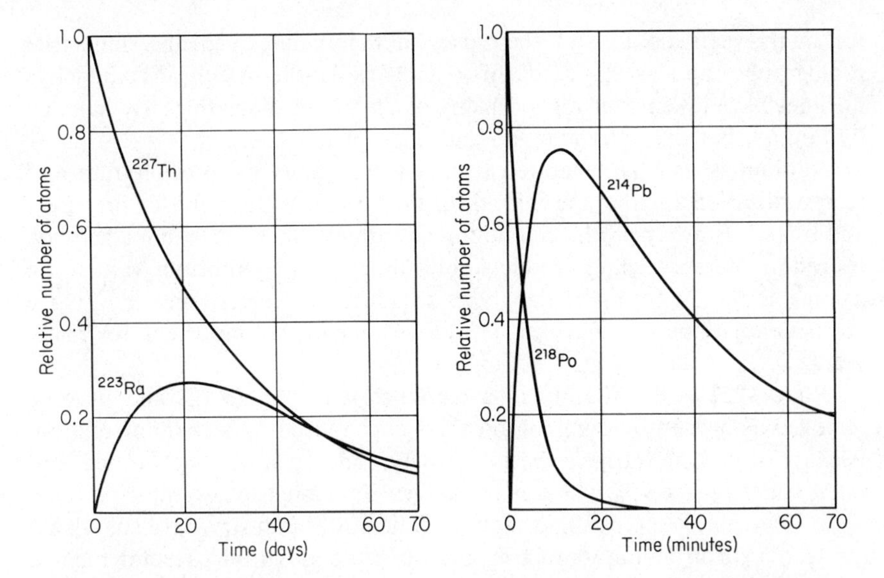

Figure 9-8. Buildup and establishment of secular equilibrium of 11.2d ^{223}Ra from 18.9d ^{227}Th.

Figure 9-9. Buildup and decay of 26.8 m ^{214}Pb from 3.05 m ^{218}Po.

From the time that Eq. (9-14) becomes valid, there will be a constant ratio established between N_2 and N_1. This is not a true equilibrium, because both N_2 and N_1 will be decaying at the rate characteristic of the parent. This condition is known as *secular equilibrium.* Figure 9-8 illustrates the attainment of secular equilibrium in the alpha transition ^{227}Th $\longrightarrow$ ^{223}Ra.

c. *Relatively short-lived parent*: $\lambda_1 > \lambda_2$. In this case the amount of the daughter rises to a maximum and then decreases, eventually with its own characteristic half-life. For short times, N_2 must be calculated from the general equation. At long times, the decay can be calculated by neglecting the factor $e^{-\lambda_1 t}$:

$$N_2 = N_{10}\frac{\lambda_1}{\lambda_1 - \lambda_2}e^{-\lambda_2 t} \qquad (9\text{-}15)$$

Figure 9-9 shows an example of this case for the alpha transformation ^{218}Po $\longrightarrow$ ^{214}Pb.

9.08 The Bateman Equations

At least two cases of practical importance require the use of serial relations between radioactive species:

1. Radium-226 and its decay products have been particularly important

for many years because of their uses in therapeutic medicine, industrial radiography, and as standards of gamma radiation. Artificial radioactive nuclides have taken over a good many of the applications formerly handled by radium, but the latter is by no means obsolete.

Radium is usually separated from its precursors by precipitation and recrystallization as either the chloride or the bromide. When freshly prepared, and free of its decay products, radium-226 shows almost no gamma activity. As radon and succeeding products build up toward equilibrium values, the gamma radiation from the daughters will increase correspondingly. These are the radiations that have made radium and radon important for many years.

Radon-222 will build up in accordance with Eq. (9-12) and may be removed as needed by pumping off the gas produced by a radium solution. Because of its high intrinsic specific activity, radon can be concentrated into small tubes or needles, which may be then implanted in malignant tissues.

2. The rapidly increasing demands of the power industry have stimulated the search for uranium, its mining, and its purification into reactor fuel. In the undisturbed ore bodies, uranium has been decaying according to the serial relations of Table 9-1, since the metal was laid down billions of years ago. Before mining operations started, each long-lived series head was in essential equilibrium with all of its daughter products.

When mining operations open up an ore pocket, there is a gross disturbance of the equilibrium conditions. Gaseous products will be released to the mine atmosphere and some will decay while airborne, to leave a series of radon daughters available for inhalation. A number of radioactive products are involved in the mine situation and calculations require an extension of the two-nuclide decay and buildup relation of Eq. (9-8).

The simple relation given is really the second of a series of parent–daughter relations known as the Bateman equations, after their developer. The first four of these equations are:

$$N_1 = N_{10}e^{-\lambda_1 t} \tag{9-16}$$

$$N_2 = N_{10}(a_1 e^{-\lambda_1 t} + a_2 e^{-\lambda_2 t}) \tag{9-17}$$

$$N_3 = N_{10}(a_3 e^{-\lambda_1 t} + a_4 e^{-\lambda_2 t} + a_5 e^{-\lambda_3 t}) \tag{9-18}$$

$$N_4 = N_{10}(a_6 e^{-\lambda_1 t} + a_7 e^{-\lambda_2 t} + a_8 e^{-\lambda_3 t} + a_9 e^{-\lambda_4 t}) \tag{9-19}$$

where

$$a_1 = \frac{\lambda_1}{\lambda_2 - \lambda_1} \qquad a_2 = -\frac{\lambda_1}{\lambda_2 - \lambda_1} \tag{9-20}$$

$$\left.\begin{aligned} a_3 &= \frac{\lambda_1\lambda_2}{(\lambda_2 - \lambda_1)(\lambda_3 - \lambda_1)} \qquad a_4 = \frac{\lambda_1\lambda_2}{(\lambda_1 - \lambda_2)(\lambda_3 - \lambda_2)} \\ a_5 &= \frac{\lambda_1\lambda_2}{(\lambda_1 - \lambda_3)(\lambda_2 - \lambda_3)} \end{aligned}\right\} \tag{9-21}$$

$$
\left.\begin{aligned}
a_6 &= \frac{\lambda_1\lambda_2\lambda_3}{(\lambda_2-\lambda_1)(\lambda_3-\lambda_1)(\lambda_4-\lambda_1)}\\
a_7 &= \frac{\lambda_1\lambda_2\lambda_3}{(\lambda_1-\lambda_2)(\lambda_3-\lambda_2)(\lambda_4-\lambda_2)}\\
a_8 &= \frac{\lambda_1\lambda_2\lambda_3}{(\lambda_1-\lambda_3)(\lambda_2-\lambda_3)(\lambda_4-\lambda_3)}\\
a_9 &= \frac{\lambda_1\lambda_2\lambda_3}{(\lambda_1-\lambda_4)(\lambda_2-\lambda_4)(\lambda_3-\lambda_4)}
\end{aligned}\right\} \tag{9-22}
$$

Relations for higher members have been developed and are available in special texts.

9.09 Nonseries Radioactive Nuclides

Careful radiochemical investigations have disclosed naturally occurring radioactivity in several nuclides that are not members of any of the four series. With two exceptions, to be considered separately, each of these non-series nuclides has an extremely long half-life and a correspondingly low intrinsic specific activity. The slow disintegration rates make detection and identification of the emitted radiations very difficult. Future measurements with improved techniques may change some of the information presented in Table 9-5 and may add new nuclides, now thought to be stable.

TABLE 9-5
NATURALLY OCCURRING, NONSERIES RADIOACTIVE NUCLEI

	Nucleus						
Symbol	*Z*	*N*	*A*	*Percent Abundance*	*Half-life (years)*	*Emission*	*Transition energy (Q)*
H	1	2	3		12.3	β^-	0.0186
C	6	8	14		5730	β^-	0.156
K	19	21	40	0.0119	1.3×10^9	β^-, EC, γ	1.5
V	23	27	50	0.25	6×10^{15}	β^-, EC, γ	2.2
Rb	37	50	87	27.8	5×10^{10}	β^-	0.27
In	49	66	115	95.8	6×10^{14}	β^-	0.49
Te	52	78	130	34.5	8×10^{20}	$\beta\beta$	?
La	57	81	138	0.09	1×10^{11}	β^-, γ	1.8
Ce	58	84	142	11.1	10^{15}	α	?
Nd	60	84	144	23.9	2.4×10^{15}	α	1.9
Sm	62	85	147	15.1	1×10^{11}	α	2.3
Lu	71	105	176	2.6	3×10^{10}	β^-, γ	1.0
Re	75	112	187		4×10^{10}	β^-	?
Pt	78	114	192	0.78	10^{15}	α	?
Bi	83	126	209	100	2×10^{18}	α	3.1

There is no obvious pattern to the distribution of the long-lived, nonseries radioactive nuclei. Some of the half-lives are surprisingly long in view of the energies available in the transitions. This presumably means that selection rules, probably related to angular-momentum requirements, are acting to hinder the decays (Secs. 10.03 and 11.06). The double-beta decay of ^{130}Te to ^{130}Xe is unique and is of particular interest in connection with the properties of the neutrino (Sec. 11-05).

Except for H-3 and C-14, all of the half-lives listed in Table 9-5 are compatible with the formation of the elements at about the time of the "big bang" some 4×10^9 years ago. The two short-lived occupants of the table must be in continuous production. In fact, each of these nuclei is being formed continuously from nuclear reactions initiated by various components of the cosmic radiation. Thus ^{14}C is formed from the bombardment of atmospheric nitrogen by neutrons:

$$^{14}\mathrm{N} + {}^{1}n \longrightarrow {}^{14}\mathrm{C} + {}^{1}p \tag{9-23}$$

Tritium can be produced by cosmic radiation in a variety of reactions.

There is some evidence that cosmic-ray activity has been relatively constant over a long period of time. If so, this would result in equilibrium concentrations of ^{3}H and ^{14}C in the atmosphere. Unfortunately, the atmospheric detonation of nuclear devices also produces substantial quantities of these two nuclei. An active weapons-testing program in the 1950s increased the atmospheric concentrations of both ^{3}H and ^{14}C, which are now slowly returning toward pretest levels.

9.10 Radioactive Dating

The relative isotopic abundances of many elements appear to be constant, almost completely independent of the origin of the sample. The principal exception to this constancy occurs in elements where one or more stable isotopes are formed by radioactive decay of another element. An outstanding example of variable abundance ratios arises in the case of lead, where the values depend upon the relative contributions from the three radioactive series as well as upon the normal isotopic ratios of lead itself.

When radiogenic isotopes do form part of the stable product, many important time intervals can be determined by accurate measurements of the isotopic ratios. These dating methods are based on the plausible but unproved assumption that the present values of the decay constants have obtained throughout the time interval being measured. Other supporting assumptions are required in particular cases.

One of the earliest applications of radioactive dating used ^{206}Pb/^{238}U or

$^{4}He/^{238}U$ ratios to establish the time at which uranium minerals were laid down. The complete decay of ^{238}U is equivalent to

$$^{238}_{92}U \longrightarrow {}^{206}_{82}Pb + 8\alpha + 6\beta + \gamma \tag{9-24}$$

The longest intermediate half-life is 1602 y, for ^{226}Ra; after a few thousands of years, the holdup at this point will be negligible compared to the amount of ^{206}Pb that has been formed. If one can be sure that no ^{206}Pb was formed by other processes, an assay for the two nuclides in question will serve to determine the time at which the ^{238}U was incorporated into the mineral. Times determined by this method agree well with those obtained from purely geological considerations.

An assay for ^{4}He may be used when there is some doubt as to the radiogenic purity of the ^{206}Pb. Helium gas can, however, diffuse out of porous structures; age determinations based on gas analyses must be accepted with caution.

Another scheme for radioactive dating utilizes ^{40}Ar, one of the decay products of ^{40}K. There are two competing decay modes:

$$^{40}K \longrightarrow {}^{40}Ca + \beta^{-} \tag{9-25}$$

$$^{40}K + e^{-} \longrightarrow {}^{40}Ar + \gamma \tag{9-26}$$

Careful measurements have established that 89 per cent of the ^{40}K decays go by beta emission, and 11 per cent by electron capture. Both daughter elements are found in nature with well-established abundances. If radiogenic ^{40}Ar is present, the mass spectrometer will show an abnormal $^{40}Ar/^{36}Ar$ ratio and the excess can be attributed to ^{40}K decays through Eq. (9-26). This method has been used successfully for samples about 2×10^{6} years old.

The dating of objects that were previously alive is based on the constancy of ^{14}C production and distribution throughout the time interval being measured. Atmospheric ^{14}C, produced by a cosmic-ray flux assumed to be constant, exists primarily as $^{14}CO_2$. This gas diffuses throughout the atmosphere and comes into equilibrium with the carbonate pool of the earth. Thus ^{14}C will be acquired by any living organism that utilizes carbon in its metabolic cycle. During life, all living organisms will contain carbon in the $^{14}C/^{12}C$ equilibrium ratio, which, before the advent of atmospheric testing of nuclear devices, was about 16 dpm g^{-1} of carbon. Upon death, ^{14}C incorporation ceases and the amount then present decays with the characteristic half-life. The specific activity will decrease at a well-established rate, and a determination of it will serve to fix the time of death. The low specific activity of contemporary samples limits the dating of organic materials to about 50,000 years.

Tritium can be used for dating in much the same way, but its application is even more limited by its relatively short half-life and its low specific activity. Both of the last two dating techniques have been disturbed by the increased specific activities that have been produced by atmospheric detonations of nuclear devices.

REFERENCES

Geyh, M. A., "Problems in Radioactive Carbon Dating of Small Samples by Means of Acetylene, Ethane, and Benzene." *Int. J. Applied Rad. and Isotopes* **20,** 463, 1969.

Hevesy, G. and F. A. Paneth, *A Manual of Radioactivity*. Oxford Univ. Press, Oxford, 1938.

Rutherford, E., *Radioactivity*. Cambridge Univ. Press, Cambridge, England, 1904.

Rutherford, E., J. Chadwick, and C. D. Ellis, *Radiations from Radioactive Substances*. Macmillan, New York, 1930.

PROBLEMS

9-1. In August, 1911, Mme. Marie Curie prepared an international standard of activity containing 21.99 mg of $RaCl_2$. Calculate the original activity of the solution, and its activity as of February, 1972, using presently accepted values of the pertinent constants.

9-2. What volume of helium gas at NTP will be produced by the complete decay of a treatment needle that originally contained 65 mCi of ^{222}Rn?

9-3. Calculate the intrinsic specific activity of ^{210}Pb. What is the specific activity of a sample that was prepared in a pure state 3 years previously?

9-4. Biologists have defined a "standard man," who contains, among other elements, 18 per cent carbon and 0.2 per cent potassium. How many microcuries of ^{14}C and of ^{40}K will be present in an 80-Kg man?

9-5. A hospital has 750 mg of ^{226}Ra which serves as a source of radon for patient treatment. The gas is pumped off and sealed into tiny seeds or needles. How many millicuries of radon will be available at each pumping if this is done at weekly intervals? How much time will be required to accumulate the 700 mCi that is needed for a massive treatment?

9-6. An assay of an equilibrium ore mixture gave a $U/^{231}Pa$ number ratio of 3.04×10^6. Calculate the half-life of ^{235}U from assay data and the known half-life of ^{231}Pa.

9-7. Radioactive sources provide power for equipment used in space missions. How many watts of power will be produced by the total absorption of the 5.5-MeV alpha particles emitted by a 4.0-Kg source of 86-y ^{238}Pu? How much petroleum with a heat of combustion of 10^4 cal g^{-1} will be required to produce energy at the same rate for a period of one year?

9-8. A short-lived radioisotope was produced in a nuclear reactor and removed at $t = 0$. A count from $t = 1.0$–1.5 min gave 11,580 after making all corrections and subtracting background. A count from $t = 2.0$–3.0 min, similarly corrected, gave a total of 746 counts. What is the half-life of the sample? Assume a counting geometry of 2.6 per cent and calculate the activity of the sample when it was removed from the reactor.

9-9. A sample of seawater contains 0.48 g of potassium and 1.8×10^{-6} g of U per liter. Assume the latter to be in equilibrium with all of its daughter products and calculate the activity of the U series and of potassium.

9-10. Relatively large volumes of H_2 and O_2 are formed in radium solutions from the decomposition of water by the alpha particles. Assume that all of the alpha-particle energy from a 5-Ci source of ^{210}Po goes into the decomposition of water with one molecule disrupted for each expenditure of 85 eV. Calculate the volume at NTP of the H_2 gas produced in 24 hours. Calculate the volume of helium produced in the same period.

9-11. Artificially produced 4.6-d ^{47}Ca is a useful tracer in studies of calcium metabolism, but its assay is complicated by the presence of its radioactive daughter, 3.4-d ^{47}Sc. Assume an assay based on beta-particle counting with equal counting efficiencies for the two nuclides, and calculate the data for plotting a correction curve that can be used to determine ^{47}Ca from a total beta count.

9-12. Tritium is replacing some alpha emitters as activators on luminous instrument dials. How much ^{3}H will be needed to replace 0.08 Ci of ^{210}Po if each is equally effective in producing scintillations, and if equal luminosities are desired 1 year after fabrication?

9-13. A piece of an ancient wooden boat shows an activity due to ^{14}C of 3.8 dpm g^{-1} of carbon. What is the age of the boat?

9-14. A counter used for dating can assay a sample of 0.1 g of carbon in 2π geometry, and has a background count of 0.08 cpm. What time will be required, on an equal-time counting schedule, to assay a sample of contemporary carbon to a coefficient of variation of 3 per cent? What times will be required under similar counting requirements if the sample is 15,000 years old?

10

Alpha-Particle Emission

10.01 Alpha-particle Spectra

For many years after alpha particles had been identified as helium-4 nuclei, it was customary to express their energy in terms of their range in air under standard conditions. Today, magnetic deflection spectrometers, Fig. 10-1, and solid-state detectors permit the rapid and accurate determination of either particle momentum or energy. Either is readily converted into the other, and in the case of the naturally radioactive elements only nonrelativistic mechanics need be used.

An examination of Tables 9-1 to 9-4 shows that alpha particles are emitted in a relatively narrow band of energies, roughly from 4–9 MeV. Each alpha emitter has a characteristic spectrum consisting of a few discrete energies. Figure 10-2A shows the decay scheme for ^{226}Ra, a typical even Z–even N alpha emitter. Only three excited states of the ^{222}Rn daughter are involved, and transitions to two of these are rare. Over 94 per cent of the transitions go to the ground state.

The decays of even–odd or odd–even nuclei are usually more complicated than an even–even decay, but here also only a relatively few lines are seen in the particle spectrum. Decays from an odd–odd nucleus are complicated by the possibility of a competing beta-particle transition. In the decay of even Z–odd N ^{235}U, Fig. 10-2B, nine excited states are involved, but an appreciable number of transitions go to each. For simplicity, gamma-ray transitions have been omitted from Fig. 10-2B.

The energy of each alpha emission is very nearly monoenergetic. Figure 3-11 shows the energy resolution that can be obtained with a solid-state

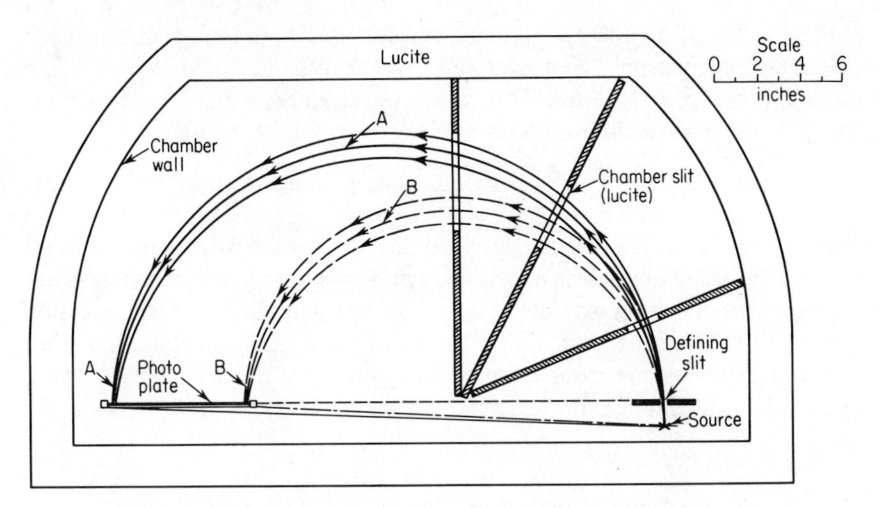

Figure 10-1. Vacuum chamber of a 180° alpha-particle spectrograph. A magnetic field directed from the plane of the paper will deflect and focus some alpha particles at *A*, less energetic ones at *B*.

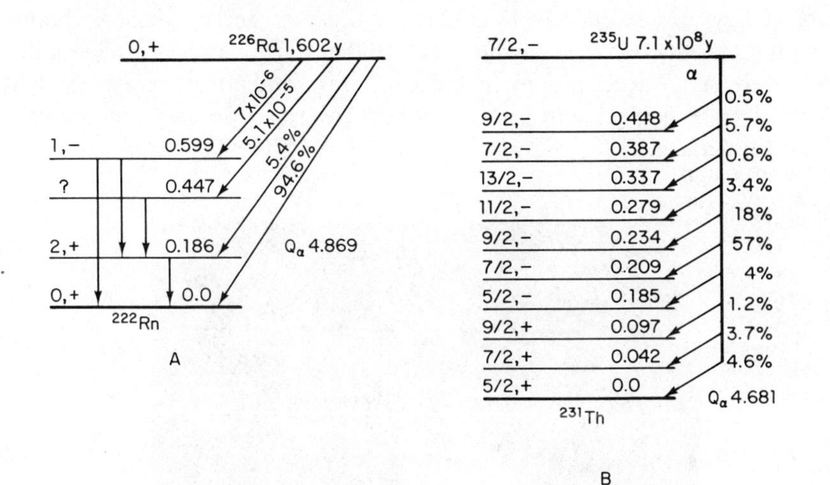

Figure 10-2. (A) The relatively simple alpha-particle spectrum of ^{226}Ra, typical of even–even nuclei. (B) The complex spectrum of alpha particles from ^{235}U is typical of even *Z*–odd *N* emitters. Gamma emission lines have been omitted for clarity.

detector but it gives a misleading idea of the width of the emission lines. Both the ^{210}Po and the ^{241}Am peaks shown in the illustration are broadened because each of the sources is so thick that some of the alphas lose an appreciable amount of energy in escaping from it.

The width of an energy level can be estimated by an application of the indeterminacy principle. Consider the extreme case of ^{212}Po, which has a mean life of 4.5×10^{-7} sec. This is the uncertainty in the lifetime of the energy state, and so the uncertainty in the energy itself will be

$$\Delta E = \frac{\hbar}{\Delta t} = \frac{6.58 \times 10^{-22}}{4.5 \times 10^{-7}} = 1.5 \times 10^{-15} \text{ MeV} \tag{10-1}$$

This extremely narrow energy level, about 10^{-16} of the level itself, is far beyond the resolving power of any known or proposed spectrometer. Since the case chosen was an extreme in Δt, it appears that alpha-particle emission is practically monoenergetic. The presence of monoenergetic alpha particles, accompanied in some cases by monoenergetic gamma rays, constitutes a powerful argument for the existence of discrete energy levels inside the nucleus.

10.02 Range and Energy

As alpha particles emitted in radioactive decay pass through matter, they lose energy almost entirely by collisions. Loss of energy through bremsstrahlung production is negligible, because the massive alpha particle undergoes only small accelerations in the collisions. Radiative energy loss may not always be negligible in the case of particles raised to very high energies in an accelerator.

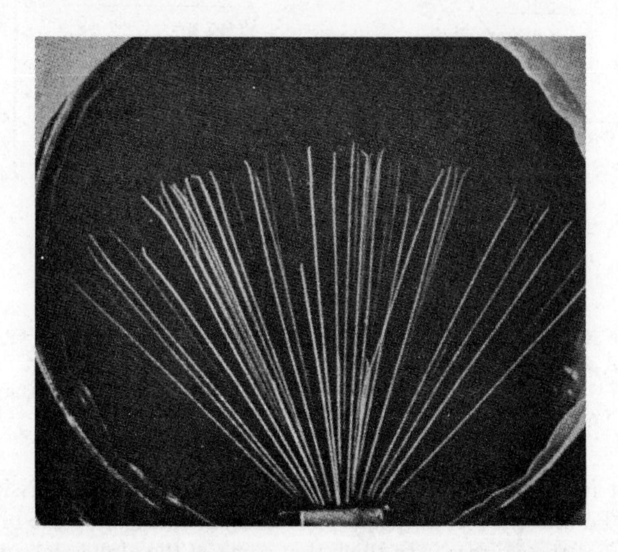

Figure 10-3. Cloud chamber photograph of alpha-particle tracks from ^{210}Po. (From F. Rasetti, *Elements of Nuclear Physics*. Prentice-Hall, Inc. Englewood Cliffs, N. J., 1947.)

The numbers of collisions and the amount of energy lost in each one of them will be subject to statistical fluctuations, and so all members of an initially monoenergetic beam will not have identical ranges as they pass through an absorber. A 5-MeV alpha particle will make, on the average, $5 \times 10^6/35 = 1.4 \times 10^5$ ionizing encounters before its initial kinetic energy is exhausted. With a normal statistical distribution, the standard deviation of this number is ± 370, which is equivalent to a standard deviation in energy of about 13 KeV.

Near the end of its path, an alpha particle may gain and lose one or two electrons several times before it finally permanently acquires two to become a neutral, thermalized helium atom. The variabilities in these latter stages of the motion add to the variabilities along the trajectory to produce a range broadening known as *straggling*. Figure 10-3 shows the nearly constant path lengths of the monoenergetic alphas from ^{210}Po in a cloud chamber. An abrupt change in the trajectory near the end of the track is the result of a collision between the slowly moving particle and a massive nucleus.

Range is customarily measured by arranging a source and a detector in a fixed geometrical relation. With gaseous absorbers the particle count rate is obtained as the gas pressure is varied. At each pressure the source–detector distance can be converted to the corresponding distance at standard conditions. Solid absorbers can be introduced into the beam and the range expressed directly in units of length. Range thus obtained is frequently converted to an *equivalent range*, usually in units of mg cm^{-2}.

Figure 10-4 shows typical results from a range determination. The count rate remains practically constant until a point 1 on the range curve is reached. Here a few of the particles have lost so much energy in collisions that they are no longer able to actuate the detector. The count rate now drops abruptly with increasing range, but shows a tailing off at the extreme range limit. The ranges will have a practically normal distribution, given by an equation of the form of Eq. (2-11), shown as B in Fig. 10-4.

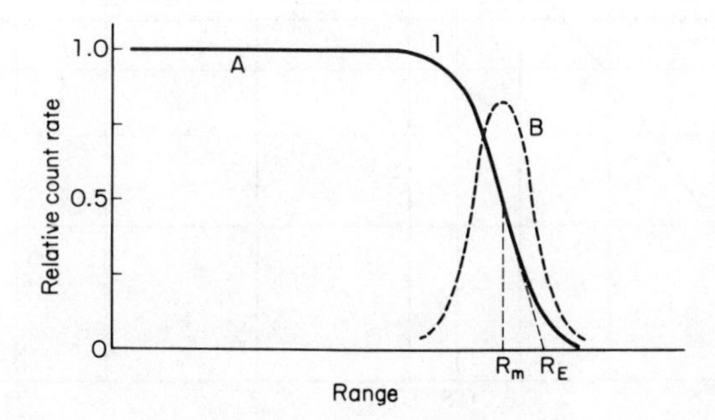

Figure 10-4. (A) Count rate and range relationship for a monoenergetic alpha-particle emission. (B) A normal probability distribution centered on the mean range R_m.

Two ranges are defined to take into account the variability due to straggling. The *mean range* R_m is that distance at which the count rate is reduced to one-half of its maximum value. A tangent drawn to the count-rate curve at the mean range will intersect the range axis at the *extrapolated range* R_E. The two range values are related by

$$R_E = R_m + \frac{\sqrt{\pi}}{2}\alpha = R_m + 0.866\alpha \tag{10-2}$$

where α is known as the *straggling parameter*. In the absence of any instrumental fluctuations, α can be expressed in terms of σ, the standard deviation of the normal distribution. As given in Eq (10-2), α expresses the range straggling from all causes.

A few materials are of particular importance in expressing alpha-particle ranges. Originally, all ranges were given in centimeters of air, and this measure is still commonly used. The upper portion of the range curve for air, Fig 10-5, is given by

$$R = 0.325E^{3/2} \quad \text{or} \quad E = 2.12R^{2/3} \tag{10-3}$$

where R is in cm of standard air and E is in MeV.

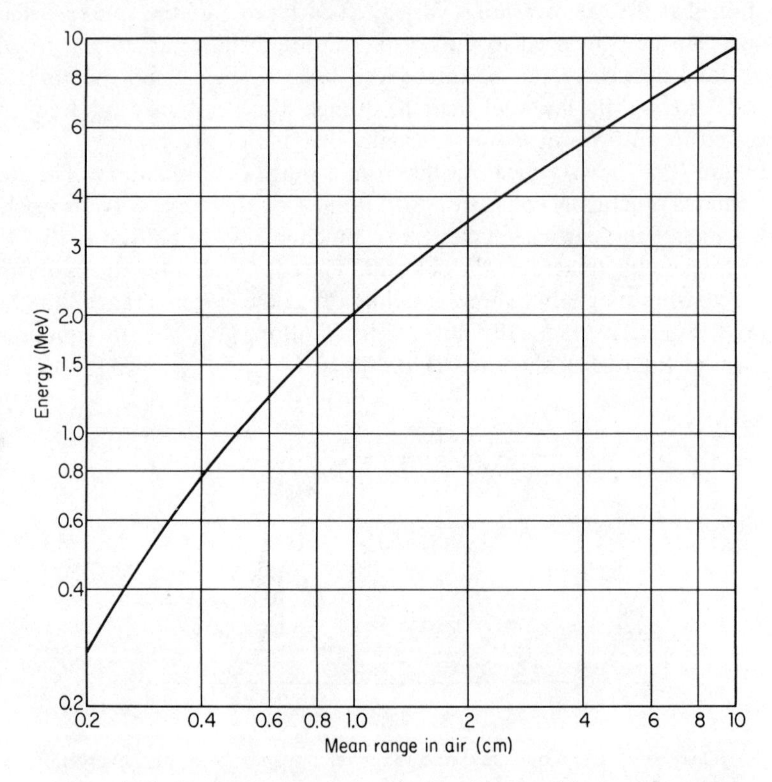

Figure 10-5. Energy–range relationship for alpha particles in air at NTP.

Another important absorber is aluminum, used as a standard for all kinds of charged particles and photons. In Fig. 10-6, the ranges are given in equivalent ranges.

It is important to know alpha-particle ranges in two kinds of particle-detector materials. Figure 10-7 shows the energy–range relations for a variety of charged particles in silicon, the most common metal used in solid-state detectors. Note that for a given range a proton needs only one-fourth the energy

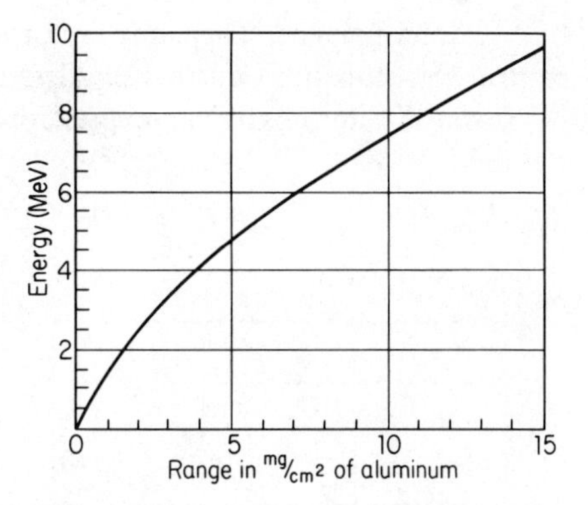

Figure 10-6. Energy–equivalent range relationship for alpha particles in aluminum.

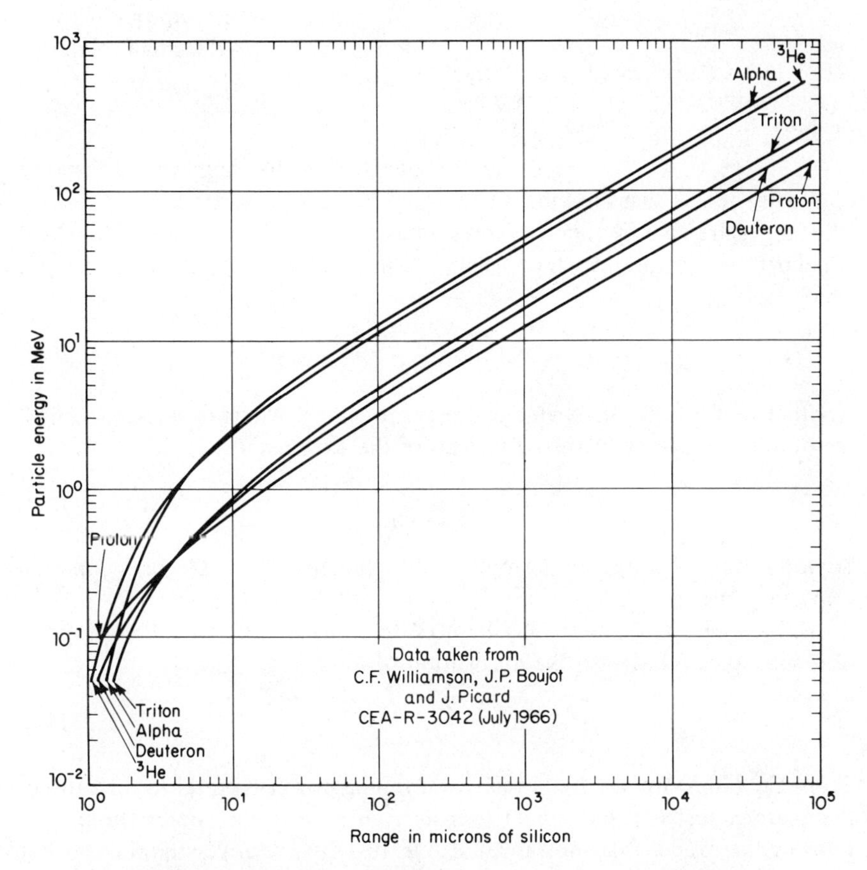

Figure 10-7. Energy–range relationships for heavy charged particles in silicon. Courtesy of Ortec, Inc.

of an alpha particle. The same factor of $\frac{1}{4}$ can be used with Fig. 10-8 if the range of protons in a nuclear emulsion is desired.

From the nonrelativistic expression for kinetic energy and Eq. (10-3) we obtain

$$v^3 = 1.03 \times 10^{27} R \text{ cm sec}^{-1} \tag{10-4}$$

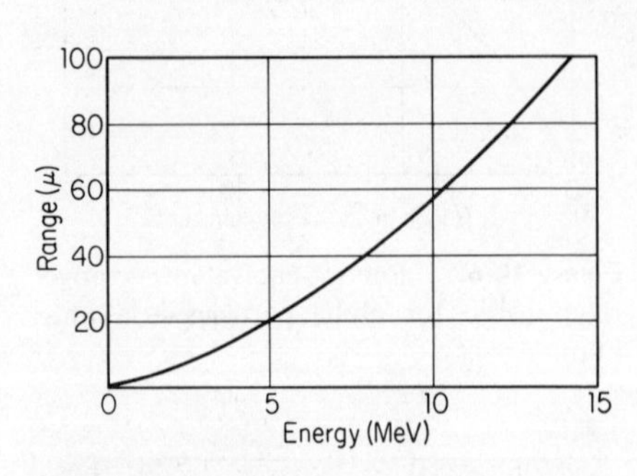

Figure 10-8. Energy–range relationship for alpha particles in a typical photographic emulsion.

Equations (10-3) and (10-4) are two forms of the *Geiger relation*. The Geiger equation holds over quite a range of particle energies. At low energies, R varies more nearly as $E^{3/4}$, and at high energies the exponent has increased to about E^2. The empirical relation is, however, useful for estimating ranges.

Another useful empirical expression is the Bragg–Kleeman relation:

$$R_s = \frac{3.2 \times 10^{-4} R_a \sqrt{A}}{\rho} \tag{10-5}$$

connecting the range in a substance of density ρ and mass number A with the range R_a in air.

Ranges are also frequently expressed as *stopping powers*. Stopping power is defined as the rate at which a particle loses energy per unit path length, which is:

$$S = \frac{dE}{dx} \tag{10-6}$$

Another useful term is *stopping power per atom*, S_A, which is assessed against the number of atoms traversed instead of the path length:

$$S_A = \frac{dE}{dn} \tag{10-7}$$

Some typical atomic stopping powers are shown in Fig. 10-9 as a function of particle energy.

Relative stopping power may be used to express the absorptive property of a substance relative to the absorption of air:

$$S_r = \frac{(dE/dx)_s}{(dE/dx)_{\text{air}}} \tag{10-8}$$

From Eq. (10-8) we see that a relative stopping power is equal to the ratio of the energies lost (substance/air) over a given path length, or to the inverse ratio of the ranges (air/substance). Table 10-1 lists relative stopping powers for a number of commonly used absorbers.

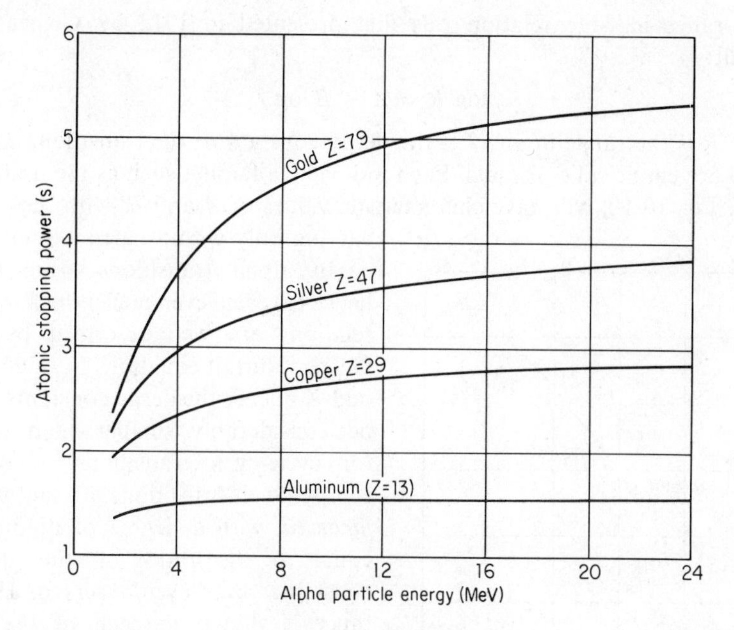

Figure 10-9. Atomic stopping powers relative to air as a function of alpha-particle energy.

TABLE 10-1
RELATIVE STOPPING POWERS

	Absorber				
	Mica	*Aluminum*	*Copper*	*Silver*	*Gold*
Relative stopping power	2000	1660	4000	3700	4800
mg cm^{-2} = 1 cm air	1.4	1.62	2.26	2.86	3.96
microns = 1 cm air	5.0	6.0	2.5	2.7	2.1

10.03 Energy and Half-life

Rutherford noted in his early studies that the most energetic alpha particles were emitted from those nuclei with the shortest half-lives. For exampie, 300-ns ^{212}Po emits an 8.95-MeV alpha particle, while the alphas from 1.4×10^{10}-y ^{232}Th have an energy of only 4.0 MeV. All of the series emitters are included between these extremes, a half-life range of 10^{24}. In contrast, the corresponding alpha energies vary by only a factor of 2. Any relation between energy (or range) and half-life must be a very strong function of the energy.

A range–half-life relation was first presented in 1912 by Geiger and Nuttall:

$$\log R = A + B \log \lambda \tag{10-9}$$

where R is the range in air, λ is the decay constant of the transition, and A and B are empirical constants. Each radioactive family, such as the uranium series, Fig. 10-10, will have characteristic values of A and B, with the latter varying only within narrow limits.

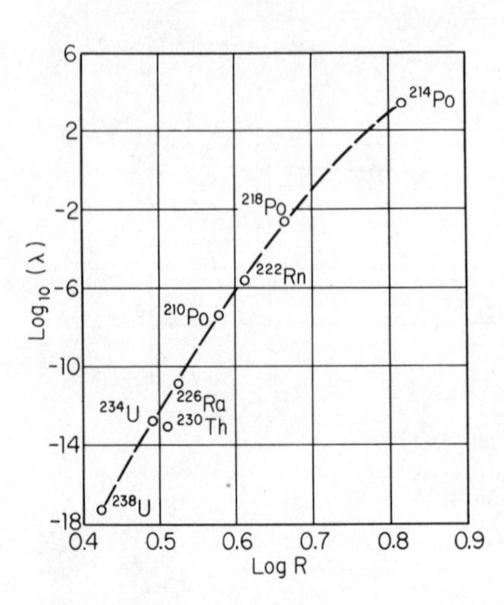

Figure 10-10. A Geiger-Nuttall plot of the alpha emissions from the even–even members of the uranium series.

In alpha transitions from (and hence to) even–even nuclei the λ–range relations are well described by the Geiger–Nuttall equation. In odd-N or odd-Z nuclei, the decay constants may be considerably smaller than those for even–even transitions of equal energy. Such transitions are said to be *hindered*, with a degree of hindrance equal to the ratio of the decay constants: even-even/observed. Hindrance is shown in each of the ten $^{235}U \longrightarrow {}^{231}Th$ transitions plotted, each with its ΔI and parity change, in Fig. 10-11.

The mass of an alpha particle is appreciable with respect to even the heaviest nucleus from which it may be ejected. An appreciable fraction of the energy available in the transition will go into the recoil of the residual nucleus and is not available to the alpha particle. From nonrelativistic mechanics

$$T_\alpha = E(A - 4)/A \tag{10-10}$$

where E is the energy available in the transition and A is the mass number of the parent nucleus. Each E value will be the Q_α for the ground-state transition, less any energy lost to gamma radiation from excited states.

As can be seen from the ordinate values in Fig. 10-12, the potential barrier is a formidable obstacle to any doubly charged alpha particle attempting to escape from the nucleus. According to classical mechanics, an escaping particle must have sufficient energy to surmount the barrier, or 25–30 MeV. After the escape the particle will acquire this amount of kinetic energy as it moves away from the barrier toward the region of zero potential. In fact, no alpha particles at or near this energy are ever observed following radioactive decay. Wave mechanics offers a ready explanation of this apparent contradiction.

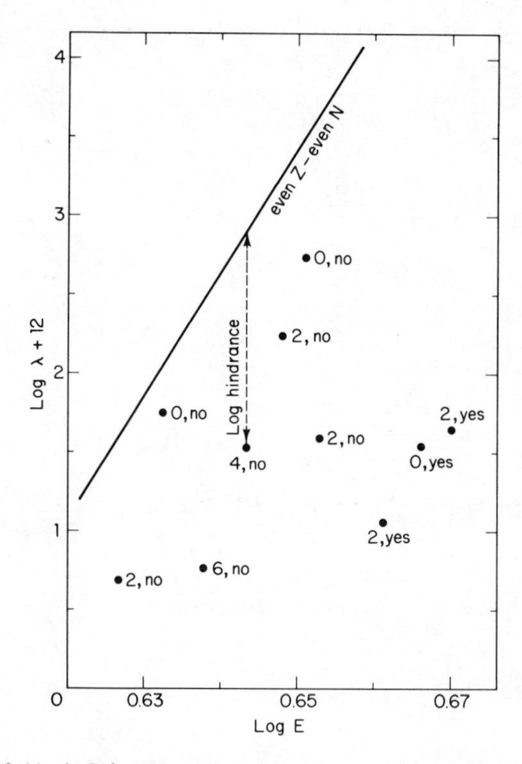

Figure 10-11. A Geiger-Nuttall plot for the members of the ^{235}U decay, with ΔI and parity change values.

Just as wave mechanics permits a linear oscillator to spend some of its time outside of the classical amplitude limits, it also permits an alpha particle to escape by *tunneling* through the barrier instead of being forced to surmount it. The possibility of tunneling permits particles with energies well below the barrier height to escape from the nucleus.

Escape probabilities would be expected to be high where the barrier is thin, and low where the barrier is thick. From the shape of the barrier, a high-energy particle, as from ^{212}Po, has a high escape probability, and the half-life is short. The low-energy particle from ^{210}Po must penetrate the barrier at a thicker portion; escape probability will be reduced and the half-life lengthened. This concept leads to an explanation for the limited range of alpha-particle energies actually observed. For transition energies greater than 9–10 MeV, penetration will take place at a very thin part of the barrier, and the half-life will be so short that the species will scarcely exist as a nucleus. At energies below 4 MeV, the barrier is very thick, and the half-lives are too long to be observable.

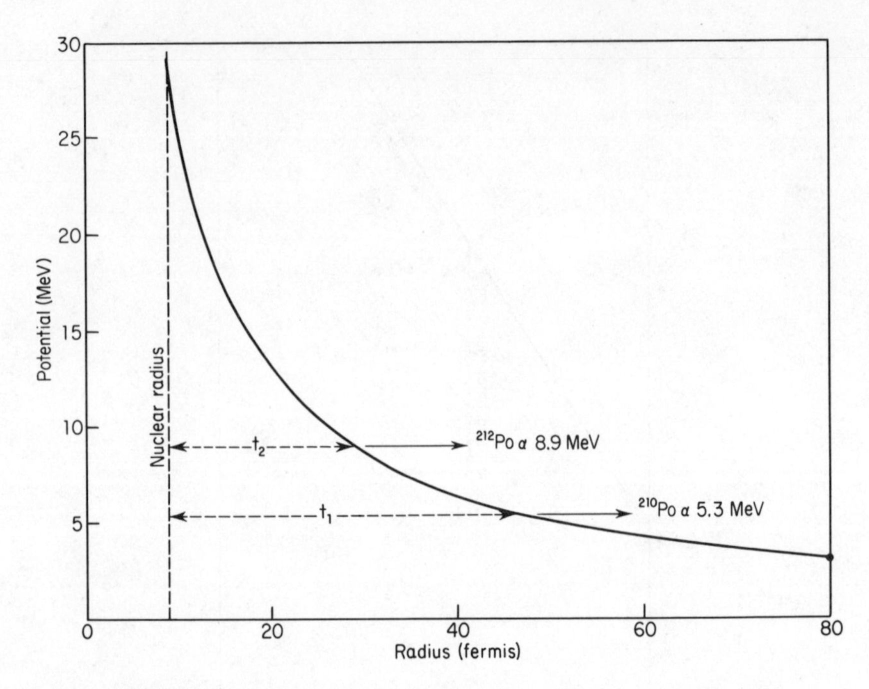

Figure 10-12. The potential barrier in ^{210}Po as seen by an escaping alpha particle.

10.04 Theory of Alpha-particle Emission

Since an alpha particle is emitted from a nucleus as a tightly bound entity, it might seem that this entity pre-existed inside the nucleus. This amounts to assuming that heavy nuclei have alpha particles as substructures. Light nuclei do exhibit periodic properties which suggest that alpha-particle groupings exist within them. In heavy nuclei, however, there is no evidence that alpha particles have an identity prior to emission.

The first wave-mechanical calculations of alpha-particle emission probabilities were remarkably successful, even though they assumed particle pre-existence. Consider an alpha emission from a heavy nucleus which will have a radius of the order of 10 F, or 10^{-12} cm. A preformed alpha particle is assumed to be in thermal motion inside the nucleus, with a mean velocity which can be estimated. From the uncertainty principle,

$$mv \simeq \frac{h}{2\pi \, \Delta x} = \frac{h}{2\pi \times 10^{-12}} \simeq 10^{-15}$$

and v is of the order of 10^9 cm sec^{-1}. According to this estimate, the alpha particle traverses the nuclear diameter 10^{21} times every second, which means that it has this number of chances to escape per second. In the case of ^{238}U with a mean life of 6.5×10^9 y, about 10^{38} attempts will be required for one escape. Only 10^{14} attempts will suffice in the case of ^{212}Po.

Gamow, and Condon and Gurney applied the concept of tunneling to the probability of alpha-particle emission from even–even nuclei, and developed an expression for the decay constant in terms of basic transition parameters. Later treatments included the probability of assembling the four-nucleon structure inside the potential barrier. When both probabilities are considered, the expression for the decay constant has the form

$$\log_{10} \lambda = \log_{10} C + \frac{8e}{2.303\hbar}(m_\alpha Zr)^{1/2} - \frac{4\pi e^2 m_\alpha}{3.256\hbar} \frac{Z}{T^{1/2}} \tag{10-11}$$

where λ is expressed in $\sec^{-1}$, C is a constant to be determined, r is the nuclear radius, and m_α is the mass of the alpha particle. All other symbols have their usual meanings.

Only a limited range of Z and A values are involved in the alpha emission from naturally radioactive nuclei, and hence the second term on the right-hand side of Eq. (10-11) is nearly constant. For most applications this term can be merged with the first constant to obtain a relation

$$\log_{10} \lambda = C_1 - C_2 \frac{Z}{T^{1/2}} \tag{10-12}$$

Theory predicts a value of 1.70 for C_2 when the energy is taken in MeV. Thus the wave mechanics offers an explanation for the form of the Geiger–Nuttall relation and permits a partial evaluation of the constants in terms of fundamental quantities. When experimental values of λ for even–even nuclei are plotted against $ZT^{-1/2}$, a linear relationship is obtained, from which both constants in Eq. (10-12) can be evaluated.

Theory provides at least some qualitative explanations for the hindrances observed in odd-N and odd-Z transitions. The four nucleons in an alpha particle are spin-paired so that there is no net spin of the particle as a whole. All nucleons in an even–even nucleus are also spin-paired, and it would seem likely that neutron pairs and proton pairs from a given shell might combine rather readily to form the alpha-particle configuration.

In an odd-N or odd-Z structure, at least one nucleon must have an unpaired spin. This nucleon will probably be the one most loosely bound, but if it is to enter into an alpha particle, it must combine with another of its own kind, with oppositely directed spin, from one of the lower energy shells. The probability of this cross-shell pairing should be less that that for the combining of two nucleon pairs that are already formed.

The degree of hindrance is also related to the angular-momentum values and the parity changes in the transition. In Fig. 10-11 the transition to the 0.279-MeV excited state is $(\frac{7}{2}, -) \longrightarrow (\frac{11}{2}, -)$ or $\frac{4}{2}$, *no*. The measured decay constant for this transition falls about 1.38 log units below the even–even line, for a degree of hindrance of 24.

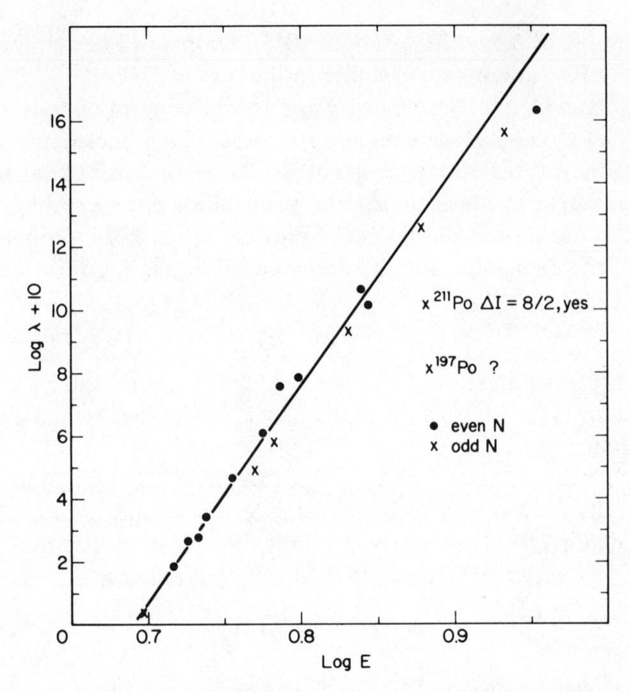

Figure 10-13. A Geiger-Nuttall plot of the alpha-emitting isotopes of Po.

Other examples of the effects of angular momentum on decay probabilities are shown in Fig. 10-13, which is a Geiger–Nuttall plot of a series of isotopes of $_{84}$Po. Most of the transitions involve only small changes in angular momentum and show only small degrees of hindrance. However, the ^{211}Po $\longrightarrow$ ^{207}Pb transition is $\frac{8}{2}$, *yes*, and this is characterized by a hindrance of about 200. The quantum states of ^{197}Po $\longrightarrow$ ^{193}Pb have not been identified, but ΔI must be large, and there is probably a parity change because the hindrance factor is about 60,000.

Even 0, *no* transitions from odd-N or odd-N nuclei may be substantially hindered, Fig. 10-11. These hindrances arise because the ground state of the parent nucleus has a large angular momentum, in this case $\frac{7}{2}$, contrasted with $I = 0$ for even–even nuclei. The centrifugal forces associated with the large values of angular momentum reduce the effective energy available for barrier penetration, and the disintegration constant is correspondingly reduced.

Angular-momentum considerations help to account for the fact that the most probable even–even alpha emissions go from ground state to ground state. In this type of transition, both the parent and the daughter nucleus will have zero angular momentum. The angular momentum of the ejected alpha particle is then also necessarily zero, and hence all of its energy is available for a direct assault on the potential barrier. Some energy is lost to photon emission in any transition to an excited state, and the escape probability may be further reduced if the alpha particle is emitted with some angular momentum, as may here be the case. There will be no parity changes in even–even transitions involving only ground states. Parity changes may be involved in transitions to excited states, and again the decay probability may be reduced. This effect may be seen in the decay scheme of ^{226}Ra, Fig. 10-2A.

Any nucleus with an unpaired nucleon will have some angular momentum, and the ground state values for parent and daughter may well be different. In addition, parity changes are quite possible in ground-state transitions involving unpaired nucleons. Ground state–ground state transitions may then be hindered on two counts. The decay scheme of ^{235}U, Fig. 10-2B, typifies this situation. In this example the most probable transition is 0, *no* to the 0.204-MeV excited state. The loss of available energy to the gamma emission is more than compensated by the fact that the transition to the ground state is $\frac{2}{2}$, *yes*. The only other 0, *no* transition in ^{235}U goes to the 0.387-MeV level, and even with this loss of available energy the transition is more probable than that to the ground state.

10.05 Binding Energy and Particle Emission

It is still necessary to explain why alpha-particle emission occurs almost exclusively from heavy nuclei. The reason for this can be seen by calculating the changes in binding energy as given by the semiempirical relation, Eq. (7-17). This can be written as

$$(M)_{Z,A} = (N)_{Z,A} - (E_B)_{Z,A} \tag{10-13}$$

In an alpha decay we are concerned with the calculation of the mass or energy change involved in the loss of a particle ($Z = 2$, $A = 4$). Then,

$$\Delta M = (N - E_B)_{Z,A} - (N - E_B)_{Z-2,A-4} - (2m_n + 2m_p - 28.3) \tag{10-14}$$

The third term in Eq. (10-14) can be written explicitly, since we know the composition of the alpha particle and its binding energy. Now, no nucleons are created or destroyed in the decay, so the N terms will cancel in the subtraction. We then have

$$\Delta E_B = 28.3 + (E_B)_{Z-2,A-4} - (E_B)_{Z,A} \tag{10-15}$$

where ΔE_B is the term denoted by Q_α in the decay schemes.

The δ term in the binding-energy expression can be neglected in alpha decay since it will have the same sign in parent and daughter and will differ in magnitude only through a small change in A. Although the calculation of Q_α involves finite differences, we may write

$$Q_\alpha = 28.3 + \left(\frac{\partial E_B}{\partial A}\right)_Z \Delta A + \left(\frac{\partial E_B}{\partial Z}\right)_A \Delta Z \tag{10-16}$$

whence

$$Q_\alpha = 28.3 + 14.1\,\Delta A - 0.92A^{-1/3}\,\Delta A + 0.197Z^2A^{-4/3}\,\Delta A$$
$$- 1.19A^{-1/3}Z\,\Delta Z + 19.3(A - 2Z)^2A^{-2}\,\Delta A \qquad (10\text{-}17)$$

The form of the last term is particularly simple, since the value of $(A - 2Z)$ is invariant under an alpha-particle decay.

Equation (10-17) can be evaluated for appropriate values of Z and A, taking $\Delta Z = -2$, $\Delta A = -4$. If the calculated value of Q_α is positive, the final state will have a greater binding energy than the initial state, and alpha-particle emission will be energetically possible. If Q_α is negative, alpha emission is energetically impossible. These results are permissive only. A positive Q_α of at least four MeV is needed to obtain a realistic probability of barrier penetration.

When the calculations are made, positive values of Q are found for all reasonable combinations of Z and A if the latter is greater than about 150. Below this mass number, no alpha emissions are possible, according to the binding-energy calculations. The nonseries alpha emitters whose mass numbers lie only slightly above the predicted limit have very long half-lives. Predictions based on the liquid-drop model are in good agreement with observations. The model cannot be expected to give accurate values at the magic numbers, since no shell-structure terms were put into it.

It is appropriate to examine at this time the possibility of a neutron or proton emission from a nucleus. Neutron emission in particular might be expected, since a neutral particle in the potential well will not encounter the coulomb potential barrier when it attempts to escape. An appropriate application of the binding-energy equation shows that the spontaneous emission of either a neutron or a proton from the ground state is energetically impossible for any value of A. Once again the predictions of the simple liquid-drop model are in accord with observations. As we shall see later, neutron emission from highly excited states of nuclei produced in fission is quite possible.

The emission of groups of nucleons larger than ^{4}He is ruled out because of the greater height of the potential barrier presented to the heavier groups. As an example, consider the decay of ^{234}U through four successive alpha-particle decays to ^{218}Po, with a total Q of about 20 MeV. The binding energy of ^{16}O is 127 MeV, to be compared to the binding energy of four alpha particles of 113 MeV. Energetically, ^{16}O emission is favored by 14 MeV. Offsetting this advantage is the fact that the potential barrier for the ^{16}O particle will be four times the height presented to an alpha. This greatly increased barrier thickness reduces the probability of tunneling to a point many orders of magnitude below our limits of detection. The spontaneous emission of a group heavier than ^{4}He must be a very rare event, if it occurs at all.

10.06 Long-range Alpha Particles

Some nuclides emit a few alpha particles with energies greater than would be expected for a ground-state-to-ground-state transition. A case in point is the α-decay of ^{212}Po, where about two particles out of 10^4 have an extralong range. The origin of these high-energy particles can be explained by considering the mode of formation of the parent.

Figure 10-14 shows the pertinent details of the decay schemes of ^{212}Bi and ^{212}Po. About 10 per cent of the ^{212}Bi decays lead to excited states of the Po daughter, rather than to the ground state. In most of these transitions the excitation will be promptly relieved by the emission of a gamma ray. However, the half-life of ^{212}Po is so short that in some cases nuclear decay to ^{208}Pb will take place while the nucleus is still in an excited state. The particle will then be ejected with the excitation energy added to the energy of the ground-state transition.

Gamma-ray emission is usually favored over particle transitions, so the long-range alphas are seen only from short-lived nuclei, and then only rarely. In longer-lived species, gamma emission will have taken place long before an appreciable number of alphas are ejected.

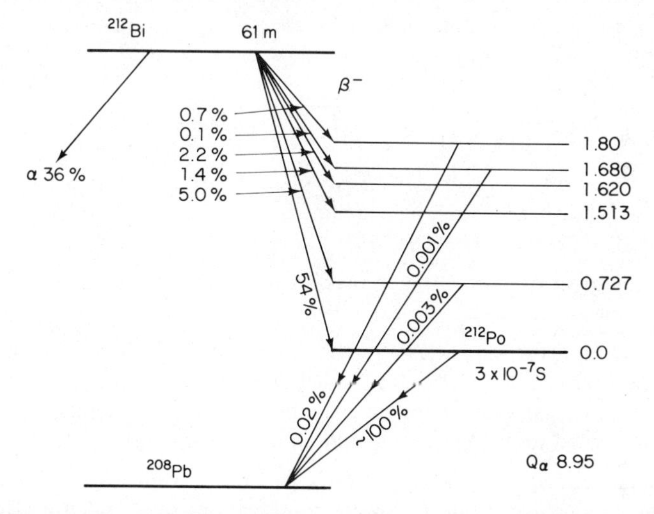

Figure 10-14. Decay of $^{212}Bi \rightarrow {}^{212}Po \rightarrow {}^{208}Pb$ with gamma rays omitted for clarity. Some alpha transitions to ^{208}Pb take place before ^{212}Po is de-excited.

REFERENCES

Evans, R. D., *The Atomic Nucleus*. International Series in Pure and Applied Physics. McGraw-Hill, New York, 1955.

Rutherford, E., J. Chadwick, and C. D. Ellis, *Radiations from Radioactive Substances*. Macmillan, New York, 1930.

Turkevich, A. L. et al., "Alpha Radioactivity of the Lunar Surface at the Landing Sites of Surveyors 5, 6, and 7." *Science* **167,** 1722, 1970.

PROBLEMS

10-1. We have been assuming that nonrelativistic mechanics can be used without appreciable error in alpha-particle calculations. What error is introduced in calculating thus the velocity of a 10-MeV alpha particle? At what energy will a velocity calculation by Newtonian mechanics be in error by 2 per cent?

10-2. What energy must be imparted to an alpha particle to force it into the nucleus of bismuth? What energy will be required to obtain proton penetration to the same radius?

10-3. Compare the Q values with the barrier heights for the following transitions: $^{234}U \longrightarrow {}^{4}He + {}^{230}Th$; $^{234}U \longrightarrow {}^{12}C + {}^{222}Rn$; $^{234}U \longrightarrow {}^{16}O + {}^{218}Po$. The pertinent mass defects are: ^{234}U, 38.16; ^{230}Th, 30.87; ^{222}Rn, 16.39; ^{218}Po, 8.38; ^{4}He, 2.42; ^{12}C, 0.00; ^{16}O, −4.74 MeV.

10-4. Use the following data to make a Geiger–Nuttall plot for the even–even isotopes of uranium, and determine the constants in Eq. (10-9). Isotope, half-life, alpha energy, and percentage occurrence: ^{238}U, 4.51×10^9 y, 4.20 MeV, 75%; ^{236}U, 2.39×10^7 y, 4.49 MeV, 76%; ^{234}U, 2.47×10^5 y, 4.77 MeV, 72%; ^{232}U, 72 y, 5.32 MeV, 68%; ^{230}U, 20.8 d, 5.89 MeV, 67%; ^{228}U, 9.1 m, 6.69 MeV, 70%.

10-5. Use the results of Prob. 10-4 to calculate the degree of hindrance for each of the following ^{235}U alpha emissions: ^{235}U, 7.1×10^8 y, 4.58 MeV (8%), 4.40 MeV (57%), 4.37 MeV (18%).

10-6. Use the data of Prob. 10-4 to make a log λ–$ZT^{-1/2}$ plot, and determine the values of C_1 and C_2 in Eq. (10-12).

10-7. Assume that the log λ–$ZT^{-1/2}$ relationship found in Prob. 10-6 applies to the nonseries alpha emitters of Table 9-5. Calculate the degree of hindrance for those decays where data are available. Can any of the hindrances found be attributed to angular-momentum and parity changes?

10-8. Assume that a Q_α value of +4 MeV is needed to insure a half-life that is measurably short. Choose realistic combinations of Z and A and calculate the limiting value of A for alpha decay.

10-9. Derive an equation of the form of Eq. (10-17) that is applicable to proton emission.

10-10. Derive an equation of the form of Eq. (10-17) that is applicable to neutron emission. Calculate ΔE_B for a typical light, intermediate, and heavy nucleus.

11

Beta Particles

11.01 Identification

Early magnetic-deflection measurements showed that beta particles had a single negative charge and an e/m ratio identical with that of electrons produced in a variety of ways. These results were strong but incomplete evidence that the emitted negative particles were indeed electrons. As new properties of electrons were discovered and measured with precision, the identity between them and beta particles was strengthened. Except for origin, there is no known difference today between the negatron and the electron.

Perhaps the strongest evidence for the identity comes from an experiment reported by the Goldhabers in 1948. If beta particles differ from electrons in any quantized parameter, such as spin, it should be possible for a beta particle to occupy a position in one of the orbital shells, even though these shells were filled with electrons. If the two particles are identical in all respects, beta-particle occupancy would be forbidden by the Pauli principle.

Any state that might be occupied by a beta particle would have an energy somewhat different from those states occupied by the regular orbitals. As a consequence, a beta-excited fluorescent X-ray spectrum should differ from one produced by electronic excitation. No differences are observed, and so the quantum identity of the two particles is established.

Beta particles originate in the nucleus, and for a good many years it was assumed that they existed there prior to emission. Before the discovery of the neutron, nuclear electrons were postulated in order to account for the differences between the number of nuclear charges and the number of protons needed to make up the nuclear mass. It is now evident on a number of grounds that electrons as such cannot exist inside the nucleus.

According to the electron–proton model, an odd Z–even A nucleus would consist of A protons and $(A - Z)$ electrons. Then $(A - Z)$ would be an odd integer, as would $(2A - Z)$, the total number of elementary particles in the structure. A case in point is $^{14}_{7}N$, where $A - Z = 7$ and $2A - Z = 21$. According to the neutron–proton model, there are an even number of nucleons in every even-A nucleus. Each of the three particles in question is a fermion, with half-integral spin. The electron–proton model requires, therefore, that every odd Z–even A nucleus also be a fermion with a half-integral spin. The neutron–proton model leads to a nucleus that is a boson with zero or integral spin. Spectroscopic observations of band spectra can distinguish unequivocally between a boson and a fermion and in this case the evidence is clear: odd Z–even A nuclei are bosons and the electron–proton hypothesis is untenable.

Another argument against the existence of nuclear electrons comes from a comparison of nuclear magnetic-dipole moments with the Bohr magneton, which is characteristic of the electron. Nuclear magnetic moments are, at most, a few nuclear magnetons, and hence are smaller than the electron's moment by a factor of 10^3 or so. Now, a neutrino is emitted simultaneously with a beta particle, and one might postulate that the two magnetic moments are equal and oppositely directed, so that they effectively cancel inside the nucleus. This possibility was ruled out when it was found that the magnetic moment of the neutrino is, at most, a small fraction of a Bohr magneton.

We have previously mentioned that the de Broglie wavelength for an electron of any reasonable energy is incompatible with the dimensions of a nucleus.

Another argument against the existence of nuclear electrons came with the discovery of nuclei that show beta branching, emitting either a negatron or a positron. If these two particles were to be in the nucleus prior to emission, they would be expected to annihilate, and so no beta branching would be observed.

The general arguments given can be applied specifically to the neutron and the proton. Thus the neutron can not be a tight combination of a proton and an electron. Similarly, a proton can not be a positron–neutron combination, because none of the considerations given depended upon the sign of the charge on the beta particle.

11.02 Beta-particle Spectra

Beta-particle spectra can be measured with a magnetic-deflection spectrograph, using a photographic film as a detector, Fig. 11-1. More quantitative data can be obtained by replacing the film with a proportional counter fitted

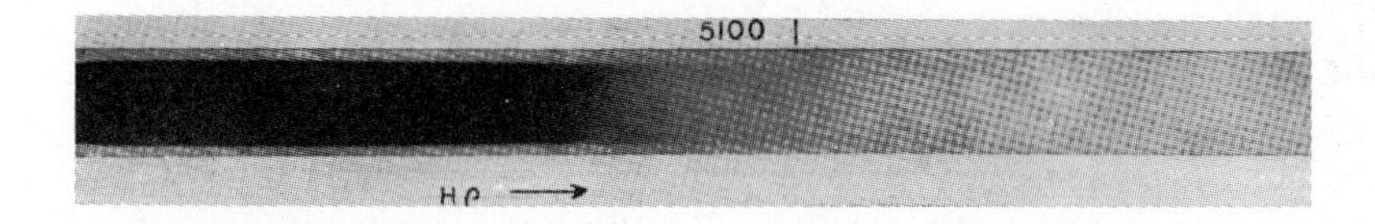

Figure 11-1. The continuous spectrum of a pure beta emitter as recorded on a photographic emulsion.

with a very thin window. Whatever the method of detection, the slit system of a magnetic spectrograph will pass a fixed $B\rho$ or momentum interval, and the instrument will provide a momentum rather than an energy spectrum.

Solid-state detectors are now available with depletion layers thick enough to absorb completely beta particles with energies up to well over 1 MeV, Fig. 10-7. The response of these detectors will be quite accurately proportional to the energy of each completely absorbed beta particle, and the output pulses can be fed into a pulse-height analyzer, which will sort them into equal energy intervals. The beta spectra thus obtained will be similar, but not identical, to those obtained by magnetic dispersion.

The striking feature of the beta spectra obtained by either method is the continuous energy and momentum distribution. Some nuclei, known as *pure beta emitters*, show only a continuous spectrum, with energies ranging from zero up to a maximum value E_m characteristic of each nuclide. Other beta decays show monoenergetic conversion electrons superposed on a continuous distribution. Careful measurements at very low energies may show numerous Auger electrons in addition to the continuous distribution.

Figure 11-2 shows the electron spectrum of ^{198}Au. The two lines result

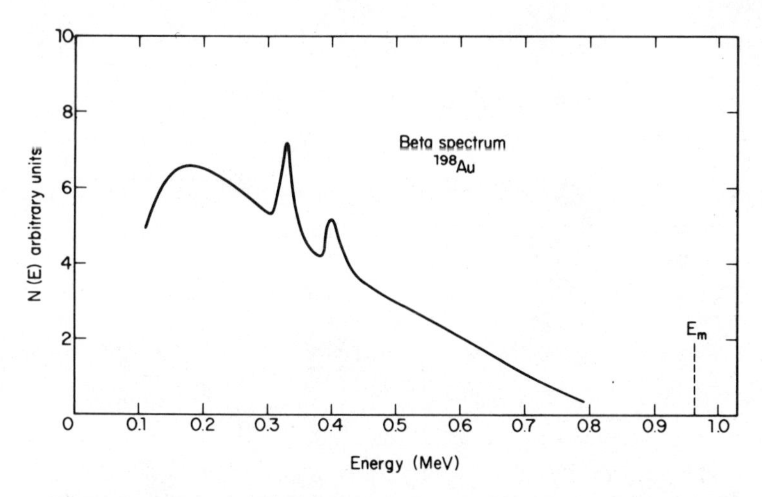

Figure 11-2. Two emission lines from internal conversion are superposed on the continuous energy distribution of the primary beta particles.

from the conversion of the 411.8-KeV gamma ray in the K and L levels of the ^{198}Hg daughter nucleus. Part of the line widths seen in the figure comes from the fact that the source had an appreciable thickness, so that some of the conversion electrons lost energy in escaping from the source. For the same reason, extreme care must be taken in preparing sources if true spectral shapes at the lower energies are desired.

11.03 Energy Relations in Beta Emission

Alpha-particle spectra and gamma-ray emissions, each consisting of a relatively few monoenergetic radiations, point strongly to the existence of a series of discrete energy levels in the nucleus. For several years, it seemed impossible to reconcile this model with the continuous energy distribution seen in beta decay.

One attempted explanation assumed that each beta particle was ejected with an energy E_m, and that part of this energy was given up to the orbitals during the escape process. Precision calorimetry showed that the amount of energy locally deposited amounted, on the average, to $\bar{E}$ per beta particle, where $\bar{E}$ is roughly $\frac{1}{3}E_m$. This result ruled out any energy transfer to the orbitals, since they would have promptly given up any energy of excitation to end up with an average energy deposition of E_m.

The energy relations in α–β branching establish beyond doubt that each nucleus loses an energy E_m and not $\bar{E}$ at beta emission. A case in point is the beta decay of ^{212}Bi, which shows 36 per cent alpha branching. Only the main decay channels are shown in Fig. 11-3. Transitions to the other states lead to the same conclusions when the gamma-ray energies are taken into account.

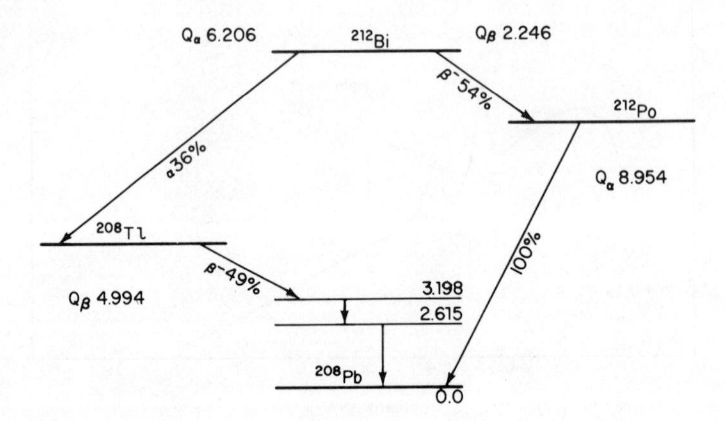

Figure 11-3. A simplified decay scheme of the branched transition at ^{212}Bi, reunited at ^{208}Pb.

The two branches separate at the ground state of ^{212}Bi and reunite at ^{208}Pb. On the left-hand side of the decay scheme, an alpha decay with Q_α 6.206 establishes the ground state of ^{208}Tl. This nucleus emits a beta particle with E_m 1.796 followed by two gamma rays. On the right-hand side, a beta decay with E_m 2.246 leads to ^{212}Po, and thence through an 8.954 alpha-decay to ^{208}Pb. The energy balance requires that each beta emission involve an energy of E_m:

α–β	β–α
6.206	2.246
1.796	8.954
3.198	11.200
11.200	

Tabular values of decay characteristics usually list values of E_m, frequently without subscript. If not otherwise indicated, tabular values will be the maximum energies associated with the transitions. In calculations of local energy deposition, as in tissue-dose estimates, $\bar{E}$ values rather than E_m are required. These will be found in some specially prepared tables. Whatever the decay scheme, tabulated values of $\bar{E}$ will include all electron emissions, properly weighted, that go with each decay.

11.04 The Neutrino Hypothesis

In 1931, Pauli suggested that the principles of conservation of energy and discrete nuclear energy levels could be retained by assuming a second particle emitted along with each beta particle. Conservation of electric charge in beta decay was already satisfied without the new particle which must, then, be electrically neutral. This requirement led to the name *neutrino*. Nuclear mass relations were also satisfied to a high degree of precision without the neutrino, whose mass must, therefore, be very small compared to that of an electron. To fit the requirements of beta emission, the neutrino must be able to carry away energy, momentum, and angular momentum, and must have an inherent spin of $\frac{1}{2}$. The neutrino must, therefore, be a fermion. The spin requirement arises from the fact that in a beta decay, a single particle of spin $\frac{1}{2}$ is converted to three particles in accordance with either reaction (11-1) or (11-2).

$$n \rightarrow p + e^- + \bar{\nu} + Q \tag{11-1}$$

$$p \rightarrow n + e^+ + \nu + Q \tag{11-2}$$

In either of these reactions, two of the product particles are known to have spins of $\frac{1}{2}$ each, so their combined spins must be either zero or integral. To maintain spin conservation, the remaining particle, the neutrino, must have a spin of $\frac{1}{2}$.

Note that neither (11-1) nor (11-2) is in energy balance. In the case of a free neutron reaction, (11-1) can take place spontaneously and then Q will represent the reaction energy, divided between the three reaction products. We are now assuming (11-1) to take place inside a large nucleus. Q now represents all of the energy available from the nuclear rearrangement, an amount that may be quite different from that resulting from the spontaneous decay of the free neutron. Reaction (11-2) is energetically impossible for a free proton and can only take place inside a nucleus. Again, Q will represent all of the energy made available by the nuclear rearrangement.

According to Pauli, the energy E_m available at each beta emission is divided between beta particle and neutrino in all possible ratios. On the average, the neutrino receives about twice the energy imparted to the beta. In 1934, Fermi developed a theory of beta decay, which included the probability of energy division between the two particles. The success of the Fermi theory has furnished additional evidence for the validity of the neutrino concept.

A neutral particle of extremely small (or zero) mass will interact very weakly with any matter through which it passes. Measurements and theory agree on a neutrino interaction cross-section of the order of 10^{-43} cm^2. This almost inconceivably small cross-section leads to, at most, a few encounters in light-year thick lead. It is understandable that the energy left by neutrinos within beta-particle calorimeters is undetectable!

According to (11-1), the emission of a negative beta particle involves the conversion of a nuclear neutron to a proton, with the ejection of two particles neither of which had a prior existence inside the nucleus. It appears to be a general rule that a matter–antimatter pair is formed whenever energy is converted into mass. Since the negative electron is a member of the matter system, we write $\bar{\nu}$ in (11-1) for the antineutrino rather than ν for the neutrino. The latter appears with positron emission, so that again a matter–antimatter pair is created, (11-2).

Particle–antiparticle pairs usually differ only in the sign of the electric charge, a condition that cannot be fulfilled by the neutrinos with zero charge. The question has been raised as to whether there are in fact two different neutrinos. A positive answer to this question seems to have been given by the requirements of a double beta-particle decay.

Consider a transition such as ${}^{130}_{52}\text{Te} \longrightarrow {}^{130}_{54}\text{Xe} + 2\beta$, where two negatron emissions are involved. Now, the emission of one type of particle is equivalent to the absorption of its antiparticle. Then (11-1) becomes

$$\left.\begin{aligned} n - \bar{\nu} &\longrightarrow p + e^- \\ n + \nu &\longrightarrow p + e^- \end{aligned}\right\} \qquad (11\text{-}3)$$

If ν and $\bar{\nu}$ are identical, reaction (11-1) will supply the particle needed for reaction (11-3), and the two beta particles will be emitted with no accompanying neutrinos. If ν and $\bar{\nu}$ are not identical, the double beta decay must be accompanied by the emission of two antineutrinos.

Theory has been quite successful in predicting disintegration constants for beta decays. In the present case, theory leads to half-lives of the order of 10^{20} years for the double-neutrino emission and to about 10^{13} years for the no-neutrino decay. Half-lives of the latter order should be readily detected, but none has been observed. The ^{130}Te decay has a half-life close to that predicted for the double-neutrino emission. Thus the experimental evidence favors the separate existence of a neutrino and an antineutrino. We shall have little need to distinguish between them and shall usually use "neutrino" to mean either particle.

The two particles in the neutrino pair can not differ in charge, but it appears that they do differ in the relative directions of their spin vectors. In the antineutrino the spin vector is aligned along with the angular-momentum vector; in the neutrino the two vectors are oppositely directed. It can be shown that this right- and left-handedness leads to the requirement that the particle masses be identically zero, with neither an electric nor a magnetic moment.

11.05 Detection of the Neutrino

First attempts to detect the neutrino utilized its ability to have momentum. If this is indeed the case, the observable tracks of an ejected beta particle and the recoiling nucleus, as seen in an expansion chamber, should not by themselves satisfy momentum conservation. Any lack of two-particle momentum conservation will be greatest in a light nucleus emitting an energetic beta particle. The reaction

$$^{6}\text{He} \longrightarrow {}^{6}\text{Li} + e^{-} + \bar{\nu} + 3.5 \text{ MeV} \tag{11-4}$$

has been used successfully. The recoil measurements not only demonstrated the participation of a third particle in beta decay, but served to establish angular relations between the directions of emission of beta and neutrino.

Although neutrino absorption cross-sections are vanishingly small, detection is possible in the intense flux associated with operation of a high-powered nuclear reactor. The absorption processes to be expected are (11-3) and

$$p + \bar{\nu} \longrightarrow n + e^{+} \tag{11-5}$$

In an experiment of heroic proportions, Cowan and Reines exposed a large volume of a hydrogen-rich scintillator solution to the intense neutrino flux of a power reactor. The reaction described in (11-5) would be expected in low yield with e^{+} producing ionization and a scintillation at the instant of formation. The neutron formed will lose energy by successive collisions with protons until its energy is reduced to about 0.18 eV. At this energy the neutron has a high probability of resonance absorption by ^{113}Cd, which was present in the scintillator solution. The reaction is

$$^{113}\text{Cd} + n \rightarrow {}^{114}\text{Cd} + \gamma \tag{11-6}$$

Some of the gamma rays from reaction (11-6) will be absorbed to produce scintillations. These will occur later than the e^+ scintillations by the amount of time required for the neutron to slow. Cowan and Reines measured the delayed coincidences between the two scintillations and thus established the existence of the neutrino beyond doubt.

All measurements are in accord with the characteristics originally postulated by Pauli. Neutrino mass appears to be identically zero, which requires that it travel with the velocity of light, since it carries energy.

11.06 Energy–Half-life Relationships

In 1933, Sargent presented a log λ–log E_m plot, now known as the *Sargent diagram*, for the beta emitters of the natural radioactive series. The points defined two straight lines, Fig. 11-4, separated by a factor of more than 100 in decay constants at equal energy. The transitions in the short-lived group

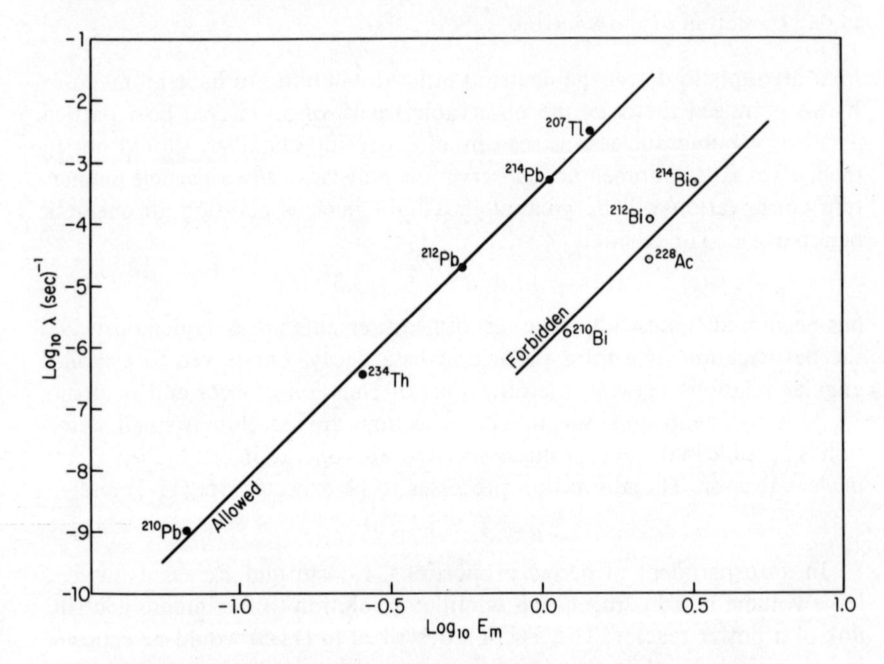

Figure 11-4. A Sargent plot of some naturally occurring beta emitters.

were said to be *allowed;* the longer half-lives came from *forbidden* transitions.

A simple two-compartment classification was soon found to be inadequate for the host of artificially produced beta emitters. Transitions are now said to be *first-*, *second-*, or *n-forbidden.* A few unusually short-lived decays are *favored.* As in alpha decay, *hindered* would be a better term than *forbidden*, since the interdiction is not absolute, but the latter is now firmly established.

It was natural to look for selection rules governing the various decay probabilities, and from the alpha-particle analog, changes in parity and in ΔI would be expected to be implicated. Two sets of selection rules governing beta decay are known, the choice depending upon the exact nature of the forces in the parent nucleus.

Most cases of beta decay obey the *Gamow–Teller rules,* and some follow the *Fermi rules,* Table 11-1. A few nuclei appear to require a mixture of the two sets of rules.

TABLE 11-1
SELECTION RULES FOR BETA DECAY

Type of transition	*Parity change*	ΔI *Gamow–Teller*	ΔI *Fermi*
Favored	*No*	0	0
Allowed	*No*	$0, \pm 1$ Not $0 \longrightarrow 0$	0
First forbidden	*Yes*	$0, \pm 1, \pm 2$	$0, \pm 1$
*n*th forbidden	*n* odd, *yes* *n* even, *no*	$\pm n, \pm(n+1)$	$\pm(n-1), \pm n$

Favored transitions are found only in light nuclei, $A < 44$. In most of these cases, parent and daughter are *mirror nuclei* characterized by $N - Z = \pm 1$. A typical mirror transition is

$$^{11}_{6}\text{C} \xrightarrow{20\text{m}} {}^{11}_{5}\text{B} + \beta^{+} \tag{11-7}$$

The short half-life of a mirror transition is explained by the presence of a single unpaired nucleon in parent and daughter. Since nuclear forces are almost independent of charge, a proton–neutron or neutron–proton conversion results in a relatively small change in nuclear arrangements. The Pauli principle permits the converted nucleon to remain in the same shell that it occupied before emission.

Favored transitions occur in a very small group of three-member isobars, each case consisting of a parent made up of two nucleons more than a core of even $N =$ even Z. A typical example is

$$^{18}_{9}\text{F} \xrightarrow{112\text{m}} {}^{18}_{8}\text{O} + \beta^{+} \tag{11-8}$$

In this example, the saturated core is $8n + 8p$; in the parent this leaves $1n$ and $1p$ in an active, unfilled shell. The Pauli principle permits not more than two like nucleons in a given shell. In the reaction described in Eq. (7-8), the tenth neutron, created by beta decay, is allowed to remain in its original shell position. This gives the transition a high probability and a correspondingly short half-life.

11.07 Allowed and Forbidden Spectra

As might be expected, the degree of forbiddenness affects the shape of the beta-particle spectrum as well as the half-life of the transition. As a rule, forbidden transitions show a greater number of low-energy particles than are seen in an allowed transition, but in many cases the differences are not great.

Figure 11-5 is an example of the spectral distribution in an allowed negatron decay. The spectrum is shown as both an energy and a momentum distribution. Characteristically, the energy plot shows a high proportion of low-energy emissions, and the momentum plot is fairly symmetrical about the maximum. In either plot, the distribution approaches the axis at the high end gradually, thus making difficult the determination of the end point. By comparison, the first forbidden spectrum shown in Fig. 11-6 exhibits an excess of low-energy betas in the energy plot, and an increase in a symmetry in the momentum plot.

As a beta particle leaves a nucleus, it will interact with the electric charge remaining. A negatron will be attracted toward the receding nucleus and hence there will be a goodly number of low-energy beta emissions. A positron, on the other hand, will be repelled by the nucleus, and many electrons will

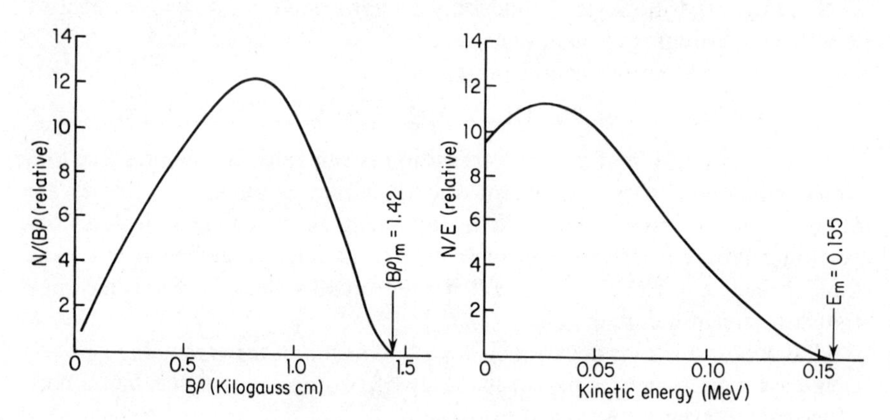

Figure 11-5. Momentum and energy spectra in the decay of ^{14}C, an allowed negatron transition.

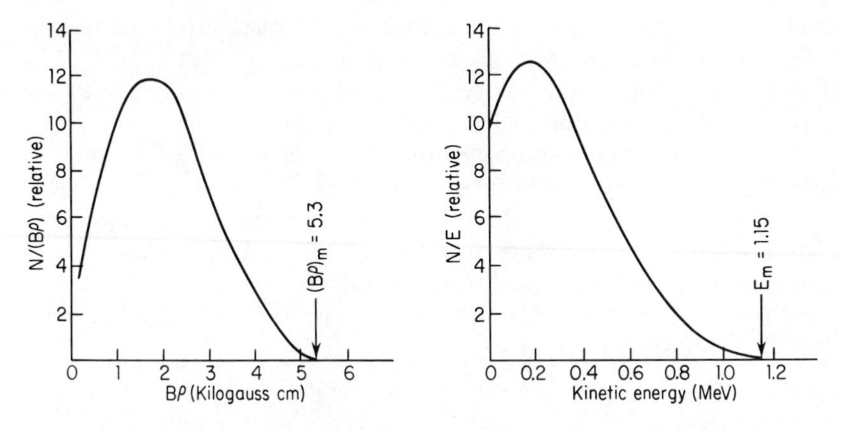

Figure 11-6. Momentum and energy spectra in the decay of ^{210}Bi, a first forbidden negatron transition.

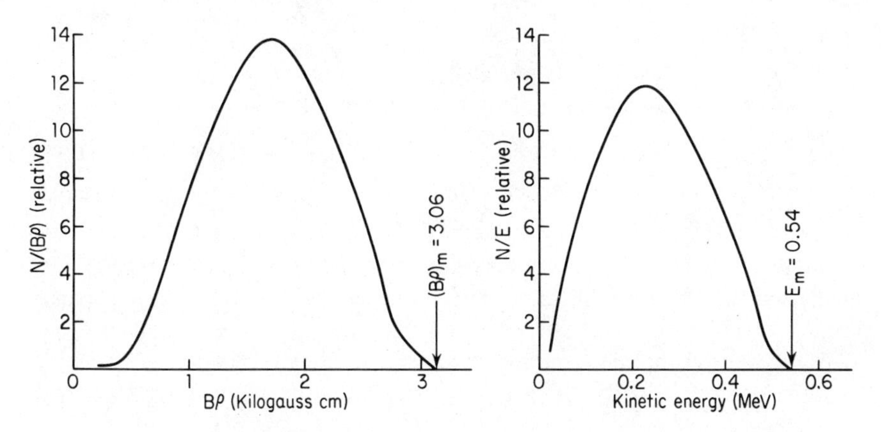

Figure 11-7. Momentum and energy spectra in the decay of ^{22}Na, an allowed-positron transition.

be pushed into higher energies, leaving a dearth at the low end of the spectrum. This effect can be seen by comparing the allowed-positron spectrum in Fig. 11-7 with the allowed-negatron spectrum of Fig. 11-5.

11.08 Fermi Theory of Beta Decay

Simple plots of $N(\beta)$ against either $B\rho$ or T are seldom used for determining E_m, because neither shape provides a sensitive end point. A linear plot is preferable and is customarily used. In 1934, Fermi presented a theory of beta decay based on reactions (11-1) and (11-2), with energy sharing between each beta particle and its accompanying neutrino. For purposes of this calculation, the recoil *energy* imparted to the daughter nucleus can be neglected. The *momentum* imparted to the recoiling nucleus will not be negligible, and consequently there will not be momentum conservation between the beta particle and its neutrino.

Fermi obtained the electron–neutrino distribution by calculating the number of electrons that could occupy a volume of *phase space* at a given electron energy, and the number of neutrinos that would be in the corresponding volume of phase space. The location of the latter volume would be determined by the requirement of energy sharing.

The results of the Fermi calculation are most simply expressed in terms of two dimensionless parameters

$$W = \frac{T + m_0c^2}{m_0c^2} = \frac{T(\text{Mev})}{0.511} + 1 \qquad (11\text{-}9)$$

and

$$\eta = \frac{mv}{m_0c} = \frac{B\rho e}{m_0c} = 5.85 \times 10^{-3}\, B\rho \qquad (11\text{-}10)$$

where $B\rho$ is in the usual units of gauss-cm.

In terms of these new variables, the Fermi relation is

$$N(W)\, dW = WF(W, Z)(W^2 - 1)^{1/2}(W_0 - W)^2\, dW \qquad (11\text{-}11)$$

where $N(W)$ = number of particles emitted at energy W in energy range dW
W_0 = maximum energy of transition.

The factor $F(W, Z)$ is the *Fermi function*, obtained by a quantum-mechanical calculation that includes the effect of the nuclear field on the escaping particles.

In the relativistic range, momentum and energy are related by

$$\eta = (W^2 - 1)^{1/2} \qquad (11\text{-}12)$$

whence Eq. (11-11) becomes

$$(W_0 - W)^2 = \frac{N(W)}{W\eta F(W)} \qquad (11\text{-}13)$$

Alternatively, all factors on the right-hand side may be expressed in terms of momentum, which gives

$$(W_0 - W)^2 = \frac{N(\eta)}{\eta^2 F(\eta)} \qquad (11\text{-}14)$$

The square root of the right-hand side of either Eq. (11-13) or (11-14) should be a linear function of $(W_0 - W)$.

Values of the Fermi function, or more conveniently $\eta^2F(\eta)$ have been tabulated.* Experimentally obtained values of $N(\eta)$ are then divided by the

* National Bureau of Standards, *Tables for the Analysis of Beta Spectra.* Applied Mathematics Series 13. For sale by the Superintendent of Documents, U.S. Government Printing Office, Washington 25, D.C.

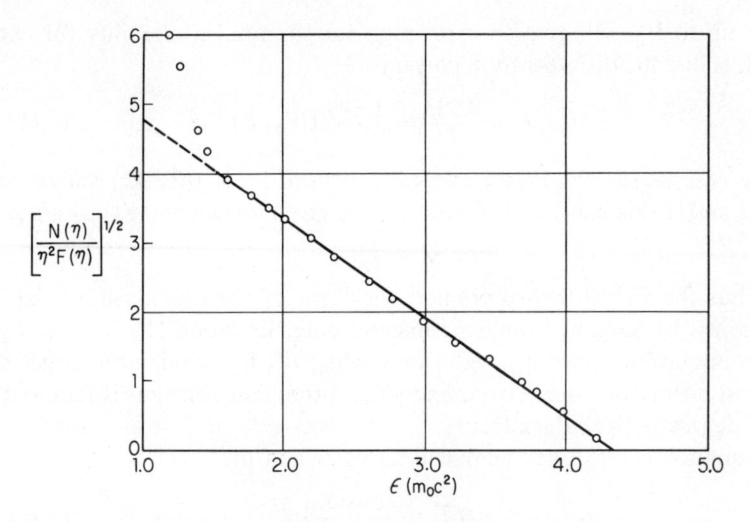

Figure 11-8. Fermi plot of the ^{15}O beta spectrum.

appropriate values from the tables and the square root of this factor is plotted against W. A linear plot will result, and this can be accurately extrapolated to give W_0.

Figure 11-8 shows a plot, called a *Fermi* or a *Kurie plot*, of Eq. (11-14). Extrapolating the linear relation to the energy axis gives a measure of W_0 and hence E_m. In the case shown, the intercept is at $4.3m_0c^2$. Subtracting m_0c^2 for the rest energy leaves $E_m = 3.3m_0c^2 = 1.68$ MeV.

Below 0.3 MeV, the experimental points depart from the predicted linear relation. This happened so frequently in Kurie plots that for a time it seemed that the Fermi theory did not hold at low energies. It now appears that the deviations are due rather to inadequate experimental conditions such as excessive self-absorption or undesired scattering.

The Fermi theory was originally developed for allowed transitions under Fermi selection rules, $\Delta I = 0$. It has been extended to forbidden transitions with complete success. With changes appropriate to the particles, it is capable of explaining muon decay, muon capture by nuclei, and pion decay.

The extension to forbidden states was made by introducing into Eq. (11-11) a term obtained from a wave-mechanical treatment of the transition probabilities. This term is

$$\frac{|P|^2}{\tau}$$

where $|P|$ is characteristic of the particular transition and τ is a time constant that is independent of the transition type.

Equation (11-11) gives the probability of emission of a particle with energy W in the small-energy range dW. Integrating this equation over all values of W up to W_0 will give an expression for the total probability for decay, which is just the disintegration constant λ:

$$\lambda = \frac{0.693}{T} = \frac{|P|^2}{\tau} f(W_0, Z) \tag{11-15}$$

where $f(Z, W_0)$ is the Fermi integral function. Since $f(Z, W_0)$ varies about as W_0^5 and $|P|^2$ is reasonably constant for a given forbiddenness classification,

$$\lambda \simeq k W_0^5 \tag{11-16}$$

Thus the Fermi theory predicts the form of the relationship originally developed by Sargent from experimental data. Equation (11-16) is not generally applicable, since it is valid only when W_0 is considerably larger than the rest energy of the electron, and since a different constant is required for each degree of forbiddenness.

Equation (11-15) may be put in terms of half-life:

$$fT = \frac{\text{constant}}{|P|^2} \tag{11-17}$$

where P will take on values depending upon the degree of forbiddenness.

TABLE 11-2
CLASSIFICATION OF BETA TRANSITIONS

Transition	*Classification*	*$\log_{10}(fT)$*
^{6}He–^{6}Li	Favored	2.76
^{11}C–^{11}B	Favored	3.62
^{18}F–^{18}O	Favored	3.57
^{35}S–^{35}Cl	Allowed	5.0
^{106}Ru–^{106}Rh	Allowed	4.2
^{45}Ca–^{45}Sc	First forbidden	5.9
^{151}Sm–^{151}Eu	First forbidden	6.9
^{32}P–^{32}S	Second forbidden	7.9
^{137}Cs–^{137}Ba	Second forbidden	9.6
^{87}Rb–^{87}Sr	Third forbidden	17.6

Experimentally determined values of T, combined with tabular values of $f(W_0, Z)$ serve to determine the classification of beta and electron-capture transitions. Because of the wide range of values, $\log_{10}(fT)$ is usually computed. Typical values are given in Table 11-2 showing the breaks in $\log_{10} fT$ as the degree of forbiddenness changes.

11.09 Beta Stability and Isobars

Since the beta particle does not exist in the nucleus prior to emission, it does not have to surmount, or tunnel through, the potential barrier. Beta-particle emissions are not, therefore, subject to the severe energy restrictions placed on alpha particles. Isobaric transitions are energetically possible under the following conditions:

Negatron emission:	$M(Z, A) > M(Z + 1, A)$
Positron emission:	$M(Z, A) > M(Z - 1, A) + 2m_e$
Electron capture:	$M(Z, A) > M(Z - 1, A) + E/c^2$

where E is the binding energy of the captured electron. To be absolutely stable against beta decay, a nucleus must have a smaller isotopic mass than either of the adjacent isobars.

We would not expect to find in nature neighboring isobars with different masses unless one was decaying to the other with a half-life at least comparable to the age of the earth. Three such isobaric pairs are known. The $^{113}Cd \longrightarrow {}^{113}In$ ground-state transition is energetically possible, but no radiations from it have been positively identified. A very long half-life is expected, since the mass difference is small and $\Delta I = 4$. Negatron emission has been detected from $^{115}In \longrightarrow {}^{115}Sn$ decay with a half-life greater than 10^{14} y. This transition is also highly forbidden with $\Delta I = 4$. An electron capture $^{123}Te \longrightarrow {}^{123}Sb$ with a half-life of more than 10^{13} y has been identified. This is the only decay channel open, since Q_β is only 60 KeV.

Two sets of triple adjacent isobars are known, but in each case the middle member is radioactive:

$$^{40}_{19}K \longrightarrow {}^{40}_{20}Ca + \beta^- \qquad {}^{40}_{19}K + e^- \longrightarrow {}^{40}_{18}Ar$$
$$^{176}_{71}Lu \longrightarrow {}^{176}_{72}Hf + \beta^- \qquad {}^{176}_{71}Lu \longrightarrow {}^{176}_{70}Yb \ ???$$

In the potassium decay there is 89:11 branching to form isobars separated by two Z units. The Lu–Hf mass difference is very small, and the transition is probably $\Delta I = 7$, *yes*, which leads to a decay rate too small to be detected.

Note that all of the double and triple isobars are even-A nuclei. Nuclei of odd mass number have no stable isobars. This marked difference in behavior can be explained by considering the semiempirical mass equation. Equation (7-17) is the most useful form. When the terms are regrouped, we have

$$M = (944.75A + 13.8A^{2/3}) - 78.5Z + (0.595A^{-1/3} + 77.2A^{-1})Z^2 \pm \delta \tag{11-18}$$

When A is constant, as for isobars, this is the equation of a parabola in M–Z coordinates.

Although Z is a discontinuous variable, Eq. (11-18) can be treated by the methods of calculus to show that a minimum of M exists for each value of A. The value of Z_m thus calculated may turn out to be fractional. In the actual case, the isobars will lie along a parabola, although none may be at the exact minimum.

We must now consider the effect of the δ term in Eq. (11-18). This term is a function of A only, and so it will have a constant value for each mass number. The simplest case is for odd A, where $\delta = 0$. The M–Z relationship is then given by a single parabola for each value of A. There will be an occupant for every value of Z, Fig. 11-9.

Proton-deficient nuclei will decay by negatron emission and the loci of the daughters will approach the atomic number of minimum mass from below. Nuclei with a proton excess will decay by positron emission or electron capture and the daughters will approach the minimum from above.

Radioactive nuclei formed by nuclear fission have large proton deficiencies and will decay by several consecutive negatron emissions before a stable energy minimum is reached. From the general shape of parabolas, the first transitions should have the greatest mass differences and the shortest half-lives. Figure 11-9 shows why odd-A nuclei have only a single stable member. Any other member will have a greater mass and will be able to decay to the more stable form.

Two possibilities must be considered when A is even: δ will be positive for odd Z–odd N and negative for even Z– even N. There will now be two mass parabolas for each value of A, Fig. 11-10, with the odd-odd parabola

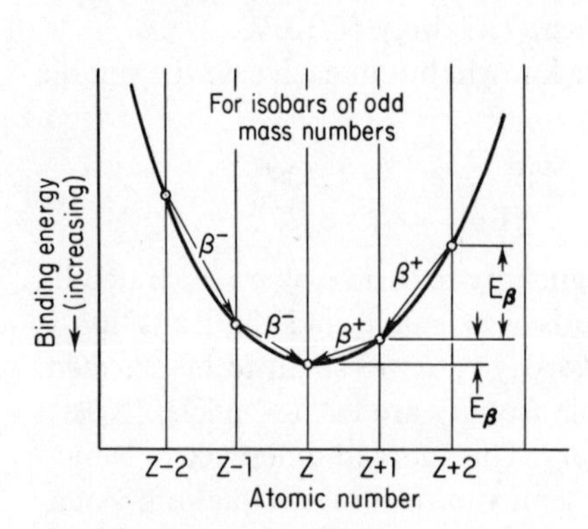

Figure 11-9. Binding energy or stability parabola for odd-A isobars.

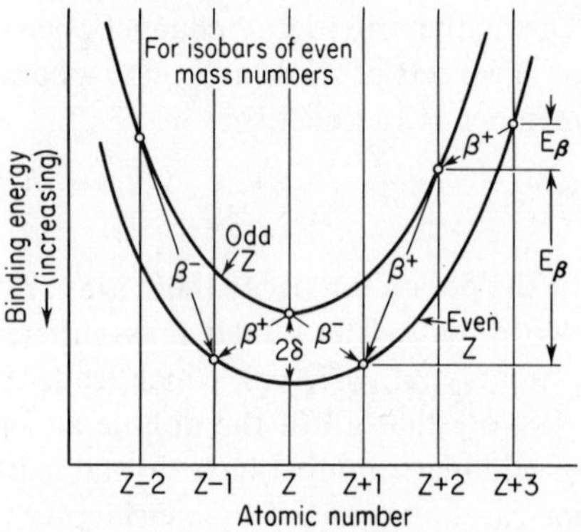

Figure 11-10. Two binding energy parabolas are required to account for even-A isobars.

2δ above the other. Each parabola will have an occupant for each 2 units of Z. Beta transitions take place as shown, with transitions from one parabola to the other. A nucleus located near the minimum of the odd–odd parabola can decay by either of two modes, since it has a choice of two positions on the even–even parabola. An example of this dual mode of decay is ^{64}Cu, whose decay scheme is shown in Fig. 11-11.

The double parabolas of Fig. 11-10 suggest that all odd–odd structures should be unstable. There are only four exceptions to this: $^{2}_{1}$H, $^{6}_{3}$Li, $^{10}_{5}$B, and

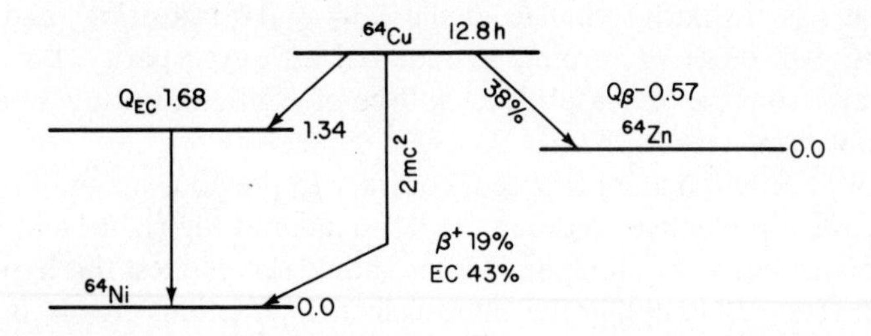

Figure 11-11. Decay scheme of ^{64}Cu, an odd–odd nucleus with two modes of decay.

$^{14}_{7}$N, all located at the low end of the periodic table. They are stable because of the exact spin-pairing associated with equal numbers of neutrons and protons. Above $Z = 8$, coulomb repulsion becomes sufficient to require more neutrons than protons, exact pairing is no longer possible, and the heavier odd–odd structures are unstable.

11.10 Parity Conservation in Beta Decay

As the concept of parity was developed, it appeared to be an important parameter that would be conserved in all nuclear reactions and radioactive decays. Although some violations of parity conservation have been found, it is still an important factor in classifying transitions and identifying subnuclear particles.

Figure 11-12 shows the parity and momentum assignments in the negatron decay of ^{11}Be. Two of the beta decays from the even-parity ground state go to states of even parity in the ^{11}B daughter. If parity is conserved, these *no* transitions will produce beta particles with an angular momentum of even parity, since the intrinsic parity of the electron is even. In the transitions to the ground state and the 2.14-MeV excited state, the angular momentum of each emitted beta particle must have odd parity.

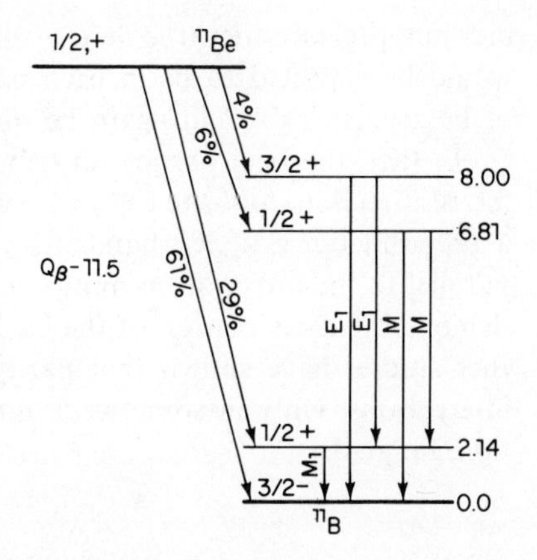

Figure 11-12. Parity and angular momentum assignments in the beta decay ^{11}Be $\rightarrow$ ^{11}B.

The gamma rays from the 8.00 and the 6.81-MeV states to the ground state must have odd parities, since these are *yes* transitions. These gamma rays are, therefore, E_1 in origin (Sec.

7.10). There is no parity change in the 2.14 → 0.0 transition, and so the gamma ray will be of M_1 origin, as required with even parity. Both of the gamma rays from the 6.81-MeV level will be of E origin, but the multipolar order is not firmly established.

In 1956, Lee and Yang proposed, contrary to previous assumptions, that nature shows a preference between the left-hand and the right-hand in reactions involving weak coupling. Beta decay should serve to test this hypothesis, since with strong *jj* coupling the internucleon interactions involved in beta decay will be weak.

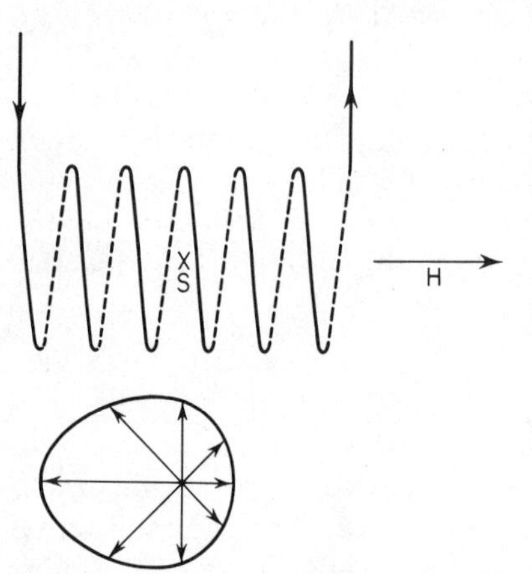

Figure 11-13. A ^{60}Co source located at S in a magnetic field shows a nonisotropic beta particle emission.

In 1957, Wu et al. described an experiment showing that at least in some beta transitions nature does prefer a left-handed system. The experiment consisted of a study of the spatial distribution of the beta particles emitted from a sample of ^{60}Co placed in a magnetic field, Fig. 11-13. To reduce disturbances due to thermal motions of the emitting nuclei, the sample was cooled to a very low temperature.

In the absence of a magnetic field, an isotropic distribution of beta particles was observed. The field H would be expected to orient the nuclear spins I, but, on the old parity concept, this should not disturb the isotropic distribution of the beta particles. If nature did not prefer either the left or the right hand, equal numbers of nuclei would be expected to be in each parity state, and an isotropic distribution of beta particles would again be observed.

In fact, the beta-particle distribution became asymmetric, with the preferred direction opposite to the direction of the magnetic field. This indicates a preponderance of left-handed states in which the spin direction is antiparallel to the direction of motion of the particles. The observed asymmetry demonstrated a violation of the basic premise of the concept of parity. Further studies have shown that parity does exist and is conserved in strong interactions. Only in some weak interactions is the law of parity conservation violated.

11.11 Internal-conversion Electrons

Section 8.13 described in detail the mechanism whereby an orbital electron may be ejected in preference to a gamma-ray emission. Figure 11-2 shows two conversion lines superposed upon the usual continuous beta-particle spectrum. Since a gamma ray has an intrinsic angular momentum different from zero, gamma radiation is absolutely forbidden from an $I = 0 \longrightarrow 0$ transition. Internal conversion provides a de-excitation channel whereby the requirements of angular-momentum conservation can be satisfied. In a $0 \longrightarrow 0$ transition, the internal-conversion coefficient will be infinite, since no photons can be emitted.

REFERENCES

Christy, R. F. et al., "The Conservation of Momentum in the Disintegration of Li^8." *Phys. Rev.* **72,** 698, 1947.

Daniel, H., "Shapes of the Beta-ray Spectra." *Rev. Modern Physics* **40,** 659, 1968.

Ellis, C. D. and W. A. Wooster, "The Average Energy of Disintegration of Radium E." *Proc. Royal Soc.* (*London*) **117,** 109, 1927.

Reines, F. and C. L. Cowan, Jr., "Detection of the Free Neutrino." *Phys. Rev.* **92,** 8301, 1953.

Widman, J. C. et al., "Average Energy of Beta Spectra." *Int. J. Applied Rad. and Isotopes* **19,** 1, 1968.

Wu, C. S. and S. A. Moszkowski, *Beta Decay.* Interscience Publishers, Wiley, New York, 1966.

PROBLEMS

11-1. Derive Eq. (11-12), starting with the relativistic relation between energy and momentum.

11-2. Show that the two variables used in the Fermi theory are related by $d\eta = [(\eta^2 + 1)^{1/2}/\eta]\, dW$, and that $dW = \beta d\eta$.

11-3. The following data were obtained with a surface-barrier detector used as an energy spectrograph. Plot the energy spectrum, and estimate the maximum beta-particle energy by extrapolation. Calculate and plot the momentum spectrum.

W:	1.0	1.2	1.4	1.6	1.8	2.0	2.2	2.4	2.6
$N(W)$:	903	1406	1658	1970	2180	2328	2395	2310	2110

W:	2.8	3.0	3.2	3.4	3.6	3.8
$N(W)$:	1527	1196	789	457	219	34

11-4. A good approximation to the maximum energy of a beta-particle spectrum can usually be obtained by plotting $\sqrt{N(W)}$ against W and extrapolating to $N(W) = 0$. Use this approximation to obtain the maximum energy of the beta particles in Prob. 11-3.

11-5. A radioactive nucleus emits a negatron from a (4, +) ground state to either of two excited states of a daughter. An 0.37-MeV beta emission leads to a (4, +) state at 2.00 MeV, from which an E_2 gamma ray leads to a lower excited state. This state is also reached directly by the emission of a 1.48-MeV negatron. A second E_2 photon emission then leads to the ground state of the daughter. Draw the decay scheme, labeling all energy levels. What is the degree of forbiddenness of each of the beta-particle emissions? What are the spin and parity designations of each daughter state? What would be the degree of forbiddenness of a beta emission directly to the ground state of the daughter?

11-6. Calculate the maximum energy available to the beta particle in the negatron decay of ^{24}Na. Why are beta particles of this energy not observed in the decay?

11-7. A radioactive nucleus is produced in either the $\frac{3}{2}$, + ground state or in an $\frac{11}{2}$, − 70-KeV excited state. Negatron emission leads either to a $\frac{5}{2}$, + ground state or to a $\frac{7}{2}$, + 37-KeV excited state. The ground state–ground state Q is 380 KeV. What beta-particle transition energies would you expect? What is the degree of forbiddenness of each transition? What photon energies would you expect to be emitted?

11-8. When a sample of 0.95 Ci of ^{210}Bi was placed in a calorimeter, the measured heat output was 0.02546 cal min^{-1}. Calculate the mean energy of the transition and the E/E_m ratio.

11-9. Phosphorous-32 emits a beta particle with $E_m = 1.71$ MeV. Calculate the maximum recoil energy imparted to the nucleus by a beta-particle emission. What will be the maximum recoil energy of the nucleus imparted by neutrino emission?

11-10. A study of the emissions from ^{181}Hf shows K- and L-conversion electrons of 68, 124, 233, 289, 414, and 470 KeV. The K- and L-shell binding energies in tantalum are known from X-ray absorption measurements to be 11.7 and 67.6 KeV, respectively. Calculate the gamma-ray energies emitted in the transition. Draw the decay scheme.

11-11. Calculate the maximum energy and momentum that can be carried away by the negatron in the reaction of Eq. (11-4). What will be the recoil energy imparted to the daughter nucleus?

11-12. The K-emission lines of gold are used to excite fluorescence in a tantalum foil. What fluorescent energies will be emitted? What will be the kinetic energies of the ejected electrons?

12

Photon Absorption and Scattering

12.01 Absorption Geometries

When X- or γ-ray photons traverse matter, some are absorbed, some pass through without interaction, and some are scattered, which really means that new photons are created, to move off in quite different directions from those in the primary beam. Figure 12-1 shows two arrangements for studying the attenuation of a photon beam by absorbers.

A photon source at S is collimated by heavy lead to provide a *narrow beam* of radiation at absorber C. In the arrangement of Fig. 12-1A, known as *good geometry*, practically every scattered photon will make a sufficiently large angle with the original beam to escape detection. Only those photons that have passed through the absorber without interaction will enter the detector, and each one of these photons will have all of its original energy. Readings taken with and without the absorber will give a measure of the *total absorption*, or the fraction of the photons removed from the narrow beam, by whatever process.

In *poor geometry*, Fig. 12-1B, the detector will receive a large fraction of the scattered photons, as well as those transmitted without interaction. Each scattered photon will be degraded in energy according to the angle through which it is scattered. The detector reading will now be a complex function of the energy distribution in the primary beam, the energy response of the detector, and the exact geometrical arrangement used. The concept of *true absorption*, or the amount of energy deposited in the absorber, is useful, but it is very difficult to measure experimentally. Simple measurements of the type described here will not suffice.

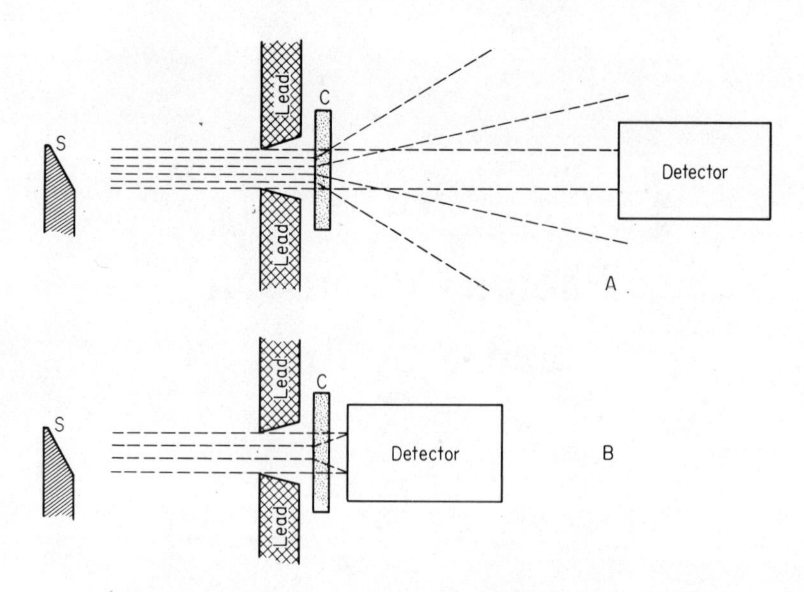

Figure 12-1. (A) "Good" geometry measures the attenuation of the photon beam. (B) In "poor" geometry the detector accepts the scattered photons and measures the coefficient of energy absorption.

In most cases of practical interest, one desires to know the beam attenuation under *broad-beam conditions.* Broad-beam irradiations will always be in poor geometry because the detector will receive some photons scattered from regions where there were no photons in a narrow beam. The difference, known as *buildup,* represents the contribution due to photons scattered from one part of the broad beam to another. In a narrow beam, all scatter is out, none in. Buildup is obviously a complicated function of beam size, photon-energy distribution, and the geometrical arrangement of the system. Practical cases are scarcely amenable to theoretical treatment and must be handled from experimentally determined attenuations.

12.02 Absorption Coefficients

The intensity of a beam of monoenergetic photons, as measured by a detector which responds selectively to the energy of the primary beam, falls off as an exponential function of the absorber thickness. Beam intensities I_0 before, and I after the introduction of an absorber of thickness x are related by

$$I = I_0 e^{-\mu_l x} \tag{12-1}$$

where μ_l is the *linear absorption coefficient.*

Equation (12-1) is identical in form to the relation governing radioactive decay. The fraction of the incident beam removed by an absorber of unit thickness is μ_l; it is related to *half-value thickness HVT* by

$$\mu_l = \frac{0.693}{HVT} \tag{12-2}$$

Mean-free-path, or the average distance traveled by a photon before absorption, is also used to describe beam penetration. By analogy with Eq. (9-5),

$$x_M = \frac{1}{\mu_l} = 1.44\, HVT \tag{12-3}$$

The linear absorption coefficient is a function of photon energy as well as of absorber material, so the simple exponential form of Eq. (12-1) will not hold in general for a heteroenergetic photon beam. Absorption will always introduce some scattered low-energy photons, so even an initially monoenergetic beam will not remain so, and measurements with thick absorbers may show an appreciable deviation from a constant value of μ_l.

Absorption coefficients depend upon the atomic composition and the amount of absorber in the beam but not upon the chemical or physical state. Thus a given mass of water, ice, or steam will produce equal beam attenuations, although the three values of μ_l will differ widely because of density differences. To avoid this it is convenient to introduce a *mass absorption coefficient* $\mu_m = \mu_l/\rho$, which will be independent of absorber density. μ_m is the beam fraction removed by unit areal density (1 g/cm^{-2}) regardless of the thickness required to obtain this density. Equation (12-1) now becomes

$$I = I_0 e^{-\mu_m (x\rho)} \tag{12-4}$$

where ρ is the absorber density in grams per cubic centimeter.

It is frequently convenient to deal with the absorption coefficient per atom, ${}_A\sigma = \mu_l/N$, where N is the number of absorbing atoms per cubic centimeter. Since μ_l has the dimensions of a reciprocal length, cm^{-1}, ${}_A\sigma$ will be in square centimeters, or area. Thus ${}_A\sigma$ is known as the *atomic absorption cross-section*. Since photon absorption takes place primarily at orbital electrons, it is also useful to define an absorption cross-section per electron ${}_e\sigma = \mu_l/NZ$.

Thus far we have considered the phenomena of photon absorption without regard to the details of their interaction with matter. Actually, each absorption coefficient is the sum of three components. For example,

$$\mu_m = \tau_m + \mu_{mc} + \mu_{mp} \tag{12-5}$$

where τ_m = mass absorption coefficient for photoelectric effect
μ_{mc} = mass absorption coefficient for Compton effect
μ_{mp} = mass absorption coefficient for pair production.

These are the three principal processes by which photons give up energy to matter and each will be discussed in some detail. Other absorptive mechanisms are known, but they are usually relatively ineffective and are neglected. Coefficients for them may be added to Eq. (12-5) in those cases where they are important.

12.03 The Photoelectric Effect

When an X-ray quantum or photon collides with an atom, it may impinge upon an orbital electron and transfer all of its energy to this particle by ejecting it from the atom. If the incident photon carried more energy than that necessary to remove the orbital electron from the atom, it imparts to the electron its additional energy in the form of kinetic energy. Figure 12-2 shows a schematic description of this interaction of the photon and electron. This process is known as the *photoelectric effect* and obeys the Einstein photoelectric equation:

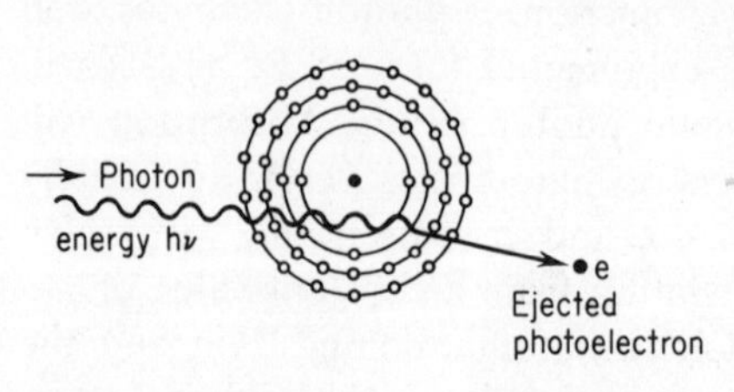

Figure 12-2. Schematic of the photoelectric process.

$$h\nu = \phi + T \tag{12-6}$$

Here, $h\nu$ represents the total energy of the incident photon, ϕ the energy required to remove the electron from its atom, and T the kinetic energy of the ejected electron.

Electrons thus ejected from atoms are called *photoelectrons*. Since these photoelectrons are produced by a process which completely absorbs the energy of the incident photon, they may carry considerable kinetic energy. This means that the photoelectrons themselves become a source of ionization, for as they pass close to neighboring atoms they strip off electrons from them.

Several attempts have been made to calculate absorption cross-sections on theoretical grounds. J. J. Thomson based his treatment on the assumption that an orbital electron was set in oscillation by the electromagnetic field of the impinging photon. The vibrating electron then reradiated energy or was set in sufficiently violent motion to be ejected from its atom. This concept led to an absorption coefficient given by

$${}_e\sigma_0 = \frac{8\pi}{3}\frac{e^4}{m_0^2 c^4} = 6.65 \times 10^{-29}\ \text{m}^2\ \text{electron}^{-1} \tag{12-7}$$

The dimensions of the Thomson coefficient are m² electron^{-1}, and so it has become customary to refer to it as a *cross-section*. It is indeed in the nature of a target area presented to the impinging photon, but in general it has no direct relation to the actual geometrical area of the target. Cross-sections are usually given in terms of the *barn*, abbreviated b, where 1 barn $= 10^{-28}$ m^2. This unusual name arose from a casual remark by a physicist that a particular cross-section was "as big as a barn."

The Thomson theory is now known to be only an approximation, valid only when zero energy is imparted to the ejected electron. The Thomson cross-section appears, however, as a factor in all quantum mechanical treatments of absorption cross-sections.

Bethe* has derived an expression for photoelectric absorption which is valid for photon energies which do not lead to highly relativistic photoelectrons. The Bethe equation may be written

$$\left.\begin{aligned} {}_A\sigma_P &= (2 \times 10^{-8})_e\sigma_0 S(Z - 0.3)^5 \left(\frac{m_0 c^2}{h\nu}\right)^{7/2} \\ &= (4 \times 10^{-27}) S(Z - 0.3)^5 (h\nu)^{-7/2} \text{ m}^2 \\ &= 40 S(Z - 0.3)^5 (h\nu)^{-7/2} \text{ b} \end{aligned}\right\} \quad (12\text{-}8)$$

where S = a factor to be evaluated below
Z = atomic number of absorber
$h\nu$ = photon energy in KeV

S is a complicated function of photon energy and atomic number. An approximation,

$$S = -0.18 + 0.28 \log_{10} \left(\frac{\text{photon energy in eV}}{Z^2}\right) \quad (12\text{-}9)$$

is good for $10 < \text{eV}/Z^2 < 10^4$.

Equation (12-8) shows a photoelectric absorption increasing as Z^5 and decreasing as $(h\nu)^{-7/2}$. For low-Z materials, such as living tissue, photoelectric absorption becomes negligible at photon energies above about 200 KeV. In high-Z materials, ${}_A\sigma_P$ is appreciable at 1–2 MeV. Figures 12-6A and 12-6B show the contribution of ${}_A\sigma_P$ to the total absorption coefficient as a function of photon energy.

The angular distribution of the photoelectrons is given by

$$dn = \frac{\sin^2 \theta}{(1 - \beta \cos \theta)^4} d\Omega \quad (12\text{-}10)$$

where dn is the number of photoelectrons ejected in the small solid angle $d\Omega$ making an angle θ with the incoming photon. $\beta = v/c$ for the ejected electron. At low-photon energies, the second term in the denominator can be neglected and we then have a $\sin^2$ distribution which shows a maximum at right angles to the photon direction. As the photon energy increases, the angular distribution tends to maximize more and more in the forward direction, Fig. 12-3.

* W. Heitler, *The Quantum Theory of Radiation*, 2nd ed., Oxford University Press, 1944, p. 123.

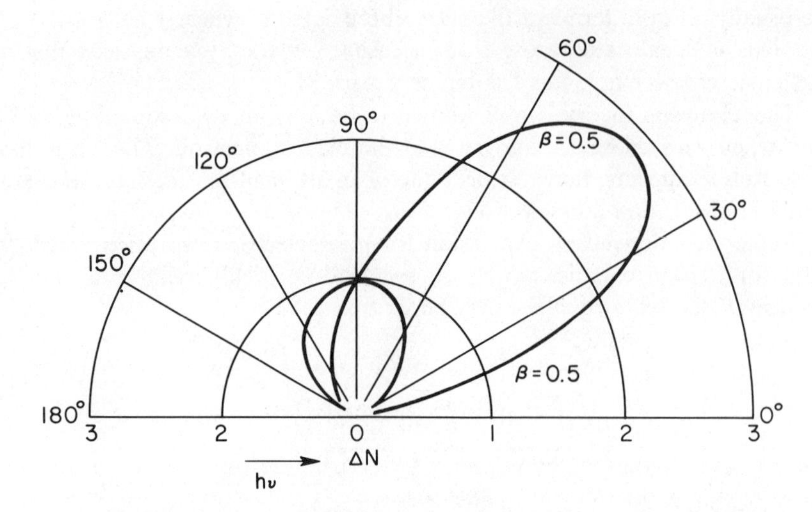

Figure 12-3. Spatial distribution of photoelectrons. The curve for $\beta = 0.5$ was calculated by the nonrelativistic relationship but is still a good approximation to the correct distribution.

12.04 Compton Scattering

In some interactions, the incident photon is absorbed, as in the photoelectric process, but only a portion of the available energy goes into kinetic energy of the ejected electron. Instead, a new photon of lower energy than the original is created, with a division of energy between it and the electron. In this process the electrons are called *recoil* or *Compton electrons* after the discoverer, A. H. Compton. In general, the new or scattered photon will not have the direction of the original, as shown in Fig. 12-4.

Compton treated the interaction as an elastic collision between two particles, one an orbital electron considered to be unbound, the other the incident photon. The latter was taken as a particle of mass $m = h\nu/c^2$ and momentum $p = h\nu/c$. From the geometrical relations shown in Fig. 12-4B, we have

conservation of momentum

$$\frac{h\nu}{c} = \frac{h\nu'}{c}\cos\theta + p\cos\phi \qquad (12\text{-}11)$$

$$\frac{h\nu'}{c}\sin\theta = p\sin\phi \qquad (12\text{-}12)$$

conservation of energy

$$h\nu = h\nu' + T \qquad (12\text{-}13)$$

The terms in ϕ are eliminated by rearranging Eqs. (12-11) and (12-12), squaring, and adding. This gives

$$p^2 = \frac{h^2}{c^2}(\nu^2 - 2\,\nu\nu'\cos\theta + \nu'^2) \qquad (12\text{-}14)$$

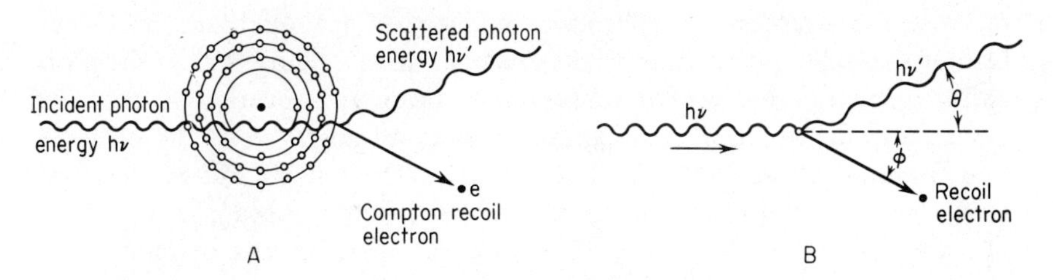

Figure 12-4. (A) Schematic of the Compton process. (B) Geometrical relations in the Compton process.

Electron energies will be high, so relativistic relations between p and T must be used. From Eqs. (4-12) and (12-13),

$$h\nu - h\nu' = m_0c^2\sqrt{1 + \frac{p^2}{(m_0c)^2}} - m_0c^2 \tag{12-15}$$

The radical can be removed by rearranging and squaring. The term in p^2 is then replaced by its value from Eq. (12-14). This gives

$$\frac{\nu - \nu'}{\nu\nu'} = \frac{h}{m_0c^2}(1 - \cos\theta) \tag{12-16}$$

or, in the more usual form,

$$\Delta\lambda = \frac{h}{m_0c}(1 - \cos\theta) \tag{12-17}$$

It is more convenient to evaluate the constant term (known as the Compton wavelength) in angstroms. The relation then becomes

$$\Delta\lambda = 0.0242(1 - \cos\theta)\ \text{Å} \tag{12-18}$$

Equation (12-17) shows that the change in wavelength depends only upon the angle of scatter and not upon the initial wavelength nor upon the material of the absorber. Low-energy photons with a long wavelength will lose only a small percentage of their energy, while high-energy photons suffer a much greater loss in scattering through the same angle. The energy relation between incident and scattered photons can be readily obtained from Eq. (12-16) (see Prob. 12-7):

$$h\nu' = \frac{h\nu}{1 + (h\nu/m_0c^2)(1 - \cos\theta)} \tag{12-19}$$

The corresponding relation for the electron is

$$T = h\nu\frac{(h\nu/m_0c^2)(1 - \cos\theta)}{1 + (h\nu/m_0c^2)(1 - \cos\theta)} \tag{12-20}$$

The Compton analysis gives no information about the probability of an interaction, nor about the angular distribution of the scattered components. These matters were treated by Klein and Nishina who, in 1929, reported one of the earliest successful applications of the then-new quantum mechanics. One of the differential forms of the Klein–Nishina equations gives the probability of photon *removal* from the primary beam as a function of scattering angle. An integration of this expression over all angles gives the crosssection for total absorption per electron, ${}_e\sigma_{CT}$. The second differential form gives the cross-section for the energy *scattered* from the beam as a function of scattering angle. Note, Fig. 12-5, how the requirements of momentum

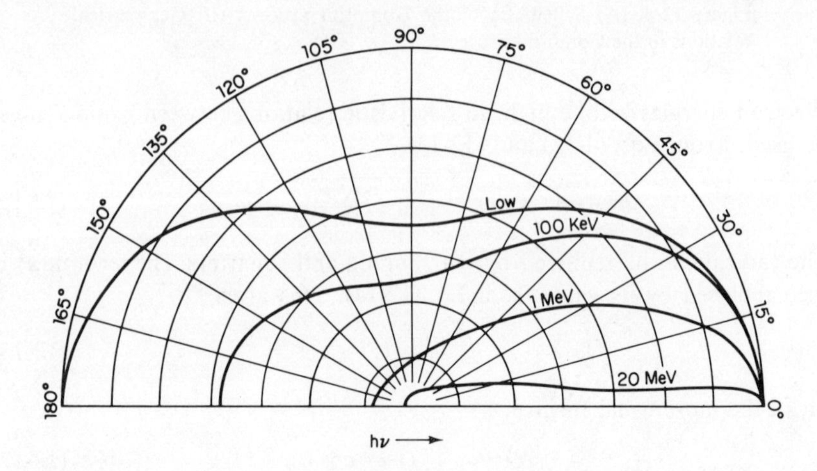

Figure 12-5. Angular distributions of Compton scattered photons.

conservation increase the probability of forward scattering at the higher photon energies.

An integration of the second form leads to the scattering cross-section per electron, ${}_e\sigma_{CS}$. The difference between the two cross-sections gives the cross-section for *true absorption*, ${}_e\sigma_{Ca}$. The three integral forms of the Klein–Nishina equations are

$${}_e\sigma_{CT} = \frac{3\sigma^0}{4\alpha^2}\left[\frac{2 + 8\alpha + 9\alpha^2 + \alpha^3}{(1 + 2\alpha)^3} - \frac{2 + 2\alpha - \alpha^2}{2\alpha}\ln(1 + 2\alpha)\right] \qquad (12\text{-}21)$$

$${}_e\sigma_{CS} = \frac{3\sigma_0}{4\alpha^2}\left[-\frac{3 + 15\alpha + 18\alpha^2 - 6\alpha^3 - 16\alpha^4}{3(1 + 2\alpha)^3} + \frac{1}{2\alpha}\ln(1 + 2\alpha)\right] \qquad (12\text{-}22)$$

$${}_e\sigma_{Ca} = \frac{3\sigma_0}{4\alpha^2}\left[\frac{9 + 51\alpha + 93\alpha^2 + 51\alpha^3 - 10\alpha^4}{3(1 + 2\alpha)^3} - \frac{3 + 2\alpha - \alpha^2}{2\alpha}\ln(1 + 2\alpha)\right] \qquad (12\text{-}23)$$

where $\alpha = h\nu/m_0c^2$ for the original photon. Numerical values obtained from the Klein–Nishina equations will be in square centimeters per electron. Values of μ_m can be obtained by multiplying the proper σ by the number of electrons per gram. Compton absorption is proportional to the first power of the electron density and is independent of the Z of the absorber. This independence results from the assumption that the orbital electrons involved are free. This assumption is acceptable because most Compton processes take place at energies which are high compared to the binding energies of the atomic electrons.

The complication in Compton absorption arises because the scattered photons leave the region in which they are produced to deposit energy at a considerable distance. This complication does not occur in photoelectric

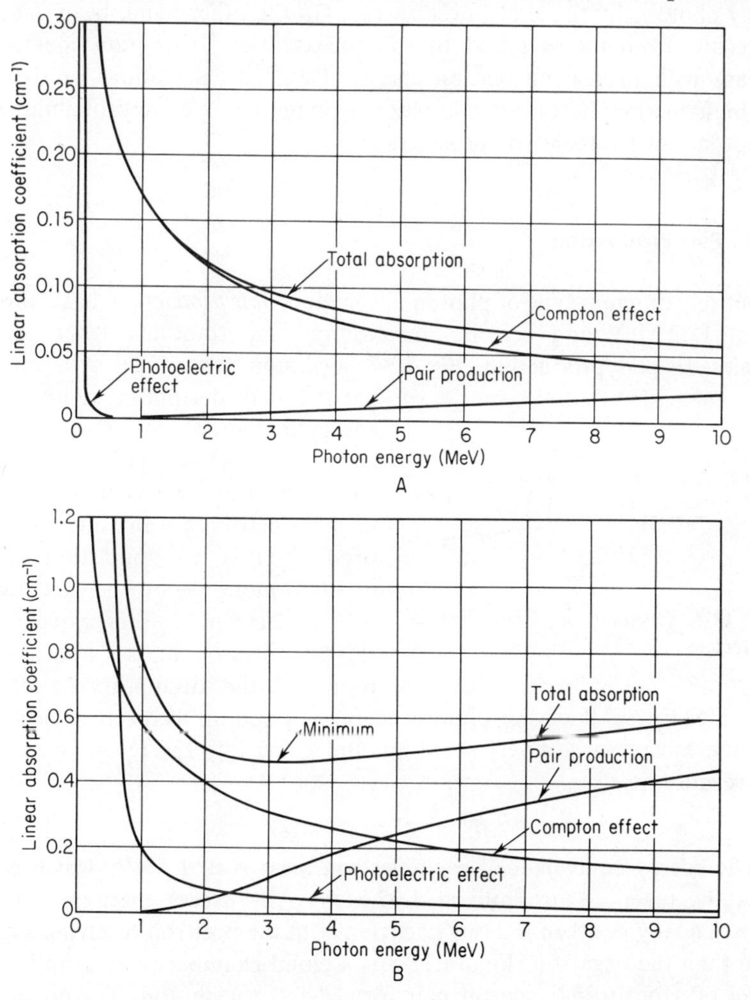

Figure 12-6. (A) Relative values of the three attenuation coefficients in aluminum. A minimum total absorption occurs at a higher energy. (B) Relative values of the three attenuation coefficients in lead. (From R. D. Evans and R. O. Evans, *Rev. Mod. Phys.*, **20**, 305, 1948.)

absorption, since at usual photon energies the range of the electron is short and the energy removed can be considered to be absorbed locally.

Compton absorption falls off with increasing energy, but at a slower rate than photoelectric absorption. Compton interactions may contribute an appreciable fraction of the total absorption out to a few MeV, Figs. 12-6A and B.

Under broad-beam conditions, which include most practical shielding situations, Compton scattering acts to *soften* the beam by replacing high-energy photons by those of lower energy. On the other hand, beam *hardening* results from the fact that the photoelectric and Compton coefficients increase with decreasing photon energy. Detailed calculations of the fate of a high-energy photon are complex, involving the relative probabilities of Compton and photoelectric processes.

12.05 Pair Production

An entirely different type of photon absorption, *pair production*, has a threshold at 1.02 MeV and becomes increasingly important at higher photon energies. In pair production, Fig. 12-7, a photon in the field of a charged particle may disappear, giving up its energy to the *creation* of a positron–negatron electron pair. Interaction with a nuclear field is preferred, but orbital electrons are also effective. Pair production is not ionization of the involved molecule by the ejection of orbital electrons. Two previously nonexistent electron masses appear as a result of the disappearance of the photon energy. All of the photon energy is given up to the two electrons, with the exception of a very small amount going into the recoiling nucleus. The reaction is

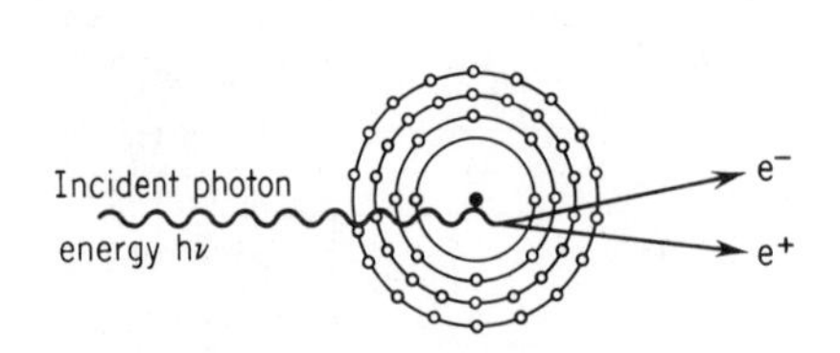

Figure 12-7. Schematic of pair production at a nucleus.

$$h\nu \longrightarrow e^+ + e^- + 2T \quad (12\text{-}24)$$

The energy equivalent of each electron mass is 0.51 MeV, which establishes the reaction threshold at 1.02 MeV. Any excess energy is divided almost equally between the two particles, with the positron receiving slightly more than the negatron. Figure 12-8 is a cloud-chamber photograph of the tracks of a positron–negatron pair formed from a photon. The equal and opposite radii produced in the magnetic field show the two particles to have equal energies but charges of opposite sign. Historically, the positron was discovered by Anderson in 1932, using a cloud chamber in connection with studies of cosmic rays.

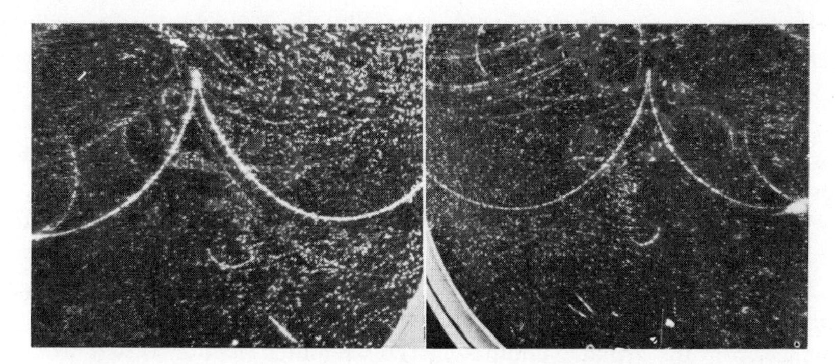

Figure 12-8. Cloud chamber photograph of pair production with track curvatures produced by a field of 1500 gauss. (Courtesy of J. A. Phillips.)

It is not possible to obtain a general expression for pair production cross-section in analytical form. In the photon-energy range usually encountered, a relation obtained by Bethe* can be used. This is

$$_{A}\sigma_{PP} = {}_{e}\sigma_{0} Z^{2}\left(\frac{28}{9}\ln 2\alpha - \frac{218}{27}\right) \qquad (12\text{-}25)$$

Equation (12-25) may be used for photon energies ranging from the threshold up to about 15 MeV in absorbers such as lead and to 30 MeV in low-Z nuclei such as those making up living tissue.

The angular distribution of the created electrons is a complicated function of the photon energy. At energies slightly above threshold, there is a tendency for the electrons to favor the forward direction (the direction of the original photon). This tendency becomes more marked at high photon energies.

Each electron of the pair loses energy by ionization as it moves off from the point of origin. When the positron energy becomes low, it combines with a negatron in a process that is the reverse of that by which it was created:

$$e^{+} + e^{-} \longrightarrow 2h\nu \qquad (12\text{-}26)$$

The two gamma rays in Eq. (12-26) are called *annihilation radiation*, since they result from the disappearance of the mass of the two electrons. Equation (12-26) is highly favored at low kinetic energies; hence in most reactions energy comes only from the electron masses. This leads to photons of 0.51 MeV. A few annihilations take place at higher kinetic energies, producing photons with energies somewhat above 0.51 MeV. The two annihilation photons move off in almost exactly opposite directions, which permits some localization measurements by simultaneous counting.

* H. A. Bethe and J. Ashkin, *Experimental Nuclear Physics*, ed. E. Segrè, John Wiley & Sons, Inc., New York, 1953, Vol. I, p. 337.

Reaction (12-18) takes place only when the two electrons have very little energy, and hence almost zero momentum. This precludes the production of a single annihilation photon in free space, since it is then not possible to account for the momentum of the photon. A single photon might be produced at a third body that is capable of absorbing some momentum. This reaction requires that the two electrons react strongly with the third body. The probability of this is small, and so the overwhelming probability is for the two-photon production.

Part of the energy initially derived from a photon absorbed by pair production goes into secondary photons and is removed to a considerable distance from the primary interaction. This requires the use of two absorption coefficients, true and total, as for Compton scattering. Total-absorption coefficients can be calculated directly from Eq. (12-25) by multiplying by the number of absorbing nuclei per gram. True absorption by pair production requires that the total coefficient be reduced by the factor

$$\frac{h\nu - 1.02}{h\nu} \tag{12-27}$$

where $h\nu$ must be in MeV. Near the threshold, there will be a substantial difference between the two coefficients.

The annihilation process is pictured as the formation, initially, of a hydrogenlike structure, *positronium*, in which the nucleus is the positron. This positron–negatron structure can be quantized exactly like the hydrogen atom. Like hydrogen, the ground state of positronium is an S state with zero angular momentum. The calculated lifetime of the singlet state (spins of positron and negatron antiparallel) is only 10^{-10} sec. *Orthopositronium*, the triplet state in which the spins are parallel, has a mean life of about 10^{-7} sec, which is long enough to permit identification and to measure some of its optical properties.

12.06 Coherent Scattering

There are several other processes, usually with small cross-sections, by which photons are absorbed. One of the more important of these processes is diffraction or coherent scattering. In previous discussions, each absorption was considered to be an independent event, with no interactions between the secondary products. Thus Compton scattering is said to be *incoherent*, with no interrelations between the scattered photons.

Interference patterns are produced, however, when a photon beam is passed through a properly oriented crystal, Sec. 5.02. This is an example of *coherent scattering*, where photons from one scattering center interfere, either constructively or destructively, with those from other centers.

Cross-sections for coherent scattering are small, and so it is unimportant as a mechanism by which energy is removed from a photon beam. Coherent scattering has proved to be an analytical tool of great importance in the determination of crystal structures.

12.07 Nuclear Photodisintegrations

Photodisintegration is another process that may be neglected in calculating the energy removed from a photon beam; on the other hand, this process has important applications. Photodisintegration is energetically possible whenever a photon has sufficient energy to remove a nucleon from a nucleus. Except for the lightest elements, binding-energy values require photons of 8 MeV or more. Two reactions,

$$^{9}\text{Be} + h\nu \longrightarrow {}^{8}\text{Be} + {}^{1}n \tag{12-28}$$

$$^{2}\text{H} + h\nu \longrightarrow {}^{1}\text{H} + {}^{1}n \tag{12-29}$$

are of importance. The thresholds for these reactions are quite sharp at 1.66 MeV and 2.22 MeV, respectively, furnishing useful calibration points for accelerators.

At energies above 20 MeV, the cross-sections for photodisintegration are sufficiently large to make useful neutron sources with the high-intensity X-ray beams available from linear accelerators and other high-voltage sources. The neutron flux is, however, accompanied by a large number of unabsorbed photons.

12.08 Absorption Edges

Absorption phenomena are more complicated than those described when photon energies are comparable to the characteristic X-ray emissions of the absorber. In optical spectra, strong resonance absorptions are observed when the incident wavelengths are the same as those which the absorber would emit upon suitable excitation. The analogous situation does not arise in X-ray absorption.

Consider a beam of photons whose energy is considerably below that of the K_α line of an absorber. There can be no K resonance absorption because of lack of energy. As the photon energy increases, absorption cross-sections will decrease due to the characteristics of both photoelectric and Compton processes. At the exact K_α energy, resonance absorption is still impossible, since this energy represents a transition from an L shell to a K vacancy. By the Pauli principle, there will be no L vacancy at the moment of photon incidence; so the transition $K \longrightarrow L$ cannot occur. Only when the incoming photon can remove a K electron from the atom will resonance absorption at the K level be possible.

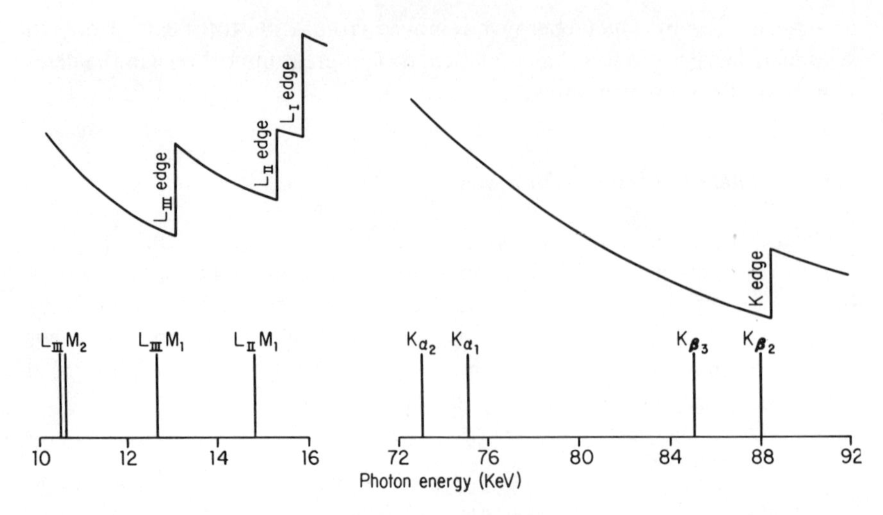

Figure 12-9. X-ray emission and absorption energies in lead. The L edge is deep enough to show substructure in the M shell.

Resonance absorption is reflected by an abrupt increase in absorption cross-section, Fig. 12-9, to form an *absorption edge* at the *critical absorption energy*. As photon energies increase above the edge, the typical decrease in cross-section is observed. Comparable absorption edges are seen in the L region. M-absorption edges exist but lie at such low energies that detection is difficult.

The absorption edge occurs at an energy determined by the atomic number of the absorber. Combinations of elements with adjacent atomic numbers will permit the transmission of a narrow band of photon energies. The energy resolution of such a filter is not as good as that which can be obtained by crystal diffraction, but the former gives a much higher transmitted intensity.

Characteristic X rays will be emitted following the absorption of photons exceeding the energy of an absorption edge. These emissions, known as *fluorescent radiation*, differ from characteristic X rays only in the mode of excitation. Fluorescent radiation is unaccompanied by a continuous spectrum, since no charged particle accelerations are involved in its production.

Fluorescent radiations lie below the absorption edge of the material in which they are produced and hence are poorly absorbed by it. These radiations can, however, be effectively removed by an absorber of lower atomic number; and the fluorescent radiation from this can, in turn, be absorbed in still lower-atomic-number material. This principle is used to harden the beams generated by X-ray tubes. The increasing absorption cross-section with decreasing energy permits the differential reduction of the usually unwanted low-energy portion of the X-ray emission spectrum. For example, an X-ray tube operating at 250 KV will have a continuous output with a maximum at about 150 KeV. A filter of 0.2–0.5 mm of copper, $Z = 29$, will effectively remove a large fraction of the lower energies and will harden the beam. A considerable amount of copper fluorescent radiation will be produced at about 8 KeV. An aluminum, $Z = 13$, filter will absorb the 8-KeV photons but will emit its own fluorescence at 1.5 KeV. If this is troublesome, it can be absorbed in a bakelite filter, $Z = 6$. To be effective, filter combinations must be inserted in the proper order.

12.09 Gamma-ray Spectra

The energy spectrum of a gamma emitter will be a reflection of the relative importance of the various absorptive processes in the detector, as well as a characteristic of the emissions themselves. Consider, for example, a solid-crystal scintillator, such as the popular sodium iodide. Light scintillations picked up by the photomultiplier tube can only come from ionizations and excitations produced by the electrons that are released by the photons that are absorbed in the crystal.

The simplest photon–electron conversion is photoelectric absorption. According to Fig. 12-10, this effect will predominate in NaI at all energies up to about 250 KeV. Most of the absorption will take place at the iodine electrons, for with $Z = 53$ this atom is much more effective than sodium, $Z = 11$. In photoelectric absorption all of the absorbed energy will go into kinetic energy of the ejected electron, except for that needed to supply the binding energy of the orbital, ϕ. In the case of iodine, $\phi = 33.2$ KeV for the *K*-shell electrons, Appendix Table 8 and Fig. 12-10.

An orbital vacancy resulting from a photon absorption will be promptly filled, characteristic iodine X rays will be emitted, and, since these are very soft, their chances of absorption in the crystal are good. If all of these processes take place within the luminous lifetime of the phosphor in the photomultiplier, the visible light produced by them will add to that produced by the original photoelectron. The total light output for the event will then be proportional to the energy of the primary photon, $h\nu$. Thus is produced the most prominent feature of the gamma-ray spectrum, the *photopeak*.

Statistical fluctuations will broaden the peak, usually according to a nearly normal distribution, Fig. 12-11. Energy resolution of the recording system is usually measured as the full width at half-maximum of the photopeak, expressed as a percentage of the photon energy. In Fig. 12-11, FWHM of the 662-KeV photopeak of ^{137}Cs is 9 per cent.

Some of the ionized iodine atoms escape electron capture and X-ray emission during the luminous period of the phosphor, and these do not contribute to the photopeak. Sometimes this escape can be seen as a slight

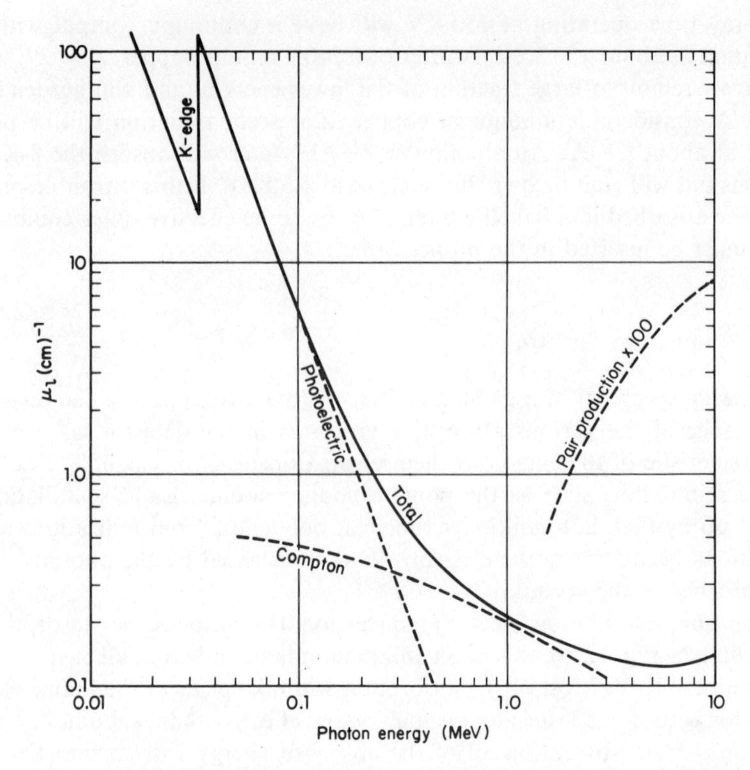

Figure 12-10. The linear absorption coefficients in sodium iodide.

depression on the low-energy side of the photopeak. When de-excitation is delayed, absorption of the soft X-ray photons will take place as an independent event, to produce a spectral peak corresponding to the energy of the *K*-absorption edge of iodine. This produces the so-called iodine *escape peak*, Fig. 12-11.

Some of the primary photons will be absorbed by Compton, instead of photoelectric, interactions. In some of these interactions the Compton-scattered photon will also be absorbed in the crystal, and again the light output will correspond to the total energy of the photon. These serve merely to enhance the photopeak.

In other interactions, the Compton-scattered photon will escape from the crystal, and now the light output will correspond only to the energy given up to the Compton electron. These electrons form an energy continuum from zero up to the maximum that the photon in question can transfer to an electron in a Compton process. From Eq. (12-20) we obtain, with $\alpha = h\nu/m_0c^2$,

$$T = h\nu \frac{\alpha(1 - \cos\theta)}{1 + \alpha(1 - \cos\theta)} \tag{12-30}$$

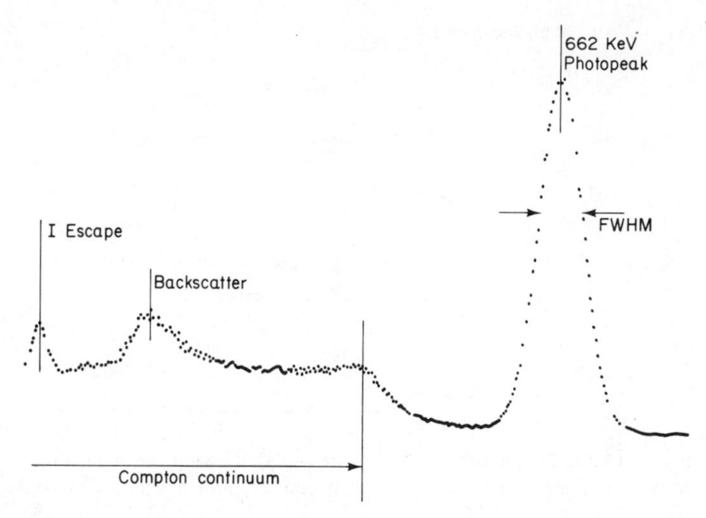

Figure 12-11. Response of a sodium iodide crystal to the 662 KeV gamma rays of ^{137}Cs.

which has a maximum value

$$T_m = h\nu \frac{2\alpha}{1 + 2\alpha} \tag{12-31}$$

The larger the crystal, the greater the probability that the scattered photon will be absorbed. Thus the ratio of the Compton continuum to the photopeak will vary with the size of the detector. A small *backscatter peak* is usually seen near the low-energy end of the continuum, resulting from the absorption in the crystal of primary photons scattered from the tube mount and nearby shielding material.

A more complicated spectrum is obtained when the energy of the primary photon is sufficient to undergo absorption by pair production. The photopeak and the other features will be seen as usual. With a photon energy of $h\nu$ available, the total kinetic energy of the created pair will be $(h\nu - 1.02)$ MeV. When all of its energy is exhausted, the positron will undergo annihilation. If both of the annihilation photons escape from the crystal, a peak will be seen at an energy of $(h\nu - 1.02)$ MeV. If only one of the 0.511-MeV photons escapes, the other will be absorbed to produce a peak at an energy of $(h\nu - 0.511)$ MeV. When both of the annihilation photons are absorbed in the crystal, the full energy of the photopeak will be realized.

Another feature of the usual pair-production spectrum is a strong peak at 0.511 MeV. This peak results from the absorption of the primary photons outside the crystal. Electron–positron annihilation will take place outside the crystal and a single one of the 0.511-MeV photons may enter the detector to form the observed peak.

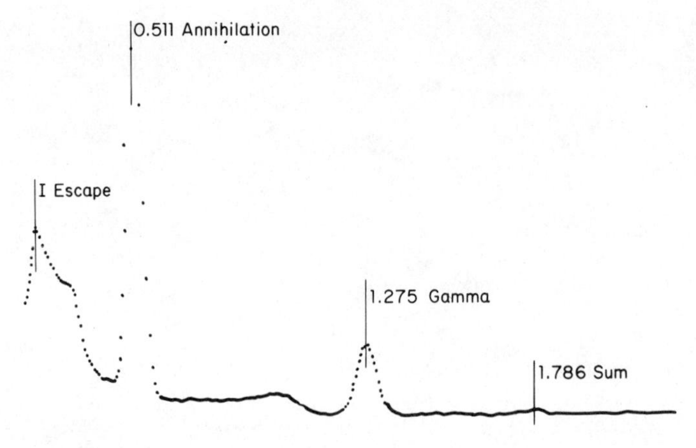

Figure 12-12. Gamma-ray spectrum of ^{22}Na, showing the typical 0.511 MeV annihilation peak of a positron emitter, and a small sum peak.

A positron emitter will always show a peak at 0.511 MeV, produced by the annihilation of positron–electron pairs, Fig. 12-12. Another feature of any complex spectrum is the presence of *sum peaks*. When two absorption processes occur simultaneously, or very nearly so, the crystal will record them as a single event with an energy equal to the sum of the two energies. Sum peaks will have much lower amplitudes than the constituent peaks, since they depend upon the chance coincidences of two emission events. Figure 12-12 shows the strong annihilation peak of the positron emitter ^{22}Na, the 1.275-MeV gamma-ray photopeak, and a small sum peak at 1.786 MeV.

12.10 Exposure and Absorbed Dose

In 1928, the Second International Congress of Radiology defined the *roentgen* (r) as the unit of quantity of photon radiation. The formal definition was:

> "The roentgen shall be the quantity of X or gamma radiation such that the associated corpuscular emission per 0.001293 gram of air produces, in air, ions carrying 1 electrostatic unit of quantity of either sign."

The 0.001293 grams of air used in the definition is the mass of dry air contained in 1 cm^3 at 0°C and 760-mm pressure. This definition has been superseded by one put in terms of MKSA units, but the two definitions are equivalent.

As sophistication in concepts and measurements increased, a greater precision was called for in many of the quantities dealing with radiation. The International Commission on Radiological Units and Measurements

(ICRU) recommended in Report 11, September, 1968, a series of new definitions. Only a few of those immediately pertinent will be given here:

"*Directly ionizing particles* are charged particles (electrons, protons, alpha particles, etc.) having sufficient kinetic energy to produce ionization by collision."

"*Indirectly ionizing particles* are uncharged particles (neutrons, photons, etc.) which can liberate directly ionizing particles or can initiate a nuclear transformation."

"*Ionizing radiation* is any radiation consisting of directly or indirectly ionizing particles or a mixture of both."

The symbol Δ precedes the symbol for a quantity that can be averaged over a volume large enough to contain many interactions and be traversed by many particles, but so small that a further reduction in size would not appreciably change the mean value of the quotient of energy by mass.

From these basic definitions the ICRU develops the following:

"The *absorbed dose* (D) is the quotient of ΔE_D by Δm, where ΔE_D is the energy imparted by ionizing radiation to the matter in a volume element, and Δm is the mass of the matter in the volume element."

$$D = \frac{\Delta E_D}{\Delta m}$$

"The special unit of absorbed dose is the *rad*:"

$$1 \text{ rad} = 100 \text{ ergs g}^{-1}$$

"The *kerma* (K) is the quotient of ΔE_K by Δm, where ΔE_K is the sum of the initial kinetic energies of all the charged particles liberated by indirectly ionizing radiation in a volume element of the specified material, and Δm is the mass of the matter in the volume element:

$$K = \frac{\Delta E_K}{\Delta m}$$

"The *exposure* (X) is the quotient of ΔQ by Δm, where ΔQ is the sum of the electrical charges on all the ions of one sign produced in air when all the electrons (negatrons and positrons) liberated by photons in a volume element of air whose mass is Δm are completely stopped in air:

$$X = \frac{\Delta Q}{\Delta m}$$

The special unit of exposure is the *roentgen* (R):

$$1 \text{ R} \equiv 2.58 \times 10^{-4} \text{ coulomb Kg}^{-1}$$

The words "charges on all the ions of one sign" should be interpreted in the mathematically absolute sense.

The ionization arising from the absorption of bremsstrahlung emitted by the secondary electrons is not to be included in ΔQ. Except for this small difference, significant only at high energies, the exposure as defined above is the ionization equivalent of the kerma in air.

The roentgen is equivalent to the production of

$$\frac{2.58 \times 10^{-4}}{1.6 \times 10^{-19}} = 1.61 \times 10^{15} \text{ ion pairs Kg}^{-1}$$

A long series of measurements has shown that an expenditure of about 34.5 eV is required to produce one ion pair in air. Thus the energy deposition associated with an exposure of 1 R will be

$$1.61 \times 10^{15} \times 34.5 = 5.6 \times 10^{16} \text{ eV Kg}^{-1}$$

A more commonly used value is

$$1 \text{ R} = 5.6 \times 10^{16} \times 10^{-3} \times 1.6 \times 10^{-12} = 89 \text{ erg g}^{-1}$$

The values for energy deposition can be expected to vary slightly with new values of W, the energy required to produce an ion pair.

A careful distinction must be made between the roentgen, the unit of exposure, and the rad, the unit of dose. Exposure characterizes the ability of a photon beam to produce ions in a standard substance, air. An electrical measurement is required in order to fulfill the basic definition. The character of an incident photon beam does not change when some substance other than air is placed in it. In general, however, the rate of energy deposition in the new substance will be quite different. The dose in rads produced by a given exposure in roentgens requires an energy measurement of some sort. It should be noted that the rad is not restricted as to the material or the type of radiation to which it applies.

The energy flow or *fluence F* required to produce an exposure of 1 R will be

$$F = \frac{89}{(\mu_m)_{\text{air}}} \text{ ergs R}^{-1} \tag{12-32}$$

For a monoenergetic beam the number fluence per roentgen, N_p, is

$$N_p = \frac{89}{h\nu(\mu_m)_{\text{air}}} \tag{12-33}$$

Now let a substance with a mass absorption coefficient $(\mu_m)_s$ be placed in the beam until it has been exposed to the energy fluence given by Eq. (12-32). Then,

$$\text{energy absorbed} = F(\mu_m)_s \text{ ergs}$$

and

$$D = \frac{\text{energy absorbed}}{100} = \frac{89(\mu_m)_s}{100(\mu_m)^{\text{air}}} \text{ rads R}^{-1} \tag{12-34}$$

Figure 12-13 shows the energy fluence per roentgen and per rad in water. In accordance with Eq. (12-32), the curve for air is essentially the reciprocal of the mass absorption coefficient for air, Fig. 12-14. The similarity of the two curves in Fig. 12-13 emphasizes the usefulness of the roentgen as a unit for biological purposes, many tissues absorbing radiation in the same fashion as does water.

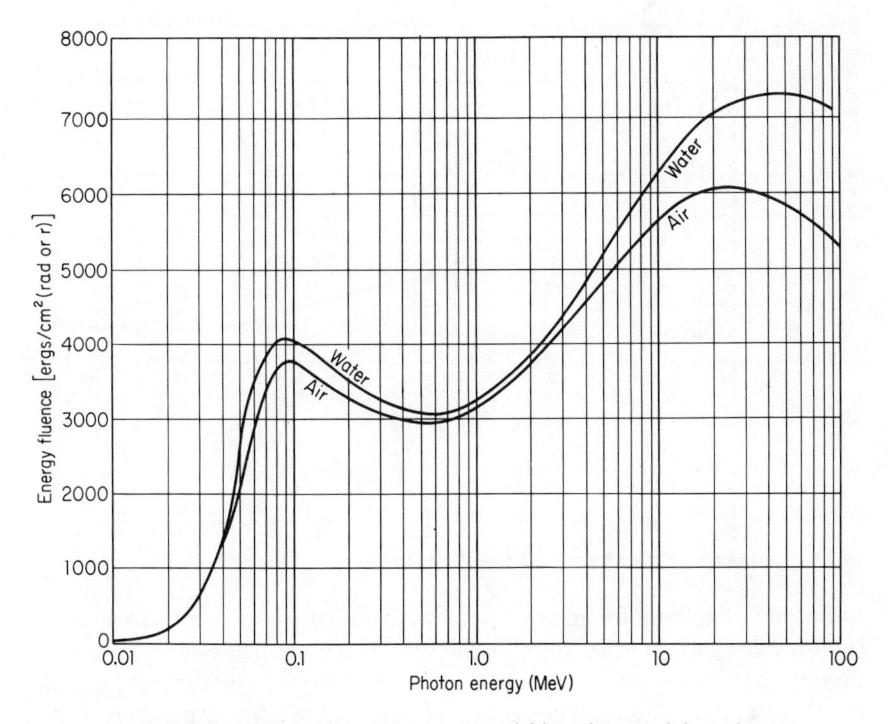

Figure 12-13. Energy fluence per roentgen (air) and per rad (water) as a function of photon energy.

Exposure is defined in electrical terms and an electrical measurement is all that is needed to determine the energy fluence in a photon beam. Strictly, the determination of absorbed dose in rads requires calorimetry in order to fulfill the basic definition. In practice, calorimetry is seldom carried out because of the difficulties in making the measurements with the small amounts of energy that are involved. Energy–ionization conversion factors have been established for a variety of substances and photon energies, and these are used to deduce dose from ionization measurements.

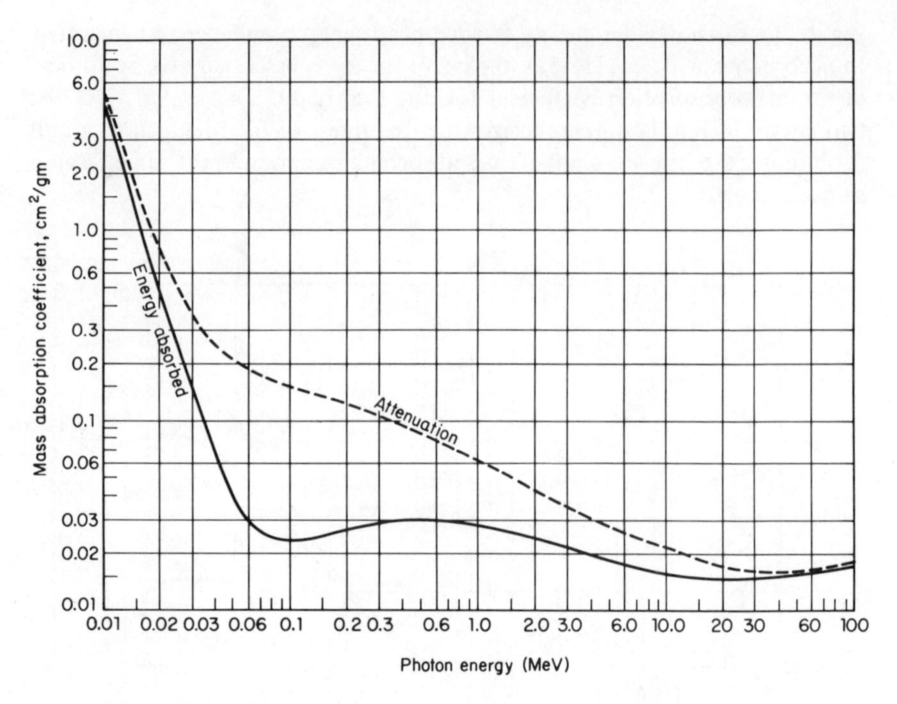

Figure 12-14. Mass attenuation and energy absorption coefficients for air.

12.11 Ionization Chambers and the Roentgen

Ionization chambers for measuring photon radiation in roentgens must be carefully designed in order to fulfill the specific requirements of the definition. Figure 12-15 shows a cross-section of a *standard air chamber* designed to fulfill the necessary conditions. The collecting potential is applied between P_1 and P_2, P_3. Points P_2 and P_3 are at ground potential until a measurement is started when P_3 is connected to the charge-measuring instrument. With this arrangement, an essentially parallel electric field can be maintained and P_3 will collect all of the negative ions from the volume whose cross-section is *EFGH*. When a narrow beam of photons is admitted to the chamber, primary ions will be formed along the beam and, in particular, in the small volume v shown in cross-section as *ABCD*. Some of the ions produced in v will leave the volume

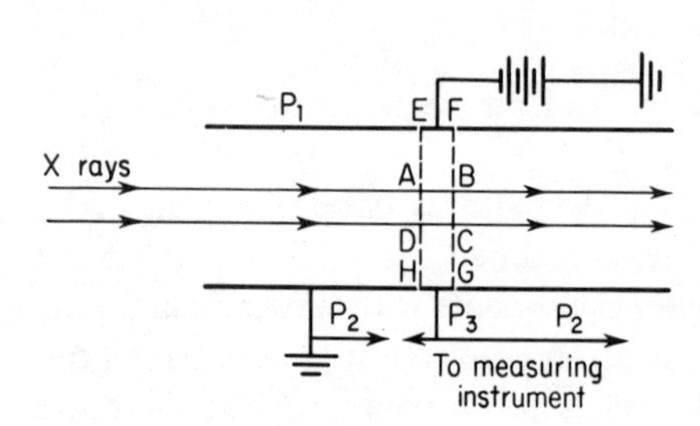

Figure 12-15. Schematic of a standard air chamber.

and produce secondary ions elsewhere, but if the chamber is so large that all primary ions formed in v give up all their energy to the gas, and not to the walls, it can be shown that the loss from v will be just compensated by gains from ions originating outside v. Under these conditions a true measurement in roentgens can be made.

The condition of equal gain and loss of ions is known as *electronic equilibrium*, or better, *charged-particle equilibrium* (CPE). Charged-particle equilibrium must exist in an ion chamber if it is to fulfill the definition of the roentgen. This requirement leads to air chambers that are far too large for routine use outside of a standardization laboratory, and so recourse must be had to a suitable substitute.

Consider a small air-filled cavity with solid walls, similar to that shown schematically in Fig. 12-16. If the walls are made of *air-equivalent* material, the energy spectrum of the electrons produced in it by the combined photon-absorptive processes will be identical with the energy spectrum in an air chamber. Air equivalence, in this case, means that the wall material must have the same atomic-number composition as air. Carbon, with an atomic number of 6, is close to the values of 7, 8, and 10 which predominate in air. Thus chamber walls are usually made of a carbon-containing plastic, frequently with some higher-Z additives. An inner coating of graphite provides the conductivity needed for ion collection. It is possible, therefore, to obtain a wall material that closely approximates air, but which will have a density about 800 times greater.

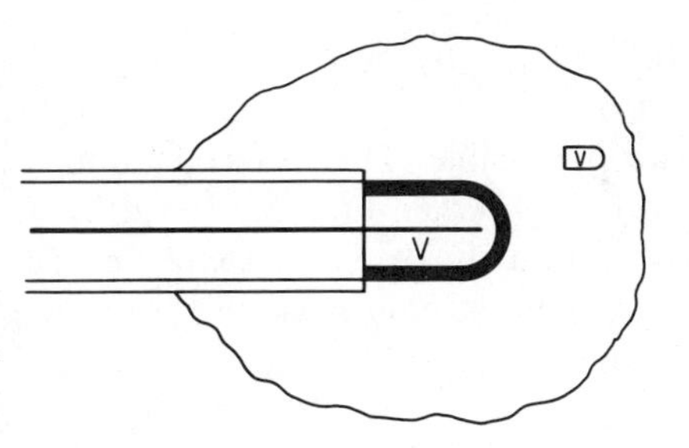

Figure 12-16. An air-filled cavity V for measuring the energy deposited in a similar volume v in a solid or liquid.

The thickness of the solid wall must be at least as great as the range of the most energetic electron released in it by the photons being measured. Under this requirement, the chamber gas will, at all times, see an electron density characterized by the walls, rather than a mixture, with some of the electrons originating outside the walls. Obviously, this restriction places an upper limit on the photon energy for which a chamber can be considered to be air-equivalent.

On the other hand, the chamber wall must not be so thick that it appreciably attenuates the primary photon beam. Should this occur, the ionization throughout the wall will not be representative of the primary beam. The thickness of the chamber wall is obviously a compromise that will be acceptable for only a limited range of photon energies.

Relative absorption coefficients of photons and electrons are such that reasonable equivalence can be obtained over an energy range of perhaps 90–300 KeV. Thinner-walled chambers must be used for lower energies, and supplementary plastic caps can be added for use with more penetrating radiation.

Electron ranges increase much more rapidly with energy than do the mean-free-paths of photons. Because of this, it is no longer possible to fulfill the definition of the roentgen for photon energies above about 3 MeV. At this energy and above, walls thick enough to contain the electrons will seriously attenuate the photon beam. At these energies, the definition of the roentgen is no longer tenable, and exposures must be expressed in terms of energy absorbed.

12.12 Cavity-ionization Measurements

It is frequently necessary to determine the energy absorbed from photon irradiation of a solid or liquid medium where direct ionization measurements are impossible. This determination can be made through the principle of *cavity ionization*, first developed by W. H. Bragg in 1912 and extended by L. H. Gray in 1936. The cavity consists of the gas-filled volume of an ionization chamber immersed in the medium as shown in Fig. 12-16, where the chamber is depicted in greatly enlarged size.

The concept of stopping power was introduced in connection with the range of alpha particles [Sec. 10.02 and Eq. (10-8)]. An identical definition holds for electrons and we use it here, focusing our attention on the relative stopping power of a solid medium S_m.

Now consider, Fig. 12-16, a chamber volume V and a geometrically similar volume v in the medium where all dimensions of v are reduced from those of V by a factor S_m. The dimensions of V must be small compared to the range of the electrons in the medium, so that the presence of V does not appreciably alter the density of electron tracks. Under these conditions, the ionization densities in both V and v will be characteristic of that in the medium.

The dimensions of V and v have been so chosen that a given electron will lose equal amounts of energy in crossing the two volumes. The cross-section of V is, however, S_m^2 times as large as that of v, so the total energy deposited in V will be S_m^2 times that in v:

$$E_a = S_m^2 E_m \tag{12-35}$$

and since

$$V = S_m^3 v$$

$$\frac{E_a}{V} = \frac{E_m}{S_m v} \tag{12-36}$$

According to Eq. (12-36), the energy deposited per unit volume is S_m times as great in the medium as in the cavity. Since S_m can be obtained by an independent measurement, a measurement of E_a/V will determine E_m/v.

Equation (12-36) is expressed in terms of energy absorption, but a direct determination of E_a can be replaced by an ionization measurement. The number of ion pairs per unit volume, J, in V is readily determined and independent measurements have established W, the energy expenditure required per ion pair. Then,

$$E_a = JWV \tag{12-37}$$

and

$$\frac{E_m}{v} = JWS_m \tag{12-38}$$

Through Eq. (12-38), an electrical measurement in a gas-filled cavity serves to determine energy deposition in a solid or liquid medium. Certain restrictions on cavity size must be observed, but in practice these are not difficult.

12.13 Specific Gamma-ray Constant

The unit of radioactivity, the curie (Ci), is defined as a disintegration rate, with no reference to the energy of any of the emitted radiations. In particular, there is nothing in the definition of activity that specifies the exposure rate from a gamma-ray emitter. The two units are related through the *specific gamma-ray constant* Γ.

Consider a source of C curies of some radioactive nuclide that emits f monoenergetic gamma rays of energy E MeV for every nuclear disintegration. Note that the *fractional occurrence* f need not be less than unity, although that is usually the case. At a distance l cm from the source, imagine an air volume 1 cm thick and 1 cm² in cross-section.

If we neglect any energy loss due to absorption along the air path l, the rate of energy flow (energy flux) F through the small volume will be

$$F = \frac{(3.7 \times 10^{10})\, CfE}{4\pi l^2}\ \text{MeV sec}^{-1} \tag{12-39}$$

The rate of energy absorption by the small volume is

$$E_a = \frac{(3.7 \times 10^{10}) CfE}{4\pi l^2} (\mu_l)_{\text{air}}\ \text{MeV sec}^{-1} \tag{12-40}$$

and the exposure rate is

$$\frac{\Delta X}{\Delta t} = \frac{(3.7 \times 10^{10}) CfE}{5.48 \times 10^7 \times 4\pi l^2} (\mu_l)_{\text{air}}\ \text{R sec}^{-1} \tag{12-41}$$

If the source emits several gamma rays, each with a characteristic energy, fractional occurrence, and absorption coefficient,

$$\frac{\Delta X}{\Delta t} = \frac{(3.7 \times 10^{10})}{5.48 \times 10^{7} \times 4\pi l^{2}} \sum_{i} f_i E_i \mu_i \text{ R sec}^{-1} \tag{12-42}$$

It is convenient to choose a standard distance for l and to lump together all of the constants and the summation into a single figure characteristic of each nuclide. This conversion factor, usually based on activity in curies, exposure rate in R hr^{-1}, and distance in meters is known as the specific gamma-ray constant Γ. Values for some nuclides will be found in the Appendix, Table 7. Other values are readily calculated from a knowledge of the decay scheme and the absorption coefficients. In making these calculations, care must be taken to use the coefficients for true absorption rather than those for beam attenuation. The former, frequently tabulated as *energy-absorption coefficients*, will be based on the "true" contributions from Compton processes and pair production.

Values of Γ are needed for calculations of doses absorbed from gamma emitters. This subject will be treated in Sec. 13.11 along with the dose calculations for beta-ray absorption.

REFERENCES

Compton, A. H. and S. K. Allison, *X-rays in Theory And Experiment*. Van Nostrand, New York, 1935.

Radiation Quantities and Units. Report No. 11 of International Commission on Radiological Units, ICRU Publications, Washington, D.C., 1968.

Segrè, E., ed., *Experimental Nuclear Physics*. Vol. I. Wiley, New York, 1953.

White, G. R., *X-ray Attenuation Coefficients from 10 kev to 100 Mev*. Nat. Bureau Standards Report 1003, 1952.

PROBLEMS

12-1. Calculate the energy limits of the Compton-scattered photons from annihilation radiation. Repeat for the 2.75-MeV gamma rays from ^{24}Na. What are the corresponding electron energies that will be recorded by a scintillation detector?

12-2. The 1.80-MeV gamma rays from ^{214}Bi are scattered from a block of graphite. What will be the energies of the photons scattered at 0°, 45°, 90°, 135°, and 180°?

12-3. The following data were obtained with lead absorbers and a monochromatic photon beam. Plot the data and determine half-value thickness, linear and mass absorption coefficients, photon mean-free-path, and the energy of the beam:

Absorber thickness (cm):	1	2	3	4	5	6
Beam intensity:	100	43	23	11	5.0	2.4

12-4. The following data were obtained with aluminum absorbers and an X-ray beam. Plot the data and determine the half-value thickness, photon mean-free-path, linear and mass absorption coefficients, and the equivalent energy of the beam:

Thickness:	0	1	2	3	4	5	6	7
Intensity:	140	103	72	51	32	28	16	14

12-5. An X-ray beam generated at 150 KVCP is hardened by a tin filter. What will be the most desirable choices of second and third filters to suppress the fluorescent radiation?

12-6. Calculate the atomic photoelectric absorption coefficients for aluminum at 20, 60, and 100 KeV. Repeat at the same energies for a lead absorber.

12-7. Derive an expression relating the energy of an incident and a scattered photon in a Compton process.

12-8. Derive Eq. (12-18) from the basic conservation laws and the relativistic energy–momentum relation.

12-9. Calculate the energy absorbed by a 80-Kg man who has received a whole-body dose of 900 rad, an amount almost certain to be fatal. Assume the body to consist entirely of water and calculate the resulting temperature rise.

12-10. Show that the old and the new definitions of the roentgen are equivalent.

12-11. Calculate the specific gamma-ray constant for ^{60}Co, in units of R hr^{-1} Ci^{-1} at a distance of 1 m.

12-12. An X-ray tube with an effective energy of 0.15 MeV delivers an exposure rate of 55 R min^{-1} at a distance of 75 cm from the target. What is the energy flux at this distance?

12-13. Thulium-170 has been used as a photon source in portable units designed for diagnostic radiography. How many curies will be required to provide an exposure rate of 35 R min^{-1} at a distance of 65 cm 1 year after the source has been prepared? (Thulium half-life 130 d, 23 per cent of the decays emit an 0.084-MeV photon.)

12-14. An X-ray beam has an effective energy of 150 KeV and delivers a dose rate of 75 R min^{-1} at a distance of 100 cm from the target. Calculate the entrance and exit dose rates for a patient 40 cm thick when the entrance surface is 50 cm from the target. Repeat for a target–patient distance of 75 cm.

12-15. A 10-Ci source of ^{60}Co is used in a foundry inspection for casting flaws. What is the minimum safe working distance from the source, without shielding, if exposure for a 40-hr working week is to be limited to 100 mR?

12-16. A needle containing 25 mg of ^{226}Ra was lost during a patient treatment. At what distance can this source be detected with a scintillation survey meter capable of responding to 0.03 mR hr^{-1}?

12-17. An X-ray therapy unit delivering 75 R min^{-1} at an effective energy of 150 KeV is adequately shielded by a concrete wall 20 cm thick. What thickness of lead must be added to the wall when the unit is replaced by a new machine delivering 100 R min^{-1} at an effective energy of 800 KeV?

12-18. In a nuclear submarine, shielding must be provided against the direct radiations from the reactor, and also against photons scattered from the surrounding water. Calculate the total attenuation in a beam of 1-MeV gamma rays that traverse 6 cm of iron and 10 m of water, are then scattered through 90°, and enter the submarine after traversing another 10 m of water and 6 cm of iron.

13

Absorption of Charged Particles

13.01 Charged-particle Collisions

From the earliest scattering experiments of Rutherford to the sophisticated high-energy experiments of today, studies of the interactions of charged particles with the matter through which they pass have been very informative. An understanding of these interactions has led to a more detailed knowledge of atomic and nuclear structure, to a better insight into the nature of the radiations themselves, and to their effects on living systems. Details of electron absorption are of particular interest since they are the particles produced in all of the common processes by which photons give up their energy.

Charged particles may be separated roughly into three classes. Electrons, the lightest of the ionizing particles, form a class by themselves. Because of their small mass, they undergo large changes of velocity at each interaction. Their paths through matter are tortuous and meandering, and they lose significant fractions of their energy through bremsstrahlung production.

Protons, alpha particles, and mesons have relatively large masses, and are characterized by nearly linear trajectories. At energies below perhaps 1 GeV, these particles will lose energy almost entirely through inelastic collisions with atomic electrons. "Collisions" is used here in its usual atomic sense, as an encounter between the two electric fields of converging particles, rather than as a mechanical contact between two rigid spheres. Bremsstrahlung production will be negligible in these collisions because the massive particles undergo only small changes in velocity.

Above 1 GeV, particles in this group interact strongly with nuclei. Here, entirely new types of reactions become important, and bremsstrahlung production becomes appreciable.

Technical developments have led to the acceleration of massive ions, even as heavy as uranium, up to very high energies. In many of these collisions, the energy of the incident particle will exceed the total binding energy of the nucleus with which it interacts. These projectiles are being used to induce a variety of nuclear reactions, and to produce transuranium elements not found in nature.

In principle, these massive particles behave much like the lighter particles. In practice, the passage of multiply charged ions is complicated by rapid variations in the net charge. These particles, originally almost completely stripped of electrons, may gain and lose orbitals many times during their flight. This is an exaggeration of the behavior already noted with alpha particles at lower energies. Figure 13-1 shows the path of a massive ion, probably a nucleus of iron, in a nuclear emulsion. Ionization at first is very dense because of the multiple charge. Near the end of its path, the nucleus succes-

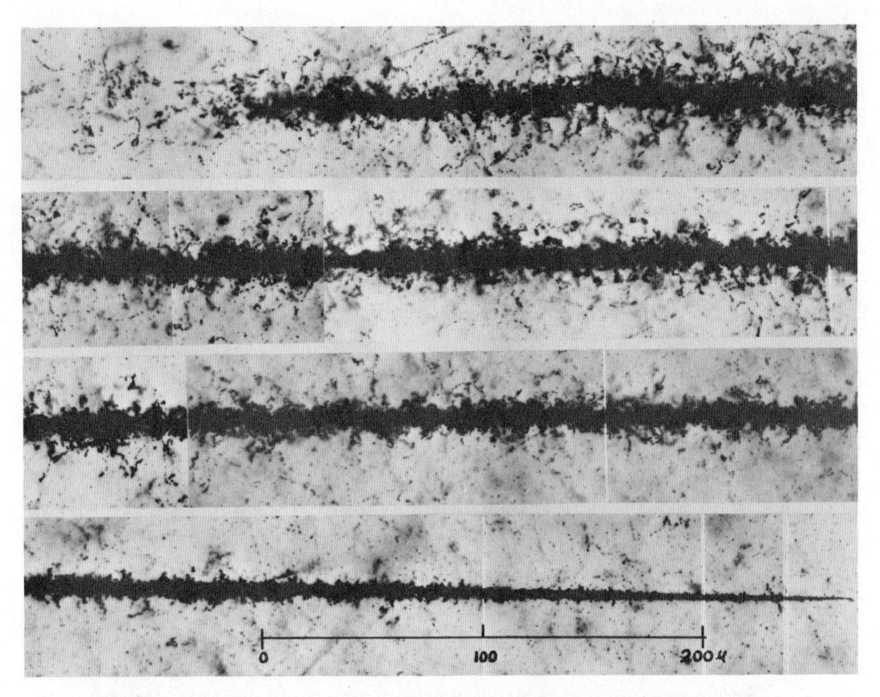

Figure 13-1. The track of a multiply-charged ion in a nuclear emulsion. As the ion slows and acquires electrons, the specific charge decreases and a typical "thindown" track results. (Courtesy of H. Yagoda.)

sively picks up electrons, the rate of ionization decreases, and we see a typical *thindown* as the particle approaches neutrality.

13.02 Some Absorption Parameters

Linear stopping power has already been defined in connection with alpha-particle emission as

$$S_l = \frac{dE}{dx} \tag{13-1}$$

where dE is taken as the loss of energy from the particle over the path length dx. From this basic definition, stopping powers based on other variables, such as mass, follow naturally.

A closely related parameter is *linear energy transfer*, L or LET. L is also defined by

$$L = \frac{dE}{dx} \tag{13-2}$$

but now dE must be taken as the energy removed from the particle and imparted to the medium at or near the site of the collision. In the absence of bremsstrahlung production, stopping power and linear energy transfer will be nearly equal. When bremsstrahlung are produced, some energy will be transferred away from the collision site, and then the two energy-loss parameters will differ substantially.

Another frequently used energy-loss term is *specific ionization*. Specific ionization is given by dN/dx, the number of ion pairs formed per unit path length.

Specific ionization and linear energy transfer are related through the *average energy expended per ion pair formed*, W. Many particle interactions transfer insufficient energy to the orbital electrons to produce an ionization. Visible and ultraviolet light will be emitted as the excited atom returns to its ground state. Dissociation may occur, but in neither of these reactions will ions be detected. Because of these energy losses that are not available for ionization, values of W can be expected to be substantially greater than the ionization potentials of the absorbing atoms.

Measurements made with a variety of radiations and a variety of absorbing gases lead to an average value of W of about 35 eV per ion pair. All values of W fall within a range of 2:1 and are essentially independent of the ionization potentials. This behavior can be made plausible, but it does not rest on a solid theoretical foundation. A limited number of measurements suggest that W may have about the same value in liquids, but experimental difficulties here are great because of the high rate of ion recombination in the dense systems.

The most important parameter in stopping-power theory is the *average excitation potential I.* This term is the *geometric* mean of all of the excitation energies and the ionization energy, with each energy weighted with a quantum-mechanical factor known as the oscillator strength (effectiveness) of the transition. Theory suggests that I should be about proportional to atomic number, or $I = kZ$, but it does not predict a value of k. Many of the oscillator strengths needed in the calculation are not known, and experimental values of I are not easy to obtain.

Fortunately, I appears in the stopping-power equations only as the logarithm, so that some uncertainty in its values can be accepted. Typical values of I are given in Table 13-1.

TABLE 13-1
AVERAGE EXCITATION POTENTIALS

Material	*Z*	*Ionization potential (eV)*	*I (eV)*	$k = I/Z$
H	1	13.6	18	18
C	6	11.2	77	12.8
N	7	14.5	88	12.6
O	8	13.6	100	12.5
Al	13	6.0	164	12.3
Ar	18	15.7	184	10.2
Fe	26	7.9	300	11.5
Pb	82	7.4	820	10.0
Air	—	—	81	—
Water	—	—	70	—

13.03 Collision Energy Loss of a Heavy Charged Particle

Bethe* has developed an expression for the rate of energy loss, or stopping power, of a heavy charged particle:

$$-\left(\frac{dE}{dx}\right)_C = \frac{4\pi z^2 e^4 NZ}{m_0 v^2}\left[\ln\frac{2m_0 v^2}{I} - \ln(1-\beta^2) - \beta^2\right] \qquad (13\text{-}3)$$

where ze = charge on the particle, esu
N = number of absorbing atoms cm^{-3}
Z = atomic number of the absorber
I = average excitation potential, ergs
m_0 = rest mass of the electron
v = velocity of the particle
$\beta = v/c$ for the particle

* H. A. Bethe and J. A. Ashkin, *Experimental Nuclear Physics*, ed. E. Segrè, Vol. I, John Wiley & Sons, Inc., New York, 1953.

Stopping power depends upon the square of the net charge on the particle but not upon its mass. For a given value of v, the stopping power can be scaled from one particle to another, from proton to alpha particle to a multiply charged ion, by the factor of z^2.

Equation (13-3) assumes no energy losses due to bremsstrahlung production and nuclear interactions. It is not applicable at low energies where the particle velocity will be comparable with the velocities of the orbital velocities in the absorbing atoms.

Equation (13-3) somewhat overestimates the rate of energy loss because it assumes that each of the orbital electrons participates equally in the absorption. This is evident from the NZ factor in the numerator. In fact, inner electrons, principally those in the K shell, cannot participate in an energy exchange unless they can absorb an energy equal to their ionization potential. They are thus constrained from entering into low-energy interactions. Corrections for nonparticipation have been developed. These can be found in the original work by Bethe, or in Evans.*

13.04 Energy Loss by Electrons

Bethe has also developed expressions for the stopping power of electrons. Bremsstrahlung production now becomes important, and so two equations are needed to account for the total energy transfer:

$$-\left(\frac{dE}{dx}\right)_c = \frac{2\pi e^4 NZ}{m_0 v^2}\left[\ln\frac{m_0 v^2 E}{2I^2(1-\beta^2)} - \ln 2(2\sqrt{1-\beta^2} - 1 + \beta^2) + (1-\beta^2) + \frac{1}{8}(1-\sqrt{1-\beta^2})^2\right] \tag{13-4}$$

$$-\left(\frac{dE}{dx}\right)_r = \frac{NEZ(Z+1)e^4}{137 m_0^2 c^4}\left(4\ln\frac{2E}{m_0 c^2} - \frac{4}{3}\right) \tag{13-5}$$

where the terms have the same meanings as in Eq. (13-3).

Although the expression for collision loss differs in some details from that for a heavy particle, it still retains the factor NZ, again with a K-shell correction. Equation (13-4) is also restricted to energies above which particle and orbital velocities are comparable.

The presence of E and Z^2 in the numerator of Eq. (13-5) shows the increasing importance of radiation losses at high energies and in absorbers of high atomic number. In lead, radiation and collision losses are about equal at 15 MeV; in air, equality is reached at about 250 MeV.

An integration of Eq. (13-5) over the total path length will give the total amount of energy lost to radiation. The path length will be approximately

* R. D. Evans, *The Atomic Nucleus*, McGraw-Hill Book Company, New York, 1955, p. 637.

proportional to E/NZ and hence the total bremsstrahlung production will be approximately

$$E_r = kE^2Z \tag{13-6}$$

This crude analysis leads to the same analytical form that is obtained when the Kramers efficiency relation, Eq. (8-4), is used to calculate the total bremsstrahlung output from a thick-target X-ray tube.

13.05 Minimum Ionization

Both stopping-power equations predict a minimum value of dE/dx at an energy determined by the mass of the particle and, to a lesser extent, by the nature of the absorbing medium. Figure 13-2 is a log–log plot of specific ionization of electrons and protons in air. The two types of particles have quite comparable values of minimum ionization, although the minima occur at energy values more than one order of magnitude apart. Each type of particle shows a rapid decrease in stopping power as the energy is increased

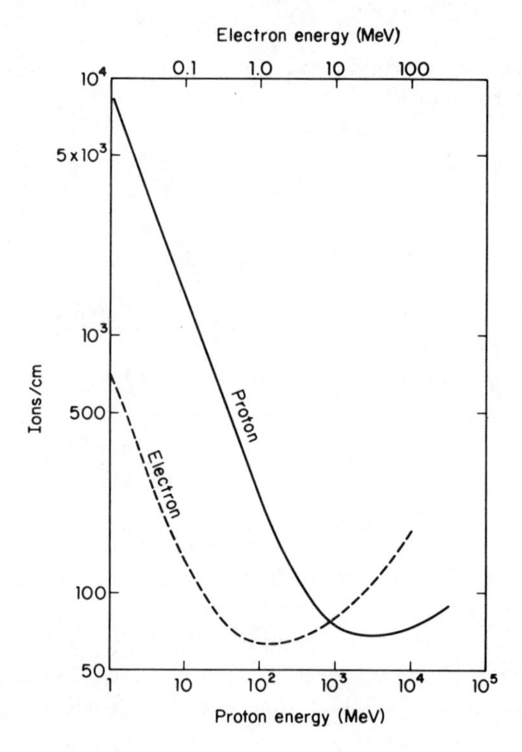

Figure 13-2. Specific ionization by electrons and protons in air. Multiplied by 4, the proton values will apply to alpha particles.

near the low end of the energy scale. Above the minimum, the stopping power increases slowly with energy.

Classical mechanics shows that the change of momentum transferred in a collision is equal to the impulse Ft. Thus the time over which a force acts is an important factor in determining its action on an object capable of undergoing a momentum change. As the velocity of a particle increases, its electric field sweeps past the fields of target atoms in a shorter and shorter time. The probability or cross-section for a transfer of energy will decrease as the velocity increases, and this will be seen as a decrease in the stopping power. This is the cause of the rapid decrease in stopping power seen at relatively low energies.

As the velocity of a particle approaches the velocity of light, it is also approaching the velocity with which its electric field is propagated. At these velocities, the normally spherical field becomes distorted, shrinking in the direction of motion and expanding laterally. The "size" of the field increases, or the value of the *impact parameter* b, Fig. 6-1, decreases. Interaction probabilities and stopping powers will then increase with further increases in particle energy.

The log–log plots in Fig. 13-2 do not bring out the details of the stopping-power maximum seen at very low particle energies. At zero energy, of course, the stopping power is zero. Above this value, but below the energy range where Eqs. (13-3) and (13-4) are valid, the stopping-power curve for a single particle passes through a sharp maximum, Fig. 13-3. A beam of particles, even though they were originally monoenergetic, will show a broad maximum because of straggling. This broader maximum is known as the *Bragg ionization peak.*

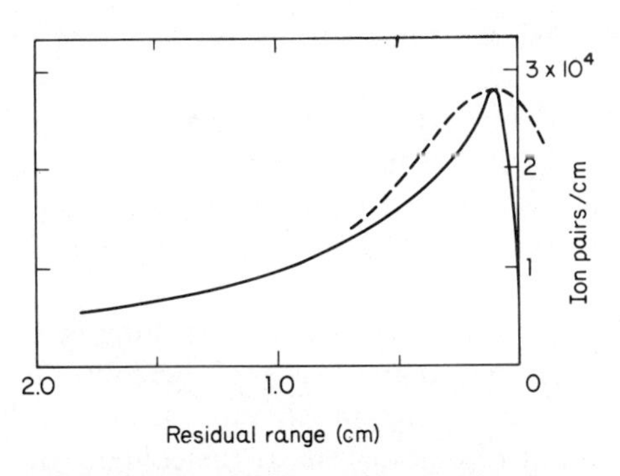

Figure 13-3. Specific ionization of a single proton has a sharp maximum value near the end of its path. A beam of monoenergetic particles will show a broader maximum because of straggling.

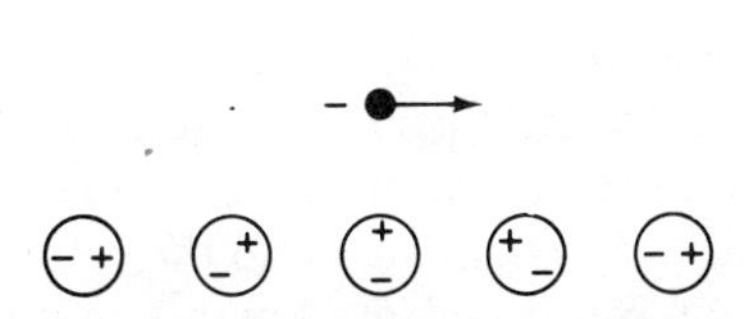

Figure 13-4. Molecular polarization along the path of an electron.

As a charged particle moves through an absorbing medium, its electric field will temporarily *polarize* the medium along its path. Charges of unlike sign will be attracted toward the moving particle, and like charges will be repelled, Fig. 13-4. Polarization reduces the effectiveness of the charge on the particle, and this acts to reduce the rate of energy loss. Polarization effects become significant only in dense media and at high energies. Thus polarization may decrease the value of dE/dx for a 100-MeV electron in water by 30 per cent.

13.06 Delta Rays

Linear energy transfer was defined as energy "imparted to the medium at or near the site of the collision." The phrase "at or near" is somewhat flexible. If one is interested in, say, the genetic effects of radiation absorption, one must be interested in the energy absorbed in a single chromosome, or a part thereof. For some other effects, a distance of a few millimeters could be considered to be near.

Most of the particle–orbital electron interactions transfer more energy than is needed to produce a single ionization. Some of the electrons will be ejected with enough kinetic energy to cause a few ionizations on their own account. Evidence for these electrons can be seen in expansion-chamber photographs as short tracks extending out from the main track of the particle. These short-range tracks are known as delta rays. A few δ-ray tracks can be seen along the thindown track in Fig. 13-1. Delta rays are more easily identified when the primary track is less heavily ionized.

Delta rays capable of making only one or two ionizing collisions probably exist, but it is usually difficult to identify a track with less than 3 or 4 events. Thus most of the δ-ray tracks observed correspond to energy transfers of 100 eV or more.

The thindown track shown in Fig. 13-1 is an extreme example of *columnar ionization*. A heavy, multiply charged particle will leave behind it an almost linear trail of very closely spaced ions. Ion densities within the trail, or column, will be much higher than the average densities, and recombination will be favored.

Some thought must be given to the geometrical relations existing in an ion chamber in which columnar ionization is taking place. If the collecting field is at right angles to the columns, or nearly so, ion separation and collection can be achieved with a minimum of recombination. If the columns are parallel to the field, recombination may seriously reduce the amount of charge collected, because now the ions must be pulled through the densely ionized column.

13.07 Cerenkov Radiation

In 1934, Cerenkov began a series of investigations into the nature of the visible light emitted by certain substances irradiated by fast particles. This visible radiation, now named after the pioneer worker, is emitted whenever a charged particle moves through a medium with a velocity greater than the velocity of electromagnetic radiation in the same medium. The effect can be thought of as an electromagnetic shock wave, analogous to the compression wave produced at the bow of a boat moving faster than the phase velocity of a compressional wave in water.

In a dielectric medium of refractive index n, photons will move with a velocity c/n, which becomes then the critical velocity for the production of Cerenkov radiation. Figure 13-5A shows the details of three interactions

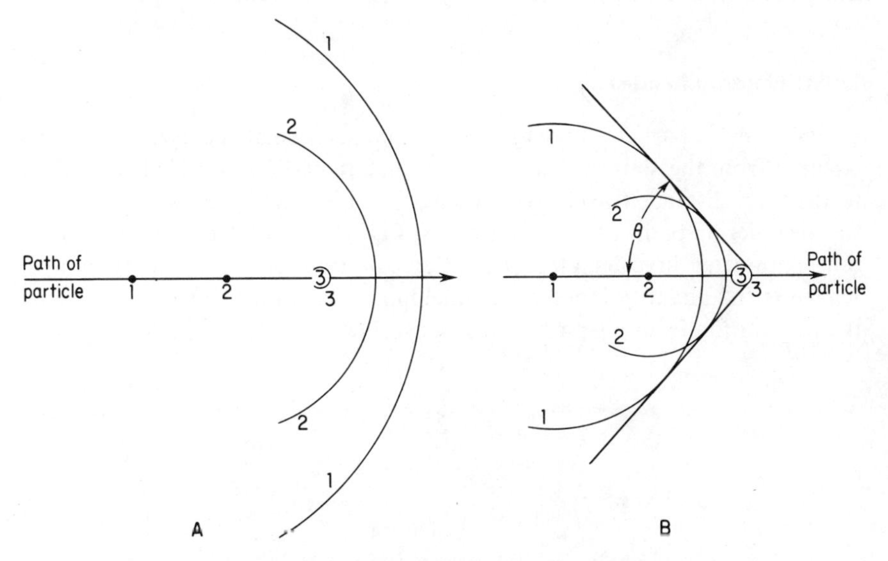

Figure 13-5. (A) Wavefronts resulting from radiation losses at 1, 2, and 3 from a particle with $v < c/n$. (B) When $v > c/n$, a tangent can be drawn to the spherical wavefronts.

between a charged particle moving with velocity $v < c/n$ and the electrons in the medium. Just after interaction 3, the electromagnetic wavefronts resulting from interactions 1 and 2 will have the positions shown. Both of the earlier fronts will have moved ahead of the point at which the third interaction occurs. There is no interrelation between the wavefronts and the three events are entirely independent.

Figure 13-5B shows the situation when $v > c/n$. The wavefronts from 1 and 2 now lie inside 3 at the time of the third interaction. According to the usual Huygens construction, a conical wavefront can be drawn tangent to the spherical surfaces. From Fig. 13-5B comes the simple relation,

$$\sin \theta = \frac{c}{nv} \tag{13-6}$$

When $v > c/n$, light will be observed out to a limiting angle determined by Eq. (13-6). The cutoff angle is quite sharp, and Cerenkov radiation can be used to determine the velocity of charged particles with considerable accuracy. If particle velocity and some other property such as momentum ($B\rho$) can be determined, an almost certain particle identification can be made.

Cerenkov radiation lies mostly in the blue portion of the visible and the near ultraviolet. The blue glow seen around nuclear reactors of the swimming-pool type is due almost entirely to Cerenkov radiation.

13.08 Electron Ranges

Heavy charged particles undergo so few and such small deflection that the distance from the source at which they come to rest (range) is almost equal to the total distance traveled (path length). This equality does not apply to the meandering paths taken by electrons, Fig. 13-6. Path length is the important parameter in calculating linear energy transfer, or stopping power. Range is the quantity involved in shielding calculations, since it represents the penetration from a source into an absorber.

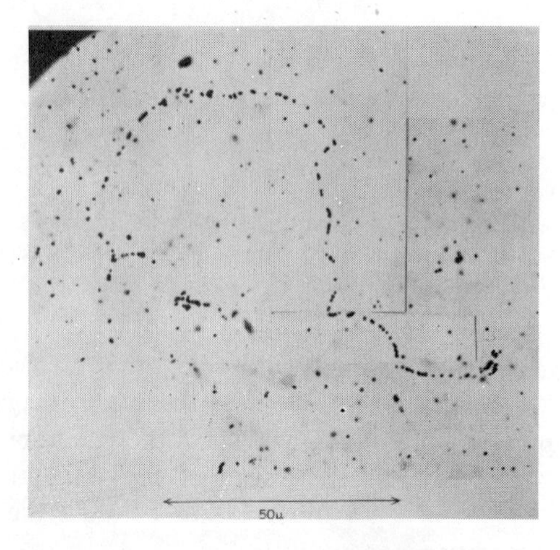

Figure 13-6. The tortuous path of an electron leads to an average range much shorter than the path length.

Aluminum has become a standard absorber for determining electron ranges, particularly those of beta particles. Results of these absorption measurements are usually expressed as *equivalent ranges* in mg cm^{-2} or g cm^{-2}, Figs. 13-7 and 13-8. For most calculations, equivalent ranges in aluminum can be transferred directly to other absorbers, since energy losses depend primarily on the number of electrons encountered and not upon the atomic number or chemical composition of the absorber. Except for hydrogen, the Z/A ratio varies only slowly with Z and so the electron density is nearly proportional to mass density.

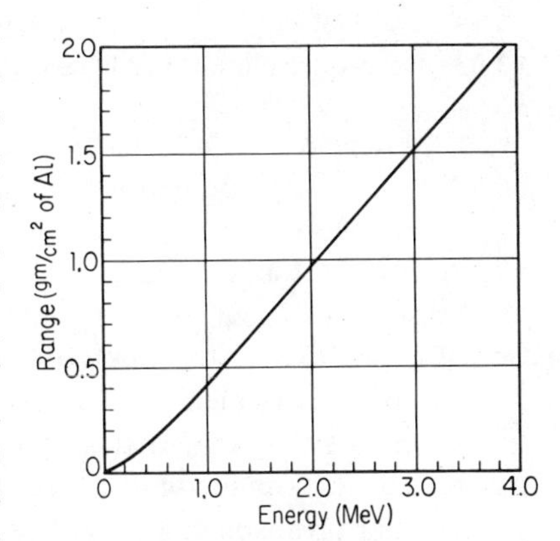

Figure 13-7. Equivalent range of electrons in aluminum.

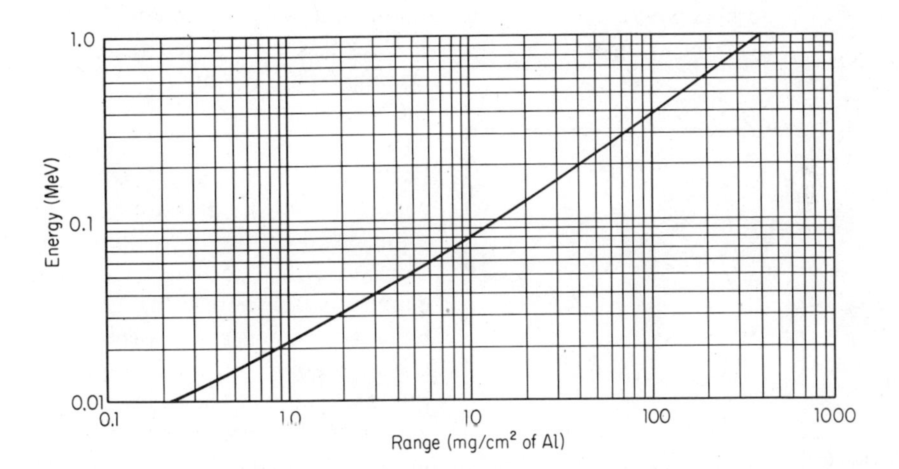

Figure 13-8. Equivalent range of low-energy electrons in aluminum.

Empirical relations have been developed from experimental data to relate range to electron energy. Two useful forms, Eqs. (13-7) and (13-8), connect range in g cm^{-2} and energy in MeV:

$$E = 1.85R + 0.245 \quad \text{for} \quad R > 0.3 \text{ g cm}^{-2} \tag{13-7}$$

$$R = 0.542E - 0.133 \quad \text{for} \quad E > 0.8 \text{ MeV} \tag{13-8}$$

Ranges for low-energy electrons can probably be obtained more accurately from the curves than from the analytical relations.

13.09 Beta-particle Energy Measurements

Whenever possible, beta-particle energies are determined from activity–energy or activity–momentum data obtained with either a solid-state or a magnetic-deflection spectrometer. A Fermi–Kurie plot, Sec 11.08, can be used to obtain the maximum beta energy from a linear extrapolation.

Reasonably accurate energy determinations can be made by applying one of a variety of extrapolation methods to an aluminum absorption curve. A series of activity measurements is made as aluminum absorbers are placed in the path of the beta particles from the unknown. A corresponding set of measurements is made with a beta emitter of known energy in an identical geometrical arrangement of source–absorber–detector. Preferably, the beta transitions of the known and the unknown should have equal degrees of forbiddenness, since then the two spectral shapes will be most similar. Equality of forbiddenness is not a requirement of the method, however. A procedure named the *Feather analysis*, after its originator, is usually applied to the absorption data in order to calculate the unknown range.

A pure beta emitter usually gives an aluminum absorption curve that is nearly linear on a log activity–absorber thickness plot. When there is an accompanying gamma-ray emission, the steep section of the absorption curve will be followed by a linear portion, with the very small slope characteristic of penetrating radiation, *A*, Fig. 13-9. To obtain the true beta-absorption curve, the linear gamma-ray portion is first extrapolated back to zero absorption. A series of points on the extrapolated line are evaluated and these values are subtracted from the corresponding points on the original curve *A*. The remainders represent activity due to beta particles alone, and are plotted to form the absorption curve *B*, Fig. 13-9. This is the curve that is used in the Feather analysis.

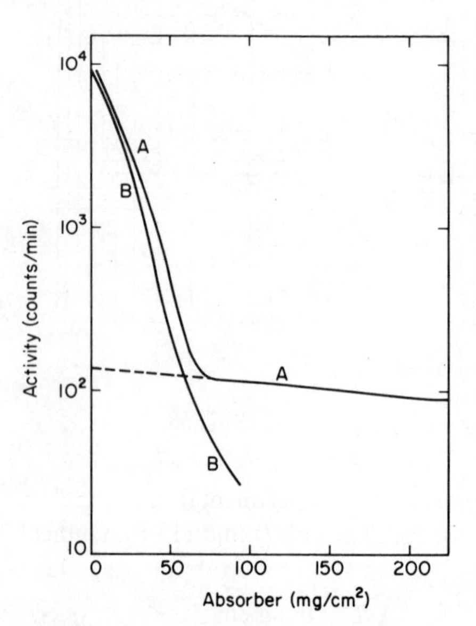

Figure 13-9. (A) Aluminum absorption curve of a beta–gamma emitter. (B) The beta particle absorption curve after the gamma contribution is subtracted.

Bismuth-210, whose beta-particle range has been carefully established as 510 mg cm^{-2}, is commonly used as the known emitter. After making any necessary corrections, as for coincidence loss or self-absorption, the known and the unknown activities are normalized to equal values at zero added absorption. Plots of log activity against equivalent absorber thickness (mg cm^{-2}) are then made, Fig. 13-10. The abscissa of this plot is then divided into deciles of the standard range (51, 102, 153, ... mg cm^{-2} for ^{210}Bi). At each range decile, a vertical line is

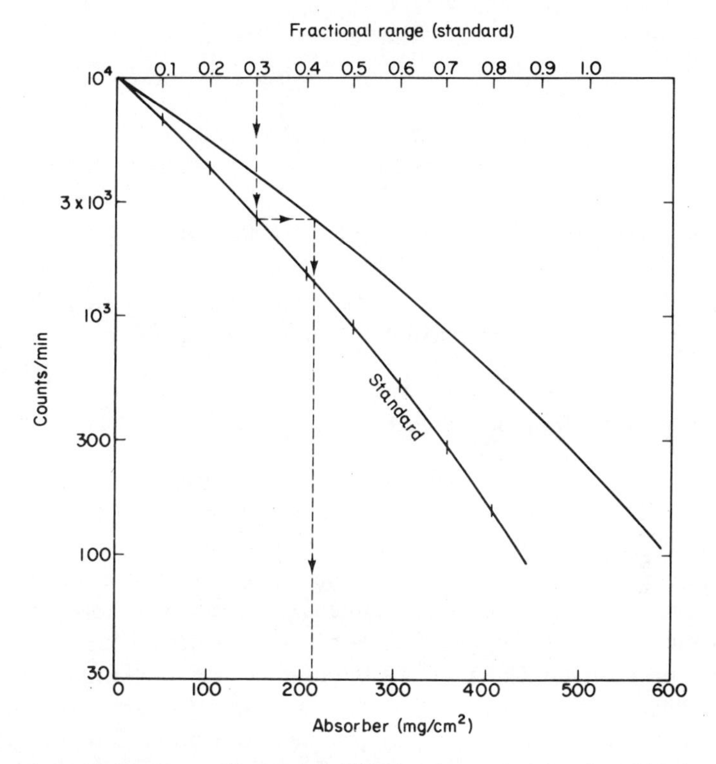

Figure 13-10. Absorption curves of ^{210}Bi and an unknown, plotted in the form for a Feather analysis.

projected to intersect with the standard absorption curve, and each of these intersections is projected horizontally to intersect with the absorption curve of the unknown. The members of each pair of intersections thus obtained correspond to the same fraction of the two ranges—standard and unknown. A projection of each intersection on the unknown curve back to the range axis will give a value of the corresponding fractional range. Each of these fractional ranges, divided by the corresponding range fraction, will yield an estimated maximum range of the unknown. These estimated ranges are then plotted against the range fraction, Fig. 13-11, and extrapolated to range 1.00, to obtain a best value of the maximum range. With the equivalent range

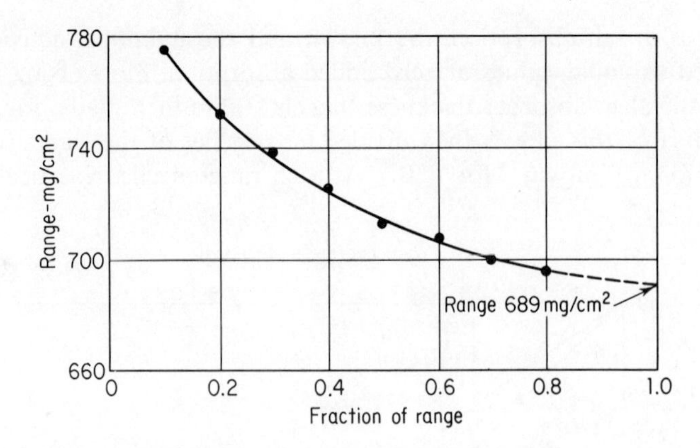

Figure 13-11. Plot of the range estimates in the Feather analysis.

determined, a range–energy plot or the empirical range–energy equation can be used to calculate the maximum beta-particle energy.

13.10 Tissue Dose from Beta Emitters

When radioactive isotopes are given to living organisms, it is usually necessary to calculate the radiation dose delivered to the tissues. When an isotope is administered for diagnostic purposes, a calculation is needed to insure that the radiation dose is within acceptable limits. In radioisotope therapy, dose calculations must be made to determine the amount of isotope needed to deliver the desired irradiation. In general, the greatest tissue dose comes from the absorption of beta particles, but in some cases the gamma dose may be governing.

Beta-dose calculations are based on the assumption that the particle energy delivered to a tissue is equal to the energy released by the isotope contained in that tissue. This assumption is not exact on a microscopic scale, but the relatively short range of beta particles makes it a good approximation for almost all tissue calculations.

Consider a tissue containing C_0 μCi g^{-1} of an isotope of *average* beta energy $\bar{E}_\beta$ MeV. Then the beta energy released by the isotope will be

$$E = C_0 \bar{E}_\beta \times 3.7 \times 10^4 \times 1.6 \times 10^{-6} \text{ erg sec}^{-1} \text{ g}^{-1} \tag{13-9}$$

From the definition of the rad,

$$\text{dose rate} = C_0 \bar{E}_\beta \times 5.92 \times 10^{-4} \text{ rad sec}^{-1} \tag{13-10}$$

The total dose delivered over any extended time can be determined by integrating Eq. (13-10), taking into account the decrease in C_0, due to both radioactive decay and biological elimination from the tissues. The exact time course of the latter process is usually complex but may be approximated by an exponential. Then, at any time t,

$$C = C_0 e^{-\lambda_B t} e^{-\lambda_P t} \tag{13-11}$$

where λ_B = biological decay constant
λ_P = physical decay constant

The decay constants may be combined into a single *effective decay constant* λ_E:

$$\lambda_E = \lambda_B + \lambda_P \tag{13-12}$$

The corresponding half-lives are connected by

$$T_E = \frac{T_B \times T_P}{T_B + T_P} \tag{13-13}$$

Integration of Eq. (13-10) gives for the dose from zero to time t

$$D_\beta = \frac{5.92 \times 10^{-4} C_0 \bar{E}_\beta}{\lambda_E} (1 - e^{-\lambda_E t}) \text{ rad} \tag{13-14}$$

where λ_E must be expressed in sec^{-1}. It is usually more convenient to put Eq. (13-14) in terms of half-lives and in days. These changes give

$$D_\beta = 73.8 \bar{E}_\beta C_0 T_E [1 - e^{-(0.693t/T_E)}] \text{ rad} \tag{13-15}$$

There will usually be some uncertainty in the choice of proper values for C_0 and T_E. For example, if iodine, say ^{131}I, is given to a patient, a substantial fraction of the isotope may be picked up by the thyroid gland, and the rest rapidly excreted. To calculate the beta dose to the thyroid gland, one must know the value of the initial uptake, and the rate of excretion from the gland. If the radiation dose to kidney or bladder is desired, some account must be taken of the fraction that was not fixed in the thyroid. Some of these values can be obtained from collateral measurements, and from past experiences, but one must be satisfied with uncertainties of 10 per cent or even more.

13.11 Tissue Dose from Gamma Rays

Calculations of radiation doses from gamma rays are complicated by the fact that we can no longer assume that the energy absorbed by a tissue is equal to the energy released in it. Radioactive materials located in any part of the body are capable of delivering a gamma-ray dose to any other part of the body. The dose delivered will be determined by the distance between emitter and absorber, the absorption characteristics of the intervening medium and at the point of dose calculation. Differential equations for the dose due to a small volume are readily set up, but the integration of these equations over the human body is impossibly complex.

An approximate solution has been obtained by Bush,* who treated the human body as a series of cylinders. Using the gamma-ray energy of radium, Bush obtained from his integrals a series of $\bar{g}$ values which give an average gamma-ray dose from a uniformly distributed isotope. These $\bar{g}$ values are put in the integrated dose equation:

$$D_\gamma = 0.346 C_0 \rho \Gamma \bar{g} T_E [1 - e^{-(0.693t/T_E)}] \quad \text{R} \qquad (13\text{-}16)$$

where C_0 = initial concentration, μCi g^{-1}
ρ = tissue density, g cm^{-3}
Γ = specific gamma-ray constant, R hr^{-1} Ci^{-1} at 1 meter
T_E = effective half-life, days
t = time, days

The $\bar{g}$ values may be used for almost all gamma-emitting isotopes, even though the energies depart somewhat from the average value for radium. The absorption curves are relatively flat over quite an energy range, so this approximation is no more serious than others that have gone into the final result.

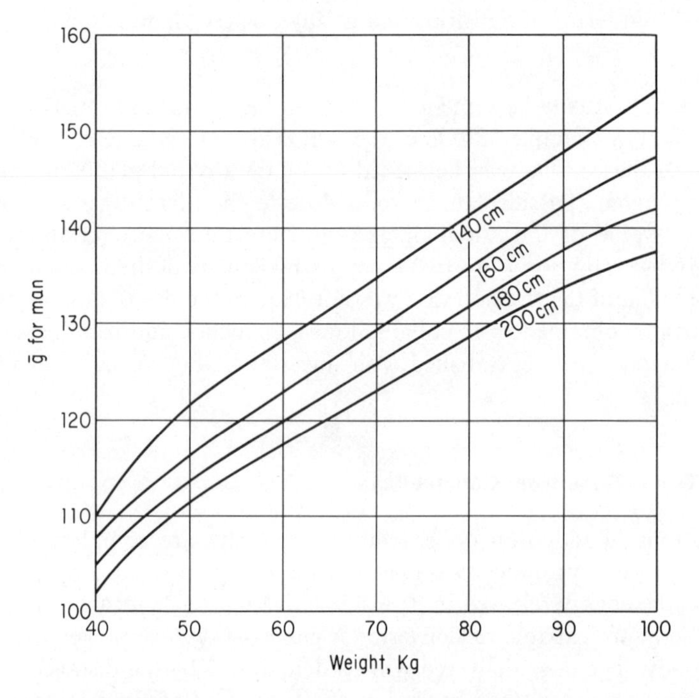

Figure 13-12. Values of $\bar{g}$ for man as a function of weight and height. From calculations of Bush, *Brit. J. Radiol.* **22,** 96, 1949.

* F. Bush, *British Journal of Radiology* **19,** 14, 1946, and **22,** 96, 1949.

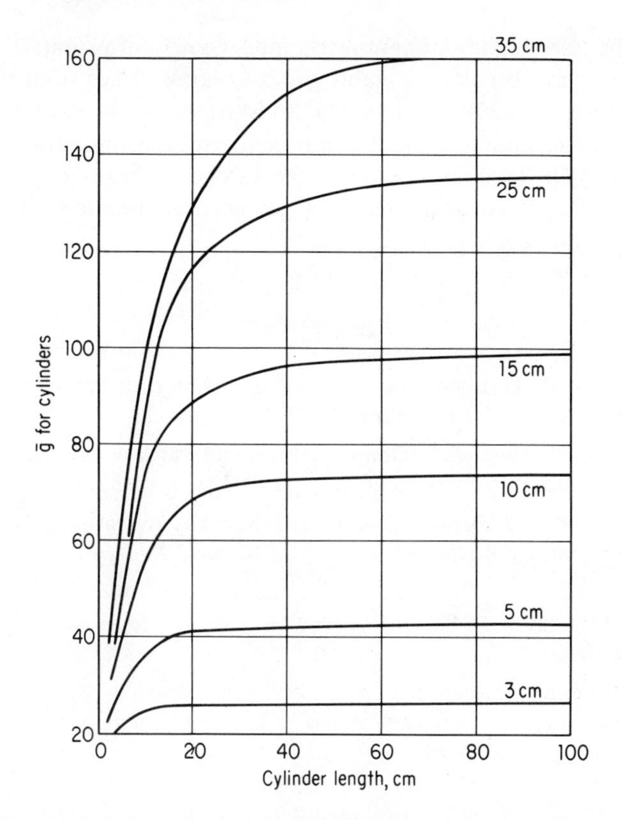

Figure 13-13. Values of $\bar{g}$ for a right cylinder as a function of length and radius. From Bush, *Brit. J. Radiol.* **22,** 96, 1949.

Figure 13-12 shows the $\bar{g}$ values as calculated by Bush and plotted over a series of weights and heights. Figure 13-13 gives plots of $\bar{g}$ values for a variety of cylinder dimensions. The case of a sphere is particularly simple; $\bar{g} = 3\pi r$ where r is the radius of the sphere in centimeters. This value is based on the assumption that all of the radioactive isotope in question is uniformly distributed throughout the sphere.

The total dose received by any tissue will be the sum of the contributions from each of the radiations emitted in the decay. However, the accuracy attainable with these dose calculations is not high, and there is little point in introducing meaningless mathematical complexities. A study of the decay scheme may show that some components may be omitted, or that they can be combined into a simpler system.

The soft X rays emitted in an electron-capture transition can usually be treated like beta particles, since they are strongly absorbed near the site of emission. The full energy of the X-ray emission is used in the calculation.

With the introduction of computers into the field of gamma-ray dosimetry, methods have become available for calculating doses from more complicated distributions. Most of the methods are based on the concept of an *absorbed fraction*, which is the fraction of the total gamma-ray energy that is absorbed by the tissue or organ of interest. Computer programs have been developed and applied to a variety of geometrical situations. The original literature should be consulted for details.*

REFERENCES

Brownell, G. L., W. H. Ellett, and A. R. Reddy, "Absorbed Fractions for Photon Dosimetry." *J. Nuc. Med.* **9,** Supp. 1, 1968.

Dillman, L. T., "Radionuclide Decay Schemes and Nuclear Parameters For Use In Radiation-dose Estimation." *J. Nuc. Med.* **10,** Supp. 2, 1969.

Elrick, R. H. and R. P. Parker, "The Use of Cerenkov Radiation in the Measurement of β-emitting Radionuclides." *Int. J. Applied Rad. and Isotopes* **19,** 263, 1968.

Fano, U., "Penetration of Protons, Alpha Particles, and Mesons." *Ann. Rev. Nuc. Sci.* **13,** 1, 1963.

Snyder, W. S. et al., "Estimates of Absorbed Dose Fractions For Monoenergetic Photon Sources Uniformly Distributed In Various Organs of a Heterogeneous Phantom." *J. Nuc. Med.* **10,** Supp. 3, 1969.

PROBLEMS

13-1. What thickness of aluminum is required to just completely stop all of the beta particles from ^{32}P, so that the bremsstrahlung spectrum can be studied without interference? How much will this thickness of aluminum attenuate the 50-KeV photons in the bremsstrahlung spectrum?

13-2. A beam of 2.5-GeV protons is directed into a tank of water of refractive index 1.34. What is the limiting angle for Cerenkov radiation? What is the energy of the electrons when the production of Cerenkov radiation ceases?

13-3. The following data were taken to determine the beta-particle energy of an unknown nuclide. All counts have been corrected for background, coincidence loss, and window absorption. Use a Feather analysis to obtain the range and energy of the beta emission.

Absorber (mg cm^{-2}):	0	50	100	150	200	250
Bi-210, cpm:	9,450	5,760	3,080	1,640	831	411
Unknown, cpm:	12,600	9,970	7,100	4,850	2,660	2,060

* *Journal of Nuclear Medicine*, Supplement No. 1, February, 1968; No. 2, March, 1969; No. 3, August, 1969.

Absorber ($mg\ cm^{-2}$):	300	350	400	450	500	550	600
Bi-210, cpm:	180	83	38.7	—	—	—	—
Unknown, cpm:	1,210	771	382	205	107	66.6	35.2

13-4. The following data, taken with aluminum absorbers, appear to show both a beta and a gamma component. Find the no-absorption count rate of the beta component, the half-value thickness for the gamma component, and the mass absorption coefficient.

Absorber ($mg\ cm^{-2}$):	2	5	8	10	15	20
Corrected cpm:	3,760	1,445	785	571	352	274

Absorber ($mg\ cm^{-2}$):	30	40	50	60	70
Corrected cpm:	232	217	207	193	180

13-5. A patient weighing 65 Kg, height 150 cm, is given a 12-μCi diagnostic dose of ^{131}I. It is estimated that the thyroid gland weighs 22 g and will take up 55 per cent of the administered dose. Assume a spherical thyroid gland and a biological half-life of 150 days, and calculate the beta- and gamma-ray doses to the thyroid for the complete decay of the isotope.

13-6. Sodium distributes promptly and uniformly throughout the body. Assume a tracer dose of 25 μCi of ^{24}Na given to the patient of Prob. 13-5, and calculate the beta- and gamma-ray doses for the complete disintegration of the isotope. How much of this dose will be received during the first 24 hours after administration?

13-7. Cesium distributes quickly and uniformly, in much the same manner as sodium. Assume that a 70-Kg, 140-cm man ingests 6 nCi of ^{137}Cs at the age of 20 years. Calculate the tissue dose received up to the age of 70 years. What is the dosage overestimate if the calculation is based on the complete disintegration of the isotope? $T_B = 150$ d.

13-8. Assume a patient with a thyroid gland estimated to weigh 21 g that has been shown by a tracer experiment to take up 40 per cent of a dose of iodine. Calculate the amount of ^{131}I that must be administered to deliver a therapeutic dose of 1800 rads to the thyroid in 72 hours. What total dose will be received by the thyroid?

13-9. A 68-Kg man ingested 2 μCi of ^{14}C in a laboratory accident. Excretion data for the compound in question are unknown. Assume no excretion and uniform distribution throughout the tissues and calculate the tissue dose for the 40 years of life estimated to remain for this individual.

13-10. In a laboratory accident, 1.5 μCi of ^{32}P was spilled over a 6-cm^2 area on the back of the hand. Estimate the tissue dose resulting from the complete disintegration of the nuclide.

13-11. A patient has received a therapeutic dose of 100 mCi of ^{198}Au, which remains in the body as an essentially point source, with no excretion. Work out a schedule of allowable exposure times to be allowed on the succeeding 8-hr nursing shifts. Assume a maximum allowable dose of 50 mR per shift, and an average working distance of 50 cm from the patient.

14

Nuclear Reactions

14.01 Artificial Radioactivity

Rutherford produced the first manmade nuclear reaction in 1919, when he bombarded nitrogen gas with alpha particles. The reaction product was a stable, naturally occurring nuclide whose existence could only be inferred from the relations observed between the bombarding particles and the ejected protons, Fig. 14-1. The reaction produced by Rutherford is

$$^{14}_{7}\mathrm{N} + {}^{4}_{2}\mathrm{He} \longrightarrow {}^{17}_{8}\mathrm{O} + {}^{1}_{1}p \tag{14-1}$$

In the usual notation this will be written as $^{14}\mathrm{N}(\alpha, p)^{17}\mathrm{O}$, one example of an (α, p) reaction.

In 1934, Irene Joliot-Curie and her husband, F. Joliot, produced the first nuclear reaction with a radioactive product nuclide. The Joliots were bombarding aluminum with alpha particles from polonium and were studying the radiations. They observed that positron emission continued after the polonium source was removed. The exponential decrease in the positron activity with time suggested a radioactive decay process.

The reaction initiated by the Joliots is

$$^{27}_{13}\mathrm{Al} + {}^{4}_{2}\mathrm{He} \longrightarrow {}^{30}_{15}\mathrm{P} + {}^{1}_{0}n \tag{14-2}$$

The positron activity was associated with the artificially produced radioactive nuclide $^{30}\mathrm{P}$, which decayed according to

$$^{30}\mathrm{P} \xrightarrow{2.5\ \mathrm{m}} {}^{30}\mathrm{Si} + \beta^{+} \tag{14-3}$$

The study of nuclear reactions is greatly facilitated in those cases where an easily detectable radioactive product is formed. All types of available accel-

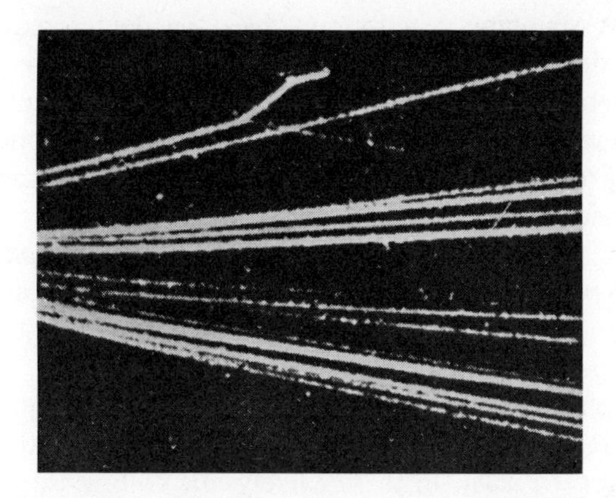

Figure 14-1. Cloud chamber photograph of a $^{14}N(\alpha, p)\,^{17}O$ reaction, showing the track of an incoming and recoiling alpha particle, and the thinner track of the ejected proton.

erators were promptly brought to bear on the problem, and by 1940 over 500 new radioactive nuclides had been produced. Today, over 1000 have been produced and identified.

14.02 Nuclear Reaction Types

The Rutherford scattering experiment yielded some most important information on atomic and nuclear structure, but this type of scattering is not a nuclear reaction. In Rutherford scattering the incident particle does not encounter the nuclear field of force. The interactions are strictly between the electric field of the bombarding particle and that of the scattering nucleus. Rutherford scattering is elastic, and the target nucleus is left in an unexcited state.

At somewhat higher energies, the electric fields may interact more strongly, energy may be transferred, and the collision becomes inelastic. When energy is transferred to a nucleus to raise it from its ground state, the collision is a *coulomb excitation.* Nuclear forces are still not involved. As a consequence the excitation levels are rather low, and de-excitation will invariably take place by gamma-ray emission.

At still higher energies, the bombarding particle will penetrate to a point where it encounters the short-range nuclear force. Now the collision may immediately pull one of the nucleons out of the target nucleus by the so-called *pickup reaction.* In the inverse process, a bombarding particle composed of more than one nucleon may lose one of them to the target by a *stripping reaction.*

The simplest stripping reaction involves the deuteron as the incident particle. As the deuteron approaches a nucleus, coulomb forces will repel the proton component, while the neutron experiences no force until it encounters the attractive nuclear force. The neutron may then be stripped off from its accompanying proton and enter the nucleus, while the proton continues on along a new trajectory.

All of the reactions described so far are *direct reactions*. They occur promptly as the bombarding particle sweeps past the target nucleus. Another class of reactions taking place through the entry of the particle well into the nucleus will be described in a following section.

With the high-energy accelerators now available it is quite easy to use bombarding particles whose kinetic energy is greater than the total binding energy of the target nuclei. In these cataclysmic collisions, a target nucleus will be raised to a very high state of excitation. Profound nuclear rearrangements can then take place. A shower of individual nucleons may leave the nucleus in a process known as *evaporation*. Again, several *groups* of nucleons may be ejected by the process of *spallation*. The nucleus may split into two pieces of approximately equal mass by *fission*. Fission is important enough to warrant a chapter all its own.

At high energies, many subnuclear particles may be ejected from the target. Nucleon–antinucleon pairs will appear at the appropriate energies, just as negatron–positron pairs are created at 1.02 MeV and above. Studies of nuclear reactions have provided important information on nuclear structure, but more questions have been raised than have been answered. The multitude of particles now known, if indeed they are separate entities, present a formidable challenge to the physicists who try to fit them into the scheme of things.

About 280 stable nuclides are available as targets for reactions with photons, deuterons, neutrons, alpha particles, and a variety of heavier projectiles. Even with a single target material and a single type of projectile, a number of different reactions may be induced. Consider the bombardment of the light isotope of copper, ^{63}Cu, with deuterons:

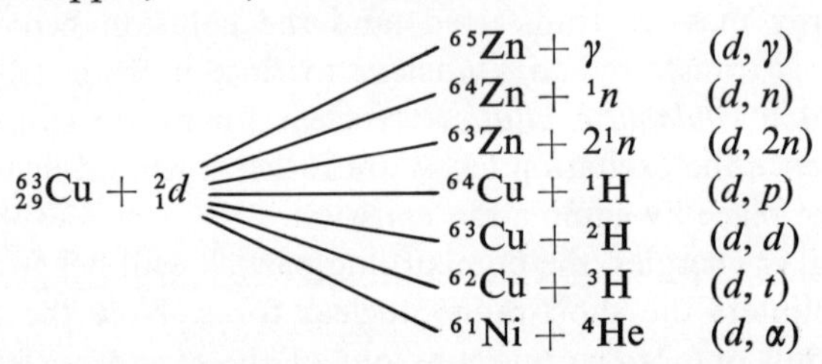

Only the (d, n) and the (d, α) reactions end in stable nuclides; all the other products are radioactive. Relative yields will depend upon the energy of the bombarding particle, because all cross-sections are energy-dependent. For

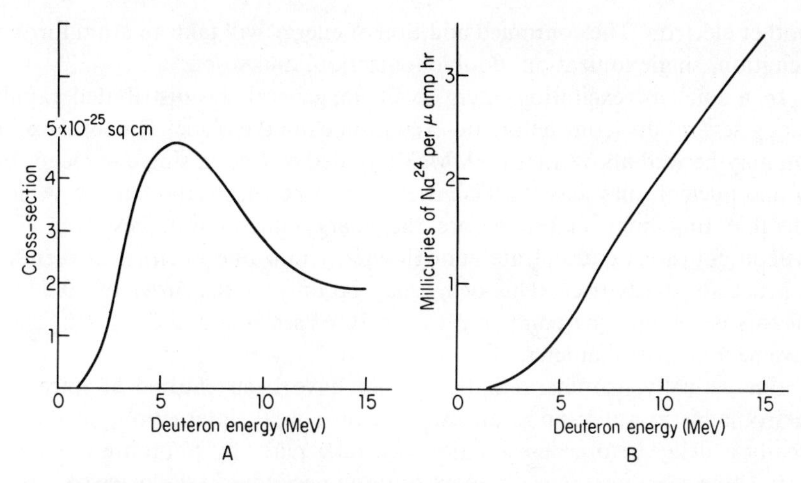

Figure 14-2. (A) Excitation function for the ^{23}Na(*d, p*) reaction. (B) Thick target yield of the ^{23}Na(*d, p*) reaction.

example, the (*d*, 2*n*) cross-section is zero at low energies but becomes appreciable at 10 MeV.

The energy at which a reaction occurs with a detectable yield is the *threshold energy*. A plot of cross-section (or reaction yield) as a function of energy is called the *excitation function*. Figure 14-2A shows the excitation function for ^{23}Na(*d, p*)^{24}Na. This yield has a threshold at 1 MeV and a maximum at 6 MeV. The decrease above 6 MeV is due to the appearance of competing reactions rather than to a decrease in the probability of the (*d, p*) process. Data for an excitation curve such as Fig. 14-2A are obtained from the bombardment of thin targets. With a thick target, yield may increase with energy, Fig. 14-2B, as the bombarding particle encounters more and more nuclei before expending all of its energy.

14.03 The Compound Nucleus

Although many direct nuclear reactions are known, most take place in a series of steps. In the first step, the incident particle penetrates the potential barrier and comes under the influence of the attractive nuclear force. At that instant the kinetic energy and the binding energy of the particle are suddenly added to the structure, which is now known as the *compound nucleus*. Because of this added energy, the compound nucleus will be formed in an excited state.

There is one important difference between atomic and nuclear excitations. We have seen how energy added to an atom may raise one of the orbital electrons into an excited state. Further additions of energy will raise the electron to higher levels, and to ionization, before energy is accepted by another electron. The continued addition of energy will take an atom through excitation, single ionization, double ionization, and so on.

In a nucleus, excitation energy will, in general, be distributed rapidly among several nucleons before de-excitation can take place. The total excitation may be well above that ($\simeq$8 MeV) needed to eject a single nucleon, but no one nucleon may have sufficient energy to penetrate the barrier. A considerable time may elapse before the energy fluctuations resulting from nucleon collisions concentrate enough energy into one nucleon to permit it to penetrate the barrier. This delay may be only of the order of 10^{-13} sec, but this is a long time compared to the 10^{-21} sec required for a particle to traverse a nuclear diameter.

Energetically, gamma radiation could have been emitted at once, but electromagnetic radiation is an example of a weak interaction, so there is usually a delay before de-excitation can take place by photon emission.

In some reactions, nuclear de-excitation does indeed take place so rapidly after particle penetration as to cast doubt that a compound nucleus is formed. These *direct interactions* do not form a class apart from those that lead to a compound nucleus; there is a gradual transition from one type to the other.

Whatever the mechanism, the excitation energy comes from the kinetic energy and the binding energy of the entering particle. Except for low-Z targets, the binding energy of a single entering nucleon will be of the order of 8 MeV. With multinucleon projectiles, some of the binding energy has already been lost in forming the particle and so is not available for excitation. In an alpha particle, for example, the energy available for excitation is about $(8 \times 4) - 28$ or only 4 MeV. The full 32 MeV would be associated with the simultaneous entry of the four individual nucleons. Even with this reduction, heavy-particle bombardment leaves the residual nucleus in a highly excited state, which leads to a variety of possible reactions.

A given compound nucleus can be formed in a variety of ways, but, no matter how formed, subsequent de-excitation will depend only upon the energy state. Thus each of the following reactions passes through the nuclide ^{7}Be, but the final products depend upon the degree of excitation present:

$$^6\text{Li}(p, \gamma)^7\text{Be}$$
$$^6\text{Li}(p, \alpha)^3\text{He}$$
$$^6\text{Li}(d, n)^7\text{Be}$$
$$^6\text{Li}(d, n)^3\text{He}^4\text{He}$$
$$^7\text{Li}(p, n)^7\text{Be}$$

14.04 Nuclear-energy Levels and Resonance

In general, the smooth excitation curve of Fig. 14-2A will have several superposed peaks corresponding to large reaction cross-sections at a series of discrete energies. At these energies, the particles emitted in nuclear de-excitation appear in high yield and we speak of these as characteristic of the reaction. Thus in $^{11}B(p, \alpha)^{8}Be$ a strong alpha emission is detected when the protons have an energy of 162 KeV. This is an example of *nuclear resonance* leading to a high reaction probability from an excited state of the compound nucleus. Note that the resonance refers to the energy of the incoming particle rather than to the outgoing particles.

Figure 14-3 represents the energy relations in a heavy nucleus. If the top of the potential well is taken as zero, the well will be filled with nucleons

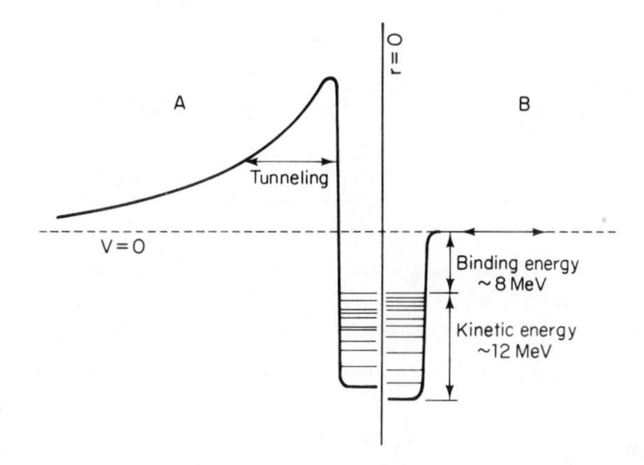

Figure 14-3. (A) Potential barrier and well in a heavy nucleus as seen by an incoming proton. (B) An incoming neutron experiences only the nuclear force.

having thermal energies up to about 8 MeV below the top, this being approximately the binding energy per nucleon. Each nucleon can exist below this only in a set of allowable levels rather than in an energy continuum.

Detailed studies of reaction yields have established nuclear-energy levels and *widths* of excited states in light nuclei ($A < 20$). In these structures, energy levels are separated by at least several KeV, permitting experimental determinations; in heavy nuclei, energy levels may be only a few eV apart.

Figure 14-4 shows some of the energy levels in the ^{14}N nucleus. Here the ground state is taken as the zero point of the energy scale. Simple mass calculations show that a proton escape according to $^{14}N \longrightarrow p + {}^{13}C$ requires that 7.54 MeV be added to the ground state of ^{14}N. Energy levels below this are *bound*, meaning that, here, de-excitation can take place only by gamma-ray emission or by internal conversion. Levels above 7.54 MeV are *virtual*. At 10.26-MeV deuteron emission, $^{14}N \longrightarrow d + {}^{12}C$ is possible, the threshold for neutron emission is at 10.54 MeV, and so on. At the higher levels of excitation, there will be many competing modes of de-excitation, for the

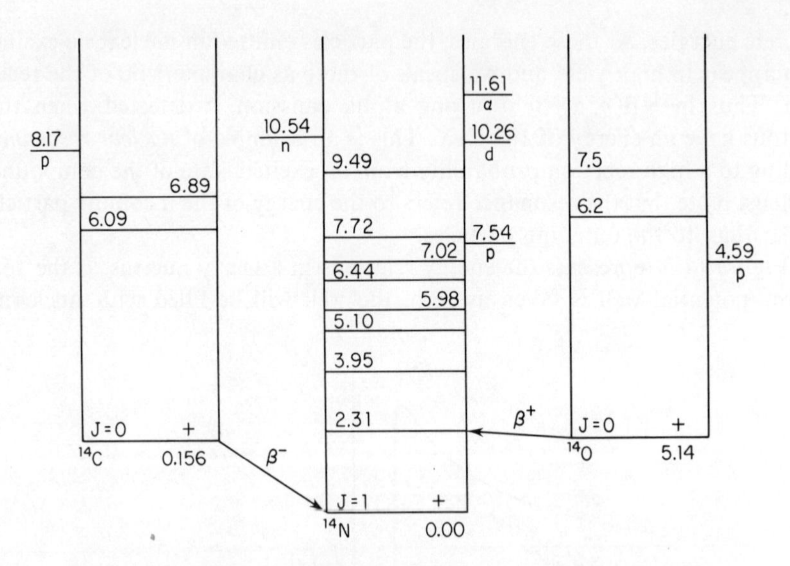

Figure 14-4. Potential energy-level diagram for ^{14}N and its neighboring isobars. Much detail has been omitted. (After T. Lauristen and F. Ajzenberg-Selove, *Am. Inst. of Physics Handbook*, p. 8–72, McGraw-Hill Co., New York, 1957.)

entrance of a new mode, or *channel*, does not exclude the use of previously existing channels.

The probability of de-excitation from a given state is obviously related to the lifetime of the state, and this, in turn, is related to the energy uncertainty or *width* of the level, through the uncertainty principle. A quantity Γ, the *width* of the energy level, or more precisely, the full-width at half-maximum, can be defined as

$$\Gamma = \hbar\lambda = \frac{\hbar}{\tau} \tag{14-4}$$

where $\tau = 1/\lambda$ is the mean lifetime of the state. It can be seen that Γ has the dimensions of energy. In our usual units of energy,

$$\Gamma = \frac{6.5 \times 10^{-16}}{\tau\ (\text{sec})}\ \text{eV} \tag{14-5}$$

If there were n_0 nuclei in an excited state at $t = 0$, these would be de-excited according to

$$n = n_0 e^{-(\Gamma/\hbar)t} \tag{14-6}$$

The total width of an energy level will be the sum of the partial widths describing each de-excitation channel. For example, if there is competition between neutron emission and gamma emission,

$$\Gamma = \Gamma_n + \Gamma_\gamma \tag{14-7}$$

Table 14-1 is an incomplete listing of the energy levels in ^{6}Li and ^{20}Ne. Note the increased number of levels and the decreased breadth for the heavier nucleus.

TABLE 14-1
NUCLEAR EXCITED STATES

^{6}Li		^{20}Ne	
E (MeV)	Γ (KeV)	E (MeV)	Γ (KeV)
2.2	22		
3.6	—		
4.5	600	13.08	0.95
5.3	100	13.19	2.8
6.6	100	13.33	2.1
7.4	1000	13.44	35
8.4	100	13.51	7.1

In heavy elements, a less specific resonance occurs at energies of about $8A^{1/3}$ MeV. This *giant resonance* is broad, sometimes extending over 4 MeV. In lighter elements, the giant resonance is found between 10–20 MeV, with no simple dependence on A. None of the giant resonances show as pronounced absorption peaks as are seen in the resonances arising from more precisely defined nuclear-energy levels.

14.05 Conservation Laws in Nuclear Reactions

Nuclear reactions obey a set of conservation laws which require a balance on the two sides of the reaction equation, much like the balances required in a chemical reaction.

1. *Mass number*. The total number of nucleons involved will be conserved.
2. *Charge*. Total electric charge will be conserved.
3. *Neutron number*. Requirements 1 and 2 imply that the total number of neutrons will also be conserved.
4. *Mass–energy*. Neither mass nor energy will be conserved individually, but the sum of mass and energy will be conserved.
5. *Linear momentum*. This will be conserved in any collision.
6. *Angular momentum*. Angular momentum will be conserved. In balancing angular momenta, account must be taken of any amount associated with gamma radiation.
7. *Statistics*. A nuclear reaction will not change the statistics that existed before reaction. They may be Fermi–Dirac or Bose–Einstein as the case may be.

8. *Parity.* Although parity does not appear to be conserved in weak interactions, no violation of parity conservation has been observed in a nuclear reaction.
9. *Isotopic spin.* The isotopic-spin quantum number remains constant in a nuclear reaction and may indeed control some selection rules governing transition probabilities.

Parameters that are not conserved in a nuclear reaction are those which depend upon some sort of geometrical arrangement or distribution. Thus electric charge will be conserved but electric quadrupole moments may not be, since this depends upon the spatial distribution of the charge.

14.06 Barrier Penetration

A charged particle attempting either to enter or to leave a nucleus will encounter the coulomb potential barrier, already discussed in connection with alpha-particle emission. An entering proton must be raised from zero energy to a point where there is a reasonable probability of tunneling through the barrier. A nuclear proton, normally existing at a potential about 8 MeV below the zero level, Fig. 14-3A, must acquire this energy in addition to that required to bring it well up on the barrier. Thus only energetic protons would be expected as products of a nuclear reaction.

Barrier heights appear twice as high to a doubly charged alpha particle which attempts to penetrate the barrier in either direction. Still higher energies are required to get more-highly-charged projectiles across the barrier. Very high excitation energies leading to spallation are required before any structure more highly charged than an alpha particle is ejected.

The potential barrier simply does not exist for the neutron. Neutrons with no more than thermal energy (0.025 eV) readily enter into the influence of the nuclear forces. In fact, many reactions are known where the probability of nuclear penetration is proportional to $1/v$, where v is the velocity of the approaching neutron. This is equivalent to saying that the probability of capture in these cases is proportonal to the length of time the neutron remains close to the target nucleus.

Because of the equality of the forces between nucleons, a nuclear neutron will also have an energy about 8 MeV below the zero level, Fig. 14-3B. This much energy must be supplied to permit neutron escape, but no more need be added to improve the probability of barrier penetration. Low-energy neutrons are, therefore, to be expected as reaction products.

A gamma ray also does not see the potential barrier, and in addition does not have any binding energy to overcome for emission. No restrictions are placed on gamma ray energies, which may have any value ranging upward from essentially zero.

An incoming gamma ray experiences no barrier to its entry into the nucleus. On the other hand, it brings to the nucleus only its own inherent energy $h\nu$. No binding energy is contributed to the target. To initiate a particle expulsion, a photon must supply at least 8–10 MeV, and preferably much more, to the target nucleus.

The net energy change, known as the Q of the reaction, is determined by the relative masses on the two sides of the reaction equation. A reaction may be *endoergic*, requiring a net energy input, or *exoergic*, from which energy will be released. The energy balance of Eq. (14-1) is

$$\begin{array}{llll} {}^{14}\text{N}: & 13{,}043.556 \text{ MeV} & {}^{17}\text{O}: & 15{,}834.318 \text{ MeV} \\ {}^{4}\text{He}: & \underline{3{,}728.337} & {}^{1}\text{H}: & \underline{\phantom{00{,}}938.767} \\ & 16{,}771.893 & & 16{,}773.085 \\ & & & \underline{-16{,}771.893} \\ & & & 1.192 \end{array}$$

The reaction Q is -1.192 MeV and is endoergic.

For the ${}^2\text{H}(d, n)$ reaction of Eq. (14-17),

$$\begin{array}{llll} {}^{2}\text{H}: & 1{,}876.092 \text{ MeV} & {}^{2}\text{He}: & 2{,}809.365 \text{ MeV} \\ {}^{2}\text{H}: & \underline{1{,}876.092} & {}^{1}n: & \underline{\phantom{0{,}}939.549} \\ & 3{,}752.184 & & 3{,}748.914 \\ & \underline{-3{,}748.914} & & \\ & 3.270 & & \end{array}$$

The reaction Q is $+3.270$ MeV and is exoergic.

An energy thus calculated applies to the reaction as a whole and not to the kinetic energy of the ejected particle. Conservation of momentum requires a sharing of energy between the ejected particle and the recoiling nucleus. In the case of ${}^2\text{H}(d, n)$, the mass of the recoiling nucleus is only three times that of the ejected particle, and the reaction Q will divide between the two kinetic energies in a 1 : 3 ratio.

If tables of mass defects, Sec. 7.02, are available, calculations of reaction Qs can be made without using total-mass values. Since nucleons are conserved in nuclear reactions, mass differences may be used directly without involving the total atomic masses. Tabulated values of Δ for the ${}^{14}\text{N}(\alpha, p)$ reaction are*

$$\begin{array}{llll} {}^{14}\text{N}: & 2.8637 \text{ MeV} & {}^{1}\text{H}: & 7.2890 \text{ MeV} \\ {}^{4}\text{He}: & \underline{2.4248} & {}^{17}\text{O}: & \underline{-0.8080} \\ & 5.2885 & & 6.481 \\ & & & \underline{-5.289} \\ & & & 1.192 \end{array}$$

The reaction is endoergic by 1.192 MeV, as before.

* C. M. Lederer, J. M. Hollander, and I. Perlman, *Table of Isotopes*, 6th ed., John Wiley & Sons, New York, 1968.

The amount of product formed, or the yield of the reaction, depends upon the number of incident particles, the number of accessible target nuclei, and the reaction cross-section, which is a measure of the probability of barrier penetration. Consider a beam of N particles per second incident upon a target cross-section of 1 cm^2. Let there be n target nuclei per cm^3 and an effective target thickness d. In a thick target, d will be the depth at which the incident particles still have a good chance of barrier penetration. Then in a short interval of time dt, the number of nuclear reactions will certainly be proportional to the total number of projectiles, Ndt, and to the total number of targets, nd. Then we have for the number of reactions, dR,

$$dR = \sigma Nnd\,dt \qquad (14\text{-}8)$$

The constant of proportionality σ has the dimensions of an area (cm^2), and hence is the *cross-section* for the reaction in question.

Reaction cross-section depends upon incident-particle energy and the height and thickness of the potential-energy barrier rather than upon the geometrical area of the target nuclei. In many cases, however, reaction cross-sections are of the same order of magnitude as nuclear areas (10^{-24} cm^2). Because of this, it is customary to express cross-sections in *barns* (b), where 1 barn $\equiv 10^{-24}$ cm^2. An obvious extension is the *millibarn* (mb).

Reaction cross-sections are so small that depletion of n in Eq. (14-8) can be neglected for most usual bombardment times and intensities. An integration then gives

$$R = \sigma Nntd \qquad (14\text{-}9)$$

Obviously, values of σ will depend upon the energy of the bombarding particles as well as upon the characteristics of the potential barrier presented to them.

If the reaction product is radioactive, there will be some decay during the bombardment. Then Eq. (14-8) becomes

$$dR = (\sigma Nnd - \lambda R)\,dt \qquad (14\text{-}10)$$

which integrates to give

$$R = \frac{\sigma Nnd}{\lambda}(1 - e^{-\lambda t}) \qquad (14\text{-}11)$$

where λ is the decay constant of the active product. The number of product nuclei given by Eq. (14-11) represents the situation at the end of a bombardment time of t seconds. Product formation ceases at time t, while radioactive decay continues at its regular rate. At any time t' after the end of the bombardment,

$$R = \frac{\sigma Nnd}{\lambda}(1 - e^{-\lambda t})e^{-\lambda t'} \qquad (14\text{-}12)$$

and the activity A will be, in disintegrations per second,

$$A = R\lambda = \sigma Nnd(1 - e^{-\lambda t})e^{-\lambda t'} \tag{14-13}$$

14.07 Nuclear Isomerism

De-excitation by gamma-ray emission usually takes place in less than 10^{-13} sec after creation of the excited state. In a good many cases, however, the gamma rays are delayed, the excited state decaying with half-lives ranging from 10^{-9} s to several months. The excited state is then said to be *metastable*, meaning "one with a measurable half-life." The excited and de-excited states are *nuclear isomers*. Obviously the distinction between prompt de-excitation and a metastable state is hazy, depending somewhat on the available measuring techniques.

One of the first cases of nuclear isomerism studied was that of bromine, whose decay scheme is shown in Fig. 14-5. When natural bromine was bombarded with neutrons, two isotopes, ^{80}Br and ^{82}Br, were produced by (n, γ) reactions. Three transitions with half-lives of 18 m, 4.4 h, and 35 h were observed but not identified. A bombardment of natural bromine with high-energy photons produced ^{80}Br and ^{82}Br by (γ, n). In this case, the ^{80}Br was formed from the heavier natural isotope rather than from the lighter. The 18-m and the 4.4-h half-lives were again observed, and so are attributable to ^{80}Br. The 35-h activity, not seen in the photon bombardment, must be due to ^{82}Br. The case of ^{80}Br is a specific example of the general rule that the properties of a reaction product are independent of the reaction type that produced it.

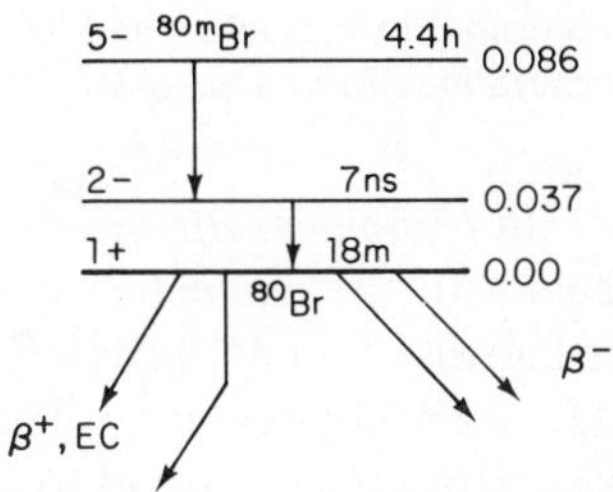

Figure 14-5. Metastable ^{80m}Br decays to the ground state by internal transition followed by two modes of decay.

14.08 Photodisintegration of the Nucleus

Photons with a quantum energy greater than the binding energy of a nucleon are capable of producing a *photodisintegration* of the absorbing nucleus. Except for ^{2}H and ^{9}Be, photodisintegration has a threshold of about 8 MeV. Cross-sections for photon-induced reactions have a sharp threshold, rise to a maximum at 3–6 MeV above threshold, and decrease slowly at high energies. Cross-sections are small under all circumstances, usually being measured in millibarns (Fig. 14-6).

At moderate energies, simple reactions like (γ, n), (γ, p), and (γ, γ) (a

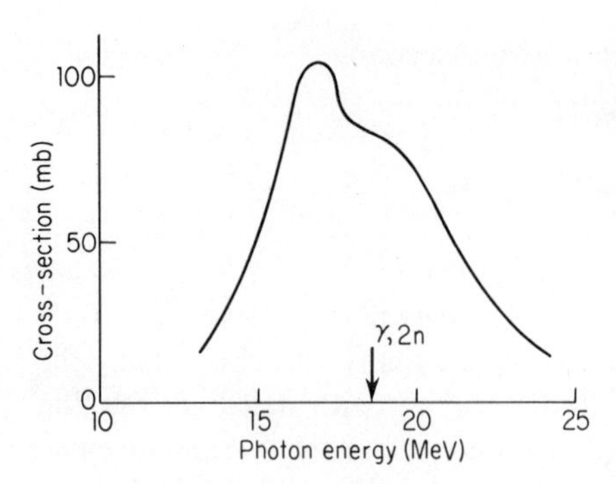

Figure 14-6. Total cross-sections for the (γ, n) and $(\gamma, 2n)$ reactions in Co. The shoulder is presumably due to a splitting of the giant resonance region. (From P. A. Fluornoy, R. S. Tinkle, and W. D. Whitehead, *Phys. Rev.*, **120**, 1424, 1960).

simple nuclear excitation) predominate. When excitation energies of 10 MeV or more are available, more complex reactions are initiated. These include (γ, np), $(\gamma, 2n)$, and $(\gamma, \alpha n)$. *Photofission*, or (γ, f), is also observed in some fissionable elements.

The (γ, n) reaction in deuterium is of particular interest because the binding energy involved is only that between two nucleons. Careful measurements have shown the threshold of ^2H (γ, n) to be 2.226 MeV, with a maximum cross-section of 2.4 mb at 5 MeV. The disruption of the deuteron takes place by either of two mechanisms.

In *photomagnetic* disintegration, the magnetic vector of the photon interacts with the magnetic dipole moments of the proton and neutron in what is known as an M_1 reaction. Interactions with quadrupole nuclear moments, M_2, are negligibly small. The cross-section for M_1 disintegration of deuterium rises from the threshold to a maximum at about 2.4 MeV and then decreases to become negligible at 4 MeV.

Photoelectric disintegration of the deuteron is pictured classically as the interaction of the electric vector of the photon with the electric field of the proton. Only the dipole interaction E_1 is large enough to be significant. The E_1 interaction is the most important component of the photodisintegration cross-section at all photon energies.

The threshold of the ^9Be (γ, n) reaction has been established at 1.666 MeV. This reaction provides a useful low-intensity source of monoenergetic neutrons if a monoenergetic source of photons is used. The sharp threshold for the reaction makes it a useful calibration point in the low-MeV region.

14.09 Proton-induced Reactions

In general, light elements emit gamma rays following proton bombardment (p, γ), and the yield curves show many sharp intensity peaks, Fig. 14-7. These peaks correspond, of course, to resonance levels in the target nucleus.

One proton reaction of unusual interest is ^7Li (p, γ) ^8Be, which has a strong resonance at 0.477 MeV and a Q of 17.2 MeV. This reaction has provided a useful source of energetic and nearly monochromatic gamma rays.

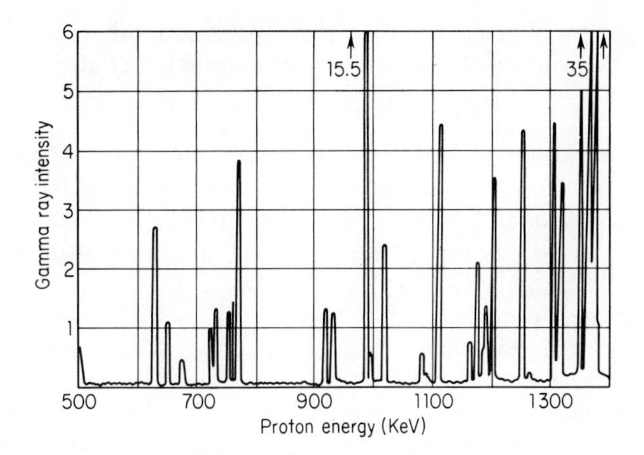

Figure 14-7. Yield of the $^{27}Al(p, \gamma)$ ^{28}Si reaction as measured by the gamma-ray energies.

Proton bombardment of the light isotope of lithium yields an interesting (p, α) reaction in which both reaction products are helium nuclei:

$$^{6}_{3}Li + ^{1}_{1}H \longrightarrow ^{4}_{2}He + ^{3}_{2}He + Q \tag{14-14}$$

Measurement of the range of the helium nuclei in a cloud chamber and conversion of the ranges to energies permits evaluation of Q in the reaction. Generally speaking, (p, α) reactions give high yield and result in nuclei which are stable, the above case being no exception, since ^{3}He occurs naturally to the extent of 1.3×10^{-4} per cent of that of ^{4}He. Proton bombardment of the light elements is most favorable to alpha emission, since, in the nuclei of higher atomic number, the coulomb repulsion presented to the incoming proton is greater. However, since both the incoming proton and the outgoing alpha particle must face a coulomb potential barrier, the (p, α) reaction is less favored than in the case where the outgoing particle is uncharged, i.e., the (p, n) reaction.

In considering the (p, n) type of reaction, we may regard it as a substitution of a proton for a neutron in the nucleus. Products of (p, n) reactions which are radioactive are unstable for the same reason that applies to (n, p) reactions. Since one pair of neighboring isobars must be radioactive, a reaction creating an isobar of a stable target isotope must result in a radioactive product. Where the target isotope is radioactive (as in the case of ^{14}C), a stable product ^{14}N results; this reaction has been observed. Reactions of the (p, n) type are characterized by a definite threshold and a high yield. These reactions, as do virtually all nuclear reactions, afford a means of measuring nuclear masses to a high degree of precision. If, for example, the mass of ^{44}Ca is measured precisely with a mass spectrograph but the ^{44}Sc mass is not so well known, we can obtain it by measuring the Q of the reaction:

$$^{44}_{20}\text{Ca} + ^{1}_{1}\text{H} \longrightarrow ^{44}_{21}\text{Sc} + ^{1}_{0}n + Q \tag{14-15}$$

Proton-induced reactions that result in neutron emission generally produce product nuclei which are positron emitters. Since the (p, n) reaction simply substitutes a proton for a neutron in the daughter nucleus and thereby increases the proton excess, we can understand why positron emission is the favored decay scheme.

Very energetic protons, such as those accelerated in synchrocyclotrons, add sufficient excitation energy to a nucleus to cause the emission of more than a single particle. Obviously, where the excitation energy is greater than, say, 50 MeV, one would expect multiple particle emission, since this value far exceeds the binding energy of a single nucleon. For such high-energy reactions, the concept of the compound nucleus has less well-defined meaning, for a high-energy particle striking a nucleus may not share its energy as we have postulated in the case of the formation of the compound nucleus. Instead, if the distance which a high-energy particle travels inside the nucleus before making a collision is comparable with the nuclear diameter, then the incoming particle may not be captured by the nucleus. It may leave the nucleus and in the encounter transfer only a fraction of its energy to it. The excited nucleus then gives off a number of particles by a "boiling-off" process, in which we may think of the nucleus as being heated up by the excitation to the point where individual nucleons evaporate from the nucleus.

14.10 Deuteron Reactions

When a deuteron approaches the field of a target nucleus, it becomes polarized from the coulomb repulsion acting on the proton component. With the polarized orientation, there is a high probability of a stripping reaction, with the neutron breaking the 2.2-MeV bond to the proton to enter the target nucleus. One would expect deuteron-induced reactions to parallel those produced by neutrons, since (d, p) and (n, γ) lead to the same product. Over 150 (d, p) reactions are known, and many of these have large cross-sections.

Both (d, p) and (n, γ) reactions yield a product which is an isotope of the target material:

$$^{31}_{15}\text{P} + ^{2}_{1}\text{H} \longrightarrow ^{32}_{15}\text{P} + ^{1}_{1}p + Q \tag{14-16}$$

In these cases, product and target cannot be separated chemically, and the product will have a low specific activity from dilution with unreacted target atoms.

Other types of deuteron reactions are known. A (d, n) reaction on a deuterium target (perhaps in the form of heavy-water ice or a hydride such as ZrD) may be used to produce neutrons by

$$^{2}_{1}H + ^{2}_{1}H \longrightarrow ^{3}_{2}He + ^{1}_{0}n + Q \qquad (14\text{-}17)$$

This reaction is useful, since it is exothermic by 3.26 MeV and since only modest deuteron energies are required to make the reaction go. Even more popular is the (*d*, *n*) reaction on a tritium target. Here the Q of the reaction is +17.6 MeV, which results in neutrons of about 14 MeV. The reaction is

$$^{2}_{1}H + ^{3}_{1}H \longrightarrow ^{4}_{2}He + ^{1}_{0}n + Q \qquad (14\text{-}18)$$

Deuterons also induce (d, α) reactions such as

$$^{37}_{17}Cl + ^{2}_{1}H \longrightarrow ^{35}_{16}S + \alpha + Q \qquad (14\text{-}19)$$

but these reactions tend to have smaller cross-sections than the (*d*, *p*)'s because of the higher potential barrier presented to the escaping alpha particle. Both (*d*, *n*) and (d, α) reactions yield products that are chemically separable from the target, so carrier-free samples can be obtained from them.

The deuteron seldom appears as the particle emitted in a reaction. Only one of the reactions listed in Sec. 14.02 shows a deuteron ejection, and this one is really a case of deuteron scattering rather than the formation of a compound nucleus with a subsequent deuteron emission.

14.11 Alpha-particle Reactions

Alpha-particle reactions were the first to be studied because of the early availability of sources of energetic particles from members of the natural radioactive series. We have already noted that an alpha-induced reaction led to the discovery of artificial radioactivity; alpha-induced reactions also led to the discovery of the neutron. The development of particle accelerators, particularly the cyclotron, made available intense sources of high-energy alphas and greatly extended the previous limits of alpha-particle reactions.

The nuclear potential barrier presents a formidable (about 16-MeV) obstacle to the doubly charged alpha particle. Because of this barrier, early studies were confined to light nuclei. For some purposes, neutron sources are conveniently made by an (α, n) reaction in a light nucleus such as Be. A well-prepared Rn–Be source will produce about 1.5×10^{7} neutrons per curie.

At energies above that binding two neutrons to the nucleus, (α, $2n$) reactions are observed; at still higher energies, (α, $3n$) becomes prominent. Figure 14-8 shows the excitation functions of alphas on indium-115. Up to about 16 MeV, the shape of the excitation curve is determined by the increasing probability of penetration through the potential barrier. Above this energy, passage over the barrier occurs, the yields then being determined by the details of the nuclear interactions.

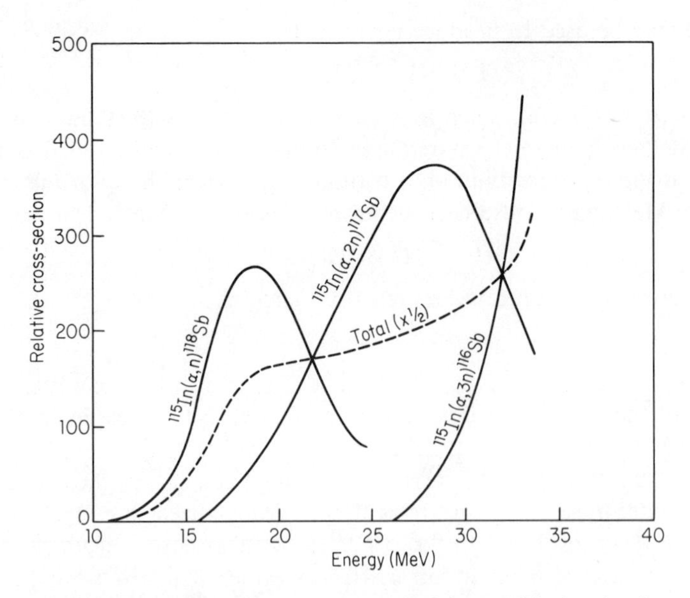

Figure 14-8. Excitation functions for the alpha-particle bombardment of ^{115}In. (After G. M. Tenner, *Phys. Rev.*, **76**, 424, 1949.)

The decrease in the ^{115}In(α, n) yield with energy is due to the onset of (α, $2n$), which decreases in turn as (α, $3n$) becomes prominent. Total cross-section, Fig. 14-8, continues to increase with energy. Note that this total-yield curve shows no resonance absorptions comparable to those seen in Fig. 14-7. In the heavy ^{115}In nucleus, the resonance levels are too closely spaced for resolution by ordinary cross-section measurements.

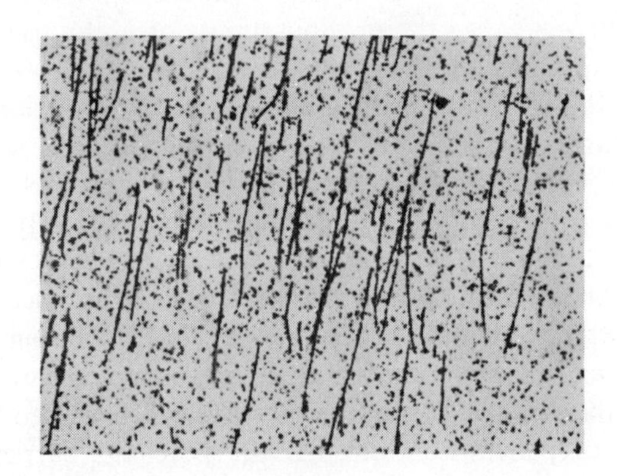

Figure 14-9. Proton tracks in a nuclear emulsion from the reaction ^{27}Al (α, p) ^{30}Si. (Courtesy J. H. Webb.)

(α, p) reactions are also observed, Fig. 14-9, since the escaping proton sees only one-half of the potential barrier presented to the alpha particle.

14.12 Heavy-particle Reactions

High-energy bombardments with heavy particles have produced new types of nuclear reactions. Not only do heavy particles raise the target or the compound nucleus to high levels of excitation, but they also impart very large angular momenta to the excited structure. This leads, among other things, to the production of metastable states with large values of angular momentum. Strong gamma radiations would be anticipated from these highly forbidden states.

Single or multiple *nucleon transfer* is observed in heavy-particle reactions usually with cross-sections of a few millibarns. Nitrogen ions are favorite projectiles, since the reaction product ^{13}N is readily identified by its 13-m positron decay. In one (n, n) transfer, ^{26}Mg(^{14}N, ^{13}N)^{27}Mg, both products are radioactive. Two radioactive products are also formed in the (p, n) transfer ^{27}Al(^{14}N, ^{14}O)^{27}Mg. Complex transfer reactions such as ($2p$, $3n$) have been observed; however, interpretation of these is somewhat uncertain.

A compound nucleus may be formed by heavy-particle bombardment, usually in a highly excited state with a large angular momentum. De-excitation may proceed by the *evaporation* of a series of nucleons until the ground state is reached or until an isomeric state can be relieved by gamma emission.

A *fragmentation* mechanism has been proposed to explain some heavy-particle reactions. According to this mechanism, the incoming projectile may break up or fragment into small clumps or individual nucleons. Some of these are absorbed into the target nucleus, while the rest proceed either as individual particles or as a coalesced unit. Experimental distinction between these unabsorbed particles and those emitted by an excited nucleus is difficult.

Fission is readily induced by heavy ions. A typical reaction might be ^{238}U(^{14}N, f), where f indicates two fission fragments of equal or nearly equal mass number. A fission reaction cannot be written explicitly, since there is no single reaction but only a distribution of fragment masses around most probable values.

Heavy ions have been most useful in producing heavy transuranium elements $Z > 100$, which are too unstable to exist in nature. Some of these reactions will be discussed in detail in a later chapter.

14.13 Reactions at High Energy

Nuclear reactions at extreme energy are of considerable interest because at these energies, the de Broglie wavelength of the bombarding particle is of the same order, or smaller than, the range of the nuclear forces. Scattering experiments done with these projectiles reveal details of nuclear force fields not determinable with particles at lower energy, and hence less localized in space.

There is no sharp distinction between high- and low-energy reactions, but for convenience we may think of a dividing line at incident particle energies of 100 MeV. In general, the higher-energy reactions are characterized by the wide range of mass numbers found in the products. At low energies, the products tend to consist of one or a few single nucleons and a residual nucleus of mass only slightly less than that of the original target. At high energies, products with all mass numbers lower than that of the target are found with substantial yields.

Four types of high-energy reactions are recognized:

1. *Spallation*, in which single nucleons or small groups are emitted from the target. With a target nucleus of mass number A, residual nuclei of 0.75–$0.99A$ are formed as a result of spallation.
2. *Fission*, where a few single nucleons are emitted and the remaining structure then splits into two approximately equal parts. The residual nuclei will have mass numbers between 0.30 and $0.65A$.
3. *Fragmentation*, where nucleon groups of $10 < A < 40$ are blasted from the target nucleus. The residual structure probably remains as a single nucleus.
4. *Secondary reactions*, initiated by particles produced in a primary interaction, usually of type 1. These secondary reactions include all of those possible with low-energy particles.

In the initial stage of a high-energy reaction, the incoming projectile undergoes one or more two-body collisions, or *knock-on processes*, with individual target nucleons. At very high bombarding energies, each struck nucleon may be considered to be free, and the collision treated as a simple two-body collision. The struck nucleons, which may themselves undergo one or more encounters before leaving the target nucleus, are known as *knock-on particles*. Following the knock-on process, de-excitation proceeds by the evaporation of one or more nucleons and by gamma-ray emission. Figure 14-10 shows *stars* produced in a cloud chamber by the products of nuclear evaporation following a high-energy interaction.

In actual fact it is quite difficult to make sharp distinctions between the various types of high-energy reactions. Not only is the available energy very high, sometimes exceeding the total binding energy of the target nucleus, but large kinetic energies can be imparted to the knock-on particles. Many of the high-energy processes appear to amount to a temporary disassembly of the original structure, with a reassembly according to microscopic details of the nucleon–nucleon interactions. A complication arises from the fact that a given nucleon does not have a fixed identity as neutron or

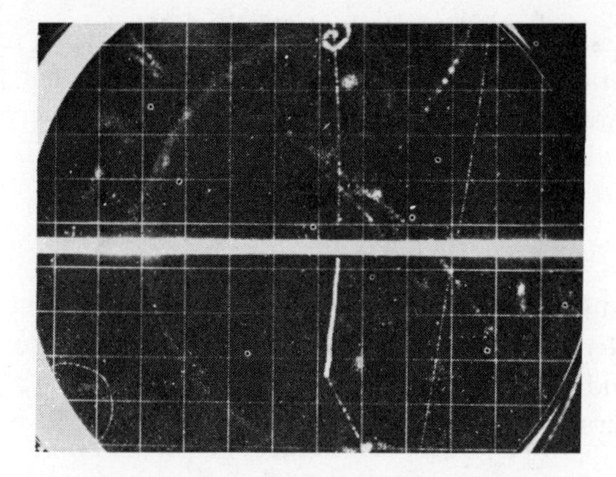

Figure 14-10. Cloud chamber photograph of a star formed in a nuclear evaporation induced by a 90-MeV neutron. (Courtesy W. M. Powell.)

proton, but changes with changes in the associated mesons. Much remains to be learned about high-energy reactions.

14.14 Particle Production at High Energies

In 1935, Yukawa predicted the existence of a particle of about 200 electron masses to serve as the nuclear-exchange particle. Such a particle was soon detected as one component of the cosmic radiation. However, it quickly became evident that this μ-meson, or muon, could not be the particle required by Yukawa. The nuclear-exchange particle must react strongly with nucleons; muons are observed deep in mines after having been formed in our atmosphere and subsequently penetrating hundreds of meters of earth without a nuclear interaction.

In 1948, Gardner and Lattes announced the production of π-mesons, or pions, from the bombardment of a carbon target with 380-MeV alpha particles. The pions originated in nucleon–nucleon reactions:

$$p + p \rightarrow d + \pi^+$$

$$p + p \rightarrow p + n + \pi^+$$

$$p + n \rightarrow n + n + \pi^+$$

$$p + n \rightarrow p + p + \pi^-$$

The pions so produced were found to have properties identical with those of pions in the naturally occurring cosmic radiation. Free charged pions decay rapidly to the corresponding muons, Table 14-2. They interact strongly with nucleons, and today there is no doubt that they are indeed the nuclear-exchange particle.

Particle masses can be determined from measurements on the magnetic deflections of tracks in expansion chambers or nuclear emulsions. Nuclear reactions also are useful for some mass determinations. Charged pion masses can be calculated from the strong nucleon reactions,

$$\pi^- + p \longrightarrow n + \gamma \tag{14-20}$$

$$\pi^+ + n \longrightarrow p + \gamma \tag{14-21}$$

from a knowledge of the nucleon masses and a measurement of the gamma-ray energies. Neutral pions must also exist, in order to account for p–p and n–n binding, and these pions are known. A competing π^-–p interaction can be used to determine the mass of π^0.

$$\pi^- + p \longrightarrow n + \pi^0 \tag{14-22}$$

As accelerator energies increased and track chamber techniques developed, more and more particles were blasted out of nuclei and were identified. Table 14-2 is an abridged listing which omits some of the less common particles and many of the infrequent modes of decay.

The particles may be divided into classes according to their masses and spins. Heavy particles with half-integral spins, and containing a nucleon, are known as *baryons:* Ω, Ξ, Σ, Λ, p, and n. Particles of intermediate mass and integral spin such as π and K are *mesons*. The lightest group, each with half-integral spin, are the *leptons:* e, ν, μ.

Particles heavier than nucleons are also known as *hyperons*. Hyperons, together with the K-mesons, form a group of *strange particles*. Although their mean lives are very short compared to ordinary standards, each of the strange particles exists for a far longer time than would be expected. Each can be classified by a *strangeness number*, which is comparable in some ways to the degrees of forbiddenness used to classify degrees of hindrance in radioactive decay. Various explanations for the strange behavior have been offered, but none of them can be considered to be firmly established.

Interparticle interactions are exceedingly complex and are beyond the scope of this text. Special mention must be made, however, of some of the activities of pions and muons.

At the present time, no important, unique role has been found for the muons in the nuclear domain. The muon seems to be a redundant particle, reacting only weakly before its decay. A positive muon can unite with a negatron to form an atom of *muonium*, μ^+e^-, which has a fleeting but detectable existence before its muon decays. Muonium has a hydrogenlike behavior, modified by the mass differences of the constituents. Since all muon production and decays are the result of weak interactions, parity is not conserved, as in beta decay.

Negative muons behave much like heavy electrons. In particular, a *μ-mesic atom* can be formed with a proton, $p\mu^-$.

TABLE 14-2
PARTICLES AND SOME OF THEIR PROPERTIES

Name	*Symbol*	*Mass (MeV)*	*Charge*	*Mean life (sec)*	*Spin*	*Decay*
Omega-minus	Ω^-	1672	1	10^{-10}	$\frac{3}{2}$	?
Xi-minus	Ξ^-	1321	1	10^{-10}	$\frac{1}{2}$	$\Xi^- \longrightarrow \Lambda^0 + \pi^-$
Xi-zero	Ξ^0	1314	0	3×10^{-10}	$\frac{1}{2}$	$\Xi^0 \longrightarrow \Lambda^0 + \pi^0$
Sigma-minus	Σ^-	1197	1	10^{-10}	$\frac{1}{2}$	$\Sigma^- \longrightarrow n + \pi^-$
Sigma-zero	Σ^0	1192	0	10^{-14}	$\frac{1}{2}$	$\Sigma^0 \longrightarrow \Lambda^0 + \gamma$
Sigma-plus	Σ^+	1189	1	8×10^{-11}	$\frac{1}{2}$	$\Sigma^+ \longrightarrow p + \pi^0$ $\Sigma^+ \longrightarrow n + \pi^+$
Lambda-zero	Λ^0	1115	0	2×10^{-10}	$\frac{1}{2}$	$\Lambda^0 \longrightarrow p + \pi^-$
Neutron	$n, \bar{n}$	939.5	0	1037	$\frac{1}{2}$	$n \longrightarrow p + e^- + \bar{\nu}$
Proton	p^-, p^+	938.3	1	∞	$\frac{1}{2}$	
K-zero	K^0	497.8	0	10^{-10} 5×10^{-8}	0	$K^0 \longrightarrow \pi^+ + \pi^-$ $K^0 \longrightarrow \pi^0 + \pi^0 + \pi^0$
K-meson	$K^\pm$	493.8	1	10^{-8}	0	$K^\pm \longrightarrow \mu^\pm + \nu$ $K^\pm \longrightarrow \pi^\pm + \pi^0$
Pion	$\pi^\pm$	139.6	1	3×10^{-8}	1	$\pi^\pm \longrightarrow \mu^\pm + \nu$
Neutral pion	π^0	135	0	10^{-16}	0	$\pi^0 \longrightarrow \gamma + \gamma$
Muon	$\mu^\pm$	105.6	1	2×10^{-6}	$\frac{1}{2}$	$\mu^\pm \longrightarrow e^\pm + \nu + \nu$
Electron	$e^\pm$	0.511	1	∞	$\frac{1}{2}$	
Neutrino*	$(\nu, \bar{\nu})_e$ $(\nu, \bar{\nu})_u$	0	0	∞	$\frac{1}{2}$	
Photon	γ	0	0	∞	1	

* Neutrinos associated with muon decays are slightly different from those that accompany beta-particle emission in ways that are not wholly understood.

If deuterium is present, the mesic atom can unite with it to form a mesic molecule. This molecule will be in ionic form, since the negatron will have been forced out of the structure by the heavier meson. The strong inter-nucleon forces now unite the proton and deuteron to form a nucleus of ^{3}He. Several MeV of binding energy become available, and the muon is ejected, ready to enter into another, similar reaction. The steps in the reaction are

$$^1p + \mu^- \longrightarrow p\mu^- + \underset{\downarrow e^-}{pne^-} \longrightarrow {}^3_2\text{He} + \underset{\downarrow \mu^-}{\mu^-} \qquad (14\text{-}23)$$

On the average, a negative muon will enter several reactions like (14-23) before it decays. It thus acts like a catalyst in the $p + d \longrightarrow$ He fusion reaction.

The final fate of muons is radioactive decay, with two choices available to μ^-.

$$\mu^\pm \longrightarrow e^\pm + \nu + \bar{\nu} \qquad (14\text{-}24)$$

$$p + \mu^- \longrightarrow n + \nu \qquad (14\text{-}25)$$

Figure 14-11 shows the tracks in a cloud chamber as a μ^+ decays to a positron.

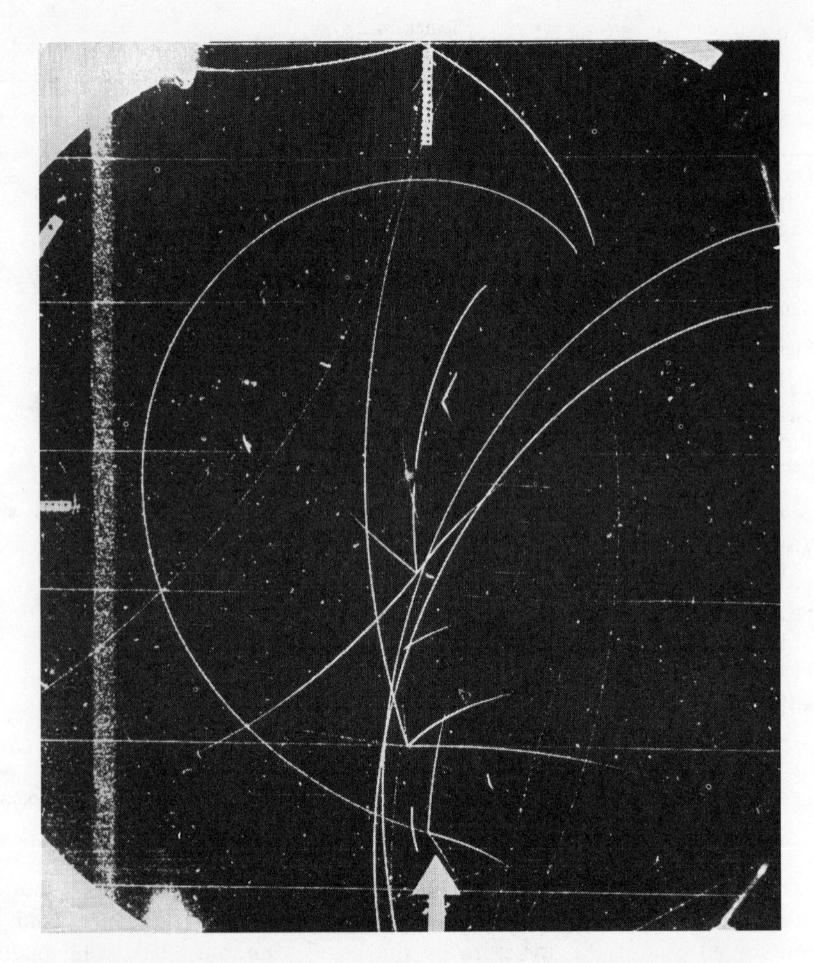

Figure 14-11. Decay of a μ^+ meson to an e^+ in a cloud chamber after passing through an aluminum plate. Note the heavy ionization along the track of the slowed meson, and the sparse ionization of the electron. (Courtesy R. W. Thompson.)

A quite different sequence of events takes place when a negative pion approaches a nucleus. This particle does not see a potential barrier but actually experiences a coulomb force of attraction as well as that from the nuclear force. It readily enters the nucleus and gives up its energy to excitation. The highly excited nucleus will then eject several heavy fragments to form an evaporation star. The heavy fragments will produce short tracks of massive ionization near the reaction site. This mechanism has been proposed as a useful technique for delivering high radiation doses to deep tissues for the radiation treatment of malignancies.

14.15 Matter and Antimatter Systems

We have seen that an electron–positron pair can be created by the destruction of a photon, provided that the latter has an energy greater than that of the mass created. This is simply a low-energy example of a general reaction whereby particle pairs are created by energy conversion. As higher energies

Figure 14-12. Bubble chamber photograph of an antiproton to antineutron conversion. The abrupt termination of the antiproton track at the arrow indicates conversion to a nonionizing particle. This is followed by a nuclear evaporation with no visible initiating particle, which must have been the antineutron. (Courtesy of E. Segrè.)

became available through the construction of mammoth particle accelerators, it became possible to create particles much heavier than the electrons.

In our experience, particles are always created in oppositely charged pairs, as would be expected if the concept of charge conservation is to be preserved. A complete duality appears to exist, with each particle having its opposite, known as its *antiparticle*. The positron is the antinegatron, the negative proton the antiproton, and so on. In most cases the particle and its antiparticle differ only in the sign of the charge. The uncharged neutrino and its antiparticle differ only in the relative spin directions, and this type of difference presumably distinguishes the neutron and the antineutron. Figure 14-12 shows the track of an antiproton, its disappearance upon conversion to an antineutron, and the nuclear evaporation that was triggered when the antineutron entered a nucleus.

Particles annihilate with their opposites when they meet at low energies, and this mutual destruction can be included as part of the definition of a particle pair. The stable particles, and hence the matter of our universe, made up of electrons, protons, and neutrons, exist only because there are no free antiparticles here. All of the antiparticles appear to be as inherently stable as those of the other system. There is no reason to believe that stable nuclei, or indeed, a stable solar system or a galaxy could not be based on the antimatter system. In such a milieu, our "stable" particles would have the fleeting existence we associate with their opposites.

If the creation of particle pairs is a universal requirement, where are the antimatter particles that must have appeared when our universe was created? If each pair was formed in close proximity, as we would expect, how did they escape annihilation before separation? Are there galaxies based on the antimatter system, and if so, where are they? At the present time we have no answers to these questions. Our space probes have found no evidence of antimatter objects, but we have probed a pitifully small portion of space as we know it.

REFERENCES

Adair, R. K. and E. C. Fowler, *Strange Particles*. Interscience Publishers, Wiley, New York, 1963.

Alfvén, H., "Antimatter and the Development of the Metagalaxy." *Rev. Modern Physics* **37,** 652, 1965.

Feinberg, G. and L. M. Lederman, "The Physics of Muons and Muon Neutrinos." *Ann. Rev. Nuc. Sci.* **13,** 431, 1963.

Segrè, E., ed., *Experimental Nuclear Physics*. Vol. II. Wiley, New York, 1953.

Swartz, C. E., *The Fundamental Particles*. Addison-Wesley, Reading, Mass., 1965.

PROBLEMS

14-1. Calculate the kinetic energy of the alpha particle ejected by a thermal neutron in $^{6}Li(n, \alpha)$.

14-2. Is the reaction $^{31}P(n, \gamma)$ feasible with thermal neutrons. If so, what is the energy of the gamma ray?

14-3. Calculate the mass defect of the neutron and the proton, expressing each in MeV and in u.

14-4. What is the minimum energy required to initiate the spallation reaction $^{109}Ag(p, 6p8n)^{96}Mo$?

14-5. Calculate the maximum kinetic energy of the electrons emitted by ^{14}C and by ^{3}H.

14-6. Calculate the maximum energy of the beta particles emitted by ^{22}Na that go to the 1.275-MeV state of ^{22}Ne.

14-7. A natural lithium target 3.8 mg cm^{-2} thick is bombarded with a 125-μA beam of 3.5-MeV protons, initiating $^{7}Li(p, n)$. Assume a reaction cross-section of 300 mb, and calculate the activity produced by a 15-minute bombardment.

14-8. Using atomic-mass values, compare the level of the excited state in the $^{12}C(d, n)$ reaction with the value shown in Fig. 14-4.

14-9. Calculate the Q value of $^{29}Si(d, p)$ from the reactions $^{30}Si(d, n)$, $^{28}Si(d, p)$, and $^{31}P(p, \alpha)$, which have Q values of +5.064, +6.251, and +1.916 MeV, respectively.

14-10. What are the gamma-ray energies associated with the charged pion reactions of Eqs. (14-20) and (14-21)?

14-11. A nuclide showing half-lives of 21 m and 5.6 d is produced by $Cr(p, n)$ and also by $Fe(d, \alpha)$. Identify the product and draw as much as you can of the decay scheme.

14-12. Which of the two possible modes of disappearance is the more energetically favorable to the μ^- meson?

14-13. The specific activity of ^{24}Na as usually made by $^{23}Na(n, \gamma)$ is sometimes too low because of the presence of unreacted ^{23}Na. Consider various particle reactions that will produce carrier-free ^{24}Na, and calculate the minimum energies needed in each case.

14-14. What is the minimum photon energy required to produce a proton–antiproton pair?

15

Charged-Particle Accelerators

15.01 The Cockcroft–Walton Circuit

Two British physicists, J. D. Cockcroft and E. T. Walton, were the first to produce a nuclear reaction with particles receiving their energy in a man-made accelerator. Their biggest obstacle was the generation of a high potential to apply to the accelerating tube. Figure 15-1 shows a diagram of the Cockcroft–Walton circuit, which was used to multiply the voltage supplied by a high-voltage transformer. The principle of operation of the circuit may be illustrated by replacing the high-vacuum tubes by switches as shown in Fig. 15-1A to C. When the high-voltage transformer swings through its positive cycle, tube T_1 is conducting (switch S_1 is closed) and condenser C_1 is charged to 100,000 volts. At each positive half-cycle, C_1 is charged, and, on each alternate negative half-cycle, switch S_1 opens and switch S_2 closes so that C_1 shares its charge with C_2. Similarly, condenser C_2 shares its charge with C_3 through the closing and opening of switches S_1 and S_3. In this way V_2 reaches a potential of +200,000 volts and V_4 becomes equal to +400,000 volts. The combination of two tubes, T_1 and T_2, together with the capacitor pair C_1–C_2, forms a single stage of the circuit and is known as a *voltage doubler*. Using a series of these voltage-doubler stages, Cockcroft and Walton produced a potential of 700,000 volts to accelerate a proton beam of about 10 μA.

Cockcroft and Walton carried out their first experiments with an accelerating voltage of about 150 KV. Hydrogen gas was admitted through a very small leak into the top end of the accelerating tube, Fig. 15-2, which was evacuated vigorously at the lower end. An arc discharge produced

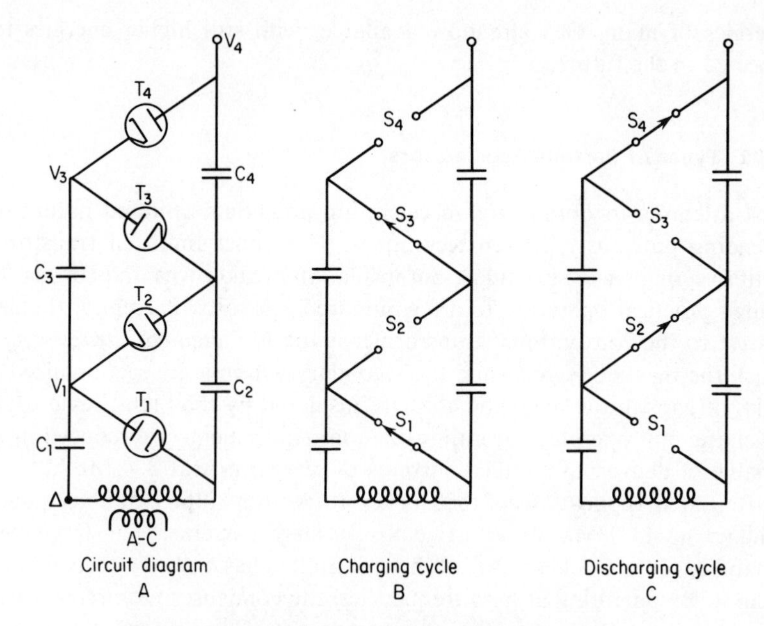

Figure 15-1. Schematic of Cockcroft–Walton circuit.

hydrogen ions, which were accelerated down the tube to strike a target of lithium oxide. The resultant nuclear reaction was

$$^{7}\text{Li} + {}^{1}\text{H} \longrightarrow {}^{4}\text{He} + {}^{4}\text{He} + Q \qquad (15\text{-}1)$$

The Q of this reaction is readily calculated from the atomic masses to be 17.347 MeV, exoergic.

Cockcroft and Walton found that the reaction yield increased with the energy of the bombarding protons. When the proton energy was 440 KeV, each alpha particle was found to have a kinetic energy of 8.8 MeV. This is in good agreement with the predicted value of $(17.35 + 0.44)/2 = 8.85$ MeV.

The success of the Cockcroft–Walton experiment brought home to physicists the great value of high, controllable energy sources for probing into the nucleus. An era of accelerator development was opened up, an era which has shown no sign of coming to a close.

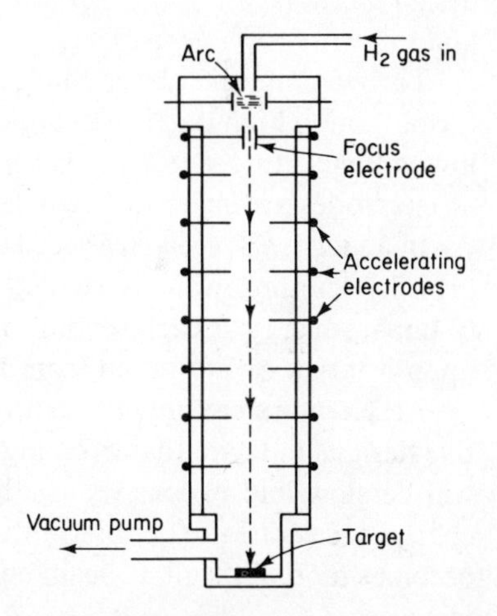

Figure 15-2. Schematic of tube for proton acceleration.

Energies of many GeV are now available, with still higher energies to be expected in the future.

15.02 Types of Particle Accelerators

First attempts to obtain high-accelerating potentials centered naturally on extending previously proven techniques. The upper limits of transformer–rectifier systems were extended, but insulation breakdowns and corona losses limited practical operation to a few hundred kilovolts. Presently, the design nearest to the conventional transformer is the *insulated-core transformer*, in which the magnetic circuit and the secondary winding are sectionalized. The series of gaps in the magnetic core are insulated by the high-pressure gas in which the unit operates. Overall insulation requirements are reduced in units capable of delivering rectified currents of several mA at 3–4 MeV.

Repeated applications of the Cockcroft–Walton voltage doubler provided voltages up to 1 MV or so at relatively large currents. The large-current capability of the Cockcroft–Walton arrangement has caused a revival of interest in it. By operating at high frequencies, the condenser requirement can be substantially reduced and the unit compacted. Currents of 10 mA appear readily attainable with this acceleration scheme.

In 1931, voltage limitations were raised by the introduction of the Van de Graaff electrostatic generator. This was a vastly improved version of the old static or "influence" machines used extensively in early days for exciting X-ray tubes. The Van de Graaff generator, like the transformer–rectifier, develops full-accelerating voltage, so large insulators are required. The design permits attaining several million volts with operation in pressure tanks.

To overcome insulator limitations, many schemes have been developed for obtaining high particle energies by the successive application of relatively low voltages. In a *linear accelerator*, the particles pass through a succession of electrodes arranged in linear fashion, acquiring energy at each electrode.

In a *magnetic-resonance* accelerator, the particles are forced into a circular orbit by an appropriately shaped magnetic field. Various methods are used to impart energy at each revolution until the particles are allowed to impinge upon a target or are taken from the machine through a thin window.

Accelerators are now operating regularly at voltages of 70 GeV. Prospective designs indicate that even higher energies can be obtained, but advances will be slow and expensive. At the high energies, radiation associated with charge acceleration (bremsstrahlung) becomes serious. Beam focusing also becomes more difficult as beam energies and, hence, acceleration path lengths increase.

15.03 Electrostatic Generators

The Van de Graaff generator has been outstanding in producing readily adjustable and highly constant energies up to about 10 MeV. The constancy of the truly *DC* output has made this machine most useful in studying nuclear-energy levels. Figure 15-3 is a schematic of the Van de Graaff principle.

A large-radius metal terminal cap is supported by an insulating column, sectionalized to obtain a uniform potential gradient along it. A fabric belt, driven by a motor at the base of the column, runs over a pulley inside the top terminal. A potential of 20–50 KV applied to inductor bar *I* causes a corona discharge at the opposing needle points *N*. This discharge "sprays" charges onto the belt, which carries them into the field-free space inside the terminal. Here the charge is led off to the terminal, which rapidly acquires a high potential. In the larger machines, a charge of opposite polarity can be sprayed onto the descending segment of the belt to increase the current-carrying capacity of the machine. The high potential thus generated can be applied to an acceleration tube, usually located inside the supporting column.

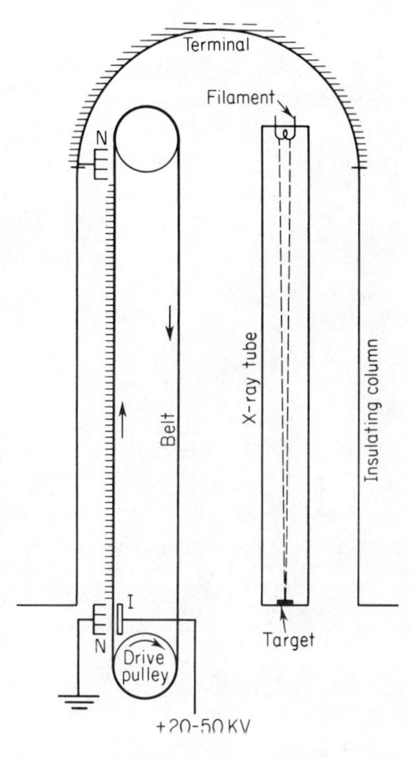

Figure 15-3. Schematic of Van de Graaff generator polarized for electron acceleration.

All surfaces at high potential must have large radii of curvature, Eq. (2-1), in order to avoid excessive corona losses or dielectric breakdown. The Van de Graaff generator is basically a constant-current device, the terminal potential being determined by the amount of charge delivered by the belt and the load imposed by the acceleration tube. Voltage control is achieved by adjusting corona points which load the generator to the desired voltage. Although Van de Graaff generators have reached 10 MV, most are designed for 5–6 MV, operating at 20 atmospheres pressure.

The development of the tandem principle abruptly doubled the potential capabilities of this type of machine. In the tandem generator, Fig. 15-4, the particles to be accelerated are injected as negative ions. These ions are accelerated to a positively charged terminal to perhaps 5 MV. At the terminal, the negative charges are stripped off, along with one or more orbital electrons. The resulting positive ions are attracted back to the negative or ground terminal, thus passing for a second time through the generator potential. Voltages of 10 MV are readily obtained; still higher values may be possible by additional charge reversals and passages through the generator potential.

Figure 15-4. Emperor tandem Van de Graaff accelerator. The accelerated ions pass through the small tube in the foreground, through an analyzing magnet, and on to the target. (Courtesy of High Voltage Engineering Corporation.)

One important feature of the Van de Graaff generator is its independence of particle mass. Accelerators based on resonances escape many of the problems associated with the insulation of extremely high potentials, but they are limited to only a few types of particles. The electrostatic generator can accelerate any charged particle, from an electron to a uranium ion. Heavy charged-particle projectiles are of particular interest in the attempts to extend the production of transuranium elements.

15.04 Linear Proton Accelerators

In the linear proton accelerator, developed by Sloan and Lawrence in 1931, the proton acquires its energy by a series of accelerations from a high-frequency oscillating voltage. A high-frequency source of several KV is applied across alternate members of a series of drift tubes. When positive ions are injected into the accelerator in proper phase, they will see a negative potential at the first drift tube and will be attracted into it. While the ions are in the field-free space inside the drift tube, the exciting voltage passes through one-half cycle. Upon emerging from the drift tube, the ions will see a negative potential on the second tube, will be accelerated into it, and so on. As the ions gain energy and velocity, longer tubes are required to maintain proper phase relations between the particles and the applied voltage.

In the original machine, Sloan and Lawrence accelerated mercury ions to 1.3 MeV in a 30-section accelerator supplied with 42 KV of radio frequency power at 10 MHz. A tube length of less than 1 meter was required.

Linear proton accelerator capabilities increased with the development of high-powered sources of radio-frequency energy. Radar transmitting tubes are capable of delivering pulses of megawatts at frequencies of 100–10,000 MHz. These pulses are applied in sequence to a series of *cavity resonators*, cylindrical tubes of proper dimensions to resonate with the driving oscillator. Under proper conditions, a traveling-wave system will be set up inside each cavity, to provide the accelerating force on the ions. Figure 15-5 shows a series of such cavities, each excited by a single-turn coil.

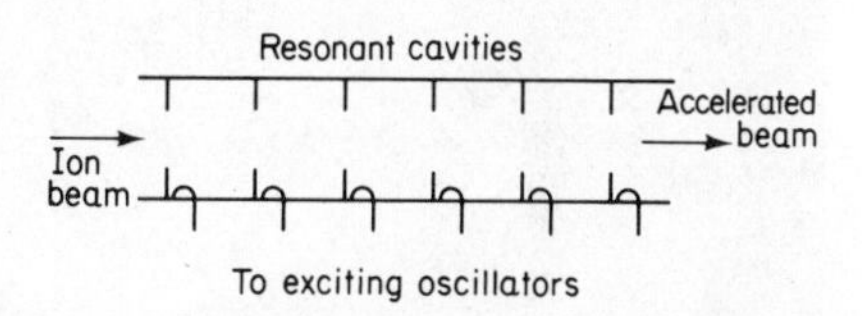

Figure 15-5. Schematic of traveling wave linear accelerator, showing the resonant cavities and the exciting coils.

One disadvantage of the linear proton accelerator is its inflexible energy output. Output energy can only be changed by adding or removing sections, which is not a simple matter. Few linear accelerators have been built for use with positive ions; in this field magnetic resonance accelerators have definite advantages.

15.05 Linear Electron Accelerators

Electron accelerators involve quite different considerations than those associated with heavy-particle acceleration. In the latter case, velocity changes substantially with energy over a wide range. A 1-MeV electron, on the other hand, has a velocity of $0.94c$, so even the largest energy increment can only increase this by less than 6 per cent. Electron masses, rather than velocities, increase with energy.

The first linear electron accelerator, or *linac*, built by Hansen in 1948, attained an energy of 6 MeV. This machine, like many of its successors, consisted of a series of wave guides adjusted to an oscillator frequency of 3000 MHz. When a pulse of r-f power is fed into a properly tuned wave guide, an electromagnetic wave will travel down the guide with a velocity determined by the guide dimensions. Electrons injected into the guide at the proper phase will be carried along by the wave, much as a surf rider is carried shoreward by an incoming water wave.

Magnetron oscillators supplied the r-f power to the early linacs. Later designs have favored the use of klystrons or Amplitrons. Tubes of this type will deliver peak powers of the order of 10 MW and average powers of 20 KW. With these powerful oscillators, pulses of 20–30-μsec duration can be obtained, if desired, and a high electron energy developed in a short wave guide. Over 600 MeV has been obtained from a design expected to produce ultimately 2 GeV.

Figure 15-6. Output end of an *L*–band linac showing the wave guides for introducing the exciting power. The accelerated beam passes through the small tube in the foreground to the target.

Early linacs operated at oscillator frequencies of about 2800 MHz in the *S* band. Operation at 1300 MHz in the *L* band permits the use of larger wave guides and acceleration tubes. The *L*-band structures will store more r-f energy and can produce higher peak currents than are possible with *S*-band linacs. Short pulses of more than 1 ampere are readily obtainable with *L*-band excitation, and average electron outputs of 30–40 KW are available (Fig. 15-6).

Linacs are becoming the machines of choice in most applications where high-energy electrons are required. The energy spread in the final beam may be as much as 10 per cent of the main-beam energy, but for many applications this is of no consequence. The energy spread can be reduced if the total-output-power requirement is not too severe.

15.06 The Cyclotron

In 1931, Lawrence and Livingston reported performance results from the first *magnetic-resonance* machine, or *cyclotron*, shown schematically in Fig. 15-7. The acceleration chamber consists of two semicylindrical *dees* (so-called because they resemble the letter D), electrically insulated from each other and separated by a small gap. This electrode system can be likened to the two halves of a pillbox, separated slightly after being cut across a diameter. The dee system is enclosed in a highly evacuated chamber with coaxial leads connecting the dees to a high-voltage, high-frequency oscillator. The dees and the vacuum chamber are mounted between the poles of a large electromagnet.

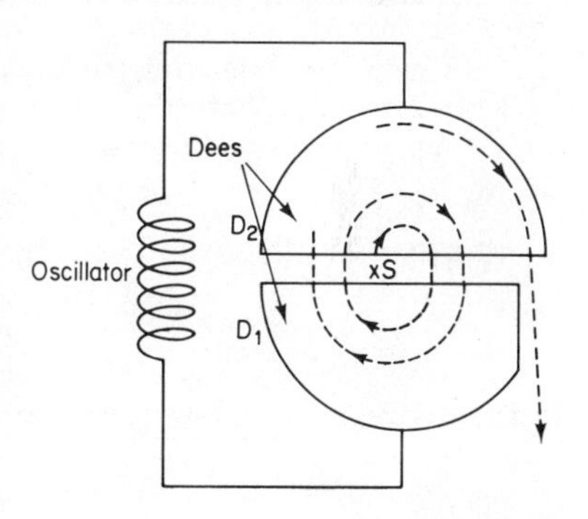

Figure 15-7. Schematic of cyclotron principle. A constant magnetic field is directed outward from the paper.

An appropriate gas such as hydrogen or deuterium is introduced through a very small leak into the center of the dees, where an arc discharge forms the corresponding ions (protons or deuterons). Consider now the fate of positive ions formed during the time when one of the dees, say D_2, is driven to a negative potential by the oscillator. Some of the ions will be pulled into the interior of D_2, which will be free of electric field because of the presence of the surrounding metal. The uniform magnetic field will exist in this region, at right angles to the velocity vector of the ions. The latter will be forced into a circular path that leads, after a deflection of 180°, into the gap between the two dees. If D_1 is now negative to D_2, the ions will be accelerated across the gap, will enter D_1, to be bent into a somewhat larger radius than before.

Repeated accelerations across the gap will cause the ions to travel in an expanding spiral path until the outer radius of the dee is reached. Here an auxiliary field can deflect the ion beam onto an internal target or onto a thin window for external bombardment (Fig. 15-8). The cyclotron is a constant-frequency device, since the extra path lengths at the outer portions of the spiral are just compensated by the increased velocities.

Figure 15-8. Ionization produced in room air by a deuteron beam emerging from a cyclotron through a thin metal foil window. (Courtesy of Radiation Laboratory, University of California.)

In usual notation, the basic cyclotron relation is Eq. (4-14):

$$Bev = \frac{mv^2}{\rho} \tag{15-1}$$

in the nonrelativistic energy range. The ion will move in a radius

$$\rho = \frac{mv}{Be} \tag{15-2}$$

For effective cyclotron action, each ion path must deflect through an angle of 180° or π radians while the high-frequency oscillator changes phase by one half-cycle. This is equivalent to requiring that the angular velocity of the ions $\omega = v/\rho$ be equal to the angular velocity of the oscillator, which is $2\pi f$. Then,

$$2\pi f = \frac{v}{\rho} \tag{15-3}$$

where f is expressed in hertz. This requirement gives

$$f = \frac{Be}{2\pi m} \tag{15-4}$$

Since f is independent of v, the cyclotron condition will be satisfied by a constant-frequency oscillator as long as the particle mass m remains constant.

The kinetic energy of the particles is

$$T = \frac{e^2(B\rho)^2}{2m} \tag{15-5}$$

or

$$T = 2\pi^2\rho^2 f^2 m \tag{15-6}$$

It is usually most convenient to use the emu system with B in gauss, e in emu, ρ in cm, and m in grams. T will then be in ergs, which are readily converted to MeV.

A choice of bombarding particle fixes the specific charge, e/m, and then the maximum energy acquired will depend upon the product $B\rho$. Because of this dependency, high-energy cyclotrons have massive yoke-type magnets, usually wound with copper tubing to permit cooling by water or oil. A typical cyclotron with a 60-in.-diameter poleface will have a magnet with 200 tons of steel. With a field of 16,000 gauss deuteron energies of 16 MeV can be realized. Beam currents up to 1 mA may be used on internal targets. External beams are more apt to be about 200 μA.

15.07 The Synchrocyclotron

At high energies, the constant frequency condition of Eq. (15-4) no longer applies because of the relativistic mass increase. The energy range may be extended by modulating either the magnetic field or the oscillator frequency. The effects of these variations can be understood in terms of some general considerations of phase stability for particles undergoing cyclic accelerations.

As we have seen, a nonrelativistic particle undergoing cyclic accelerations has a constant angular velocity and hence traverses the dee gap at precisely the same phase in each r-f cycle, as A_1 and B_1 in Fig. 15-9. Beam-focusing conditions require that these points be restricted to the last one-quarter of the acceleration cycle, as shown.

As the particle gains energy and becomes relativistic, its angular velocity will decrease so that it reaches the gap at a later phase, as B_2. Even with this late arrival, the particle will gain some energy but will continue to lag behind until it reaches the gap in the deceleration phase, as at B_3. The particle will then slow, which will result in a mass decrease and an increase in angular velocity. This will restore the particle to B_0, where it will travel in a stable orbit with no change in energy. Any displacement from the stable orbit will produce a reaction tending to restore the stable situation. Ions will bunch together and travel in this phase-stable orbit.

In the *frequency modulated-*, *FM-*, or *synchro-cyclotron*, the r-f oscillator frequency is cyclically varied, usually by a motor-driven capacitor. As the

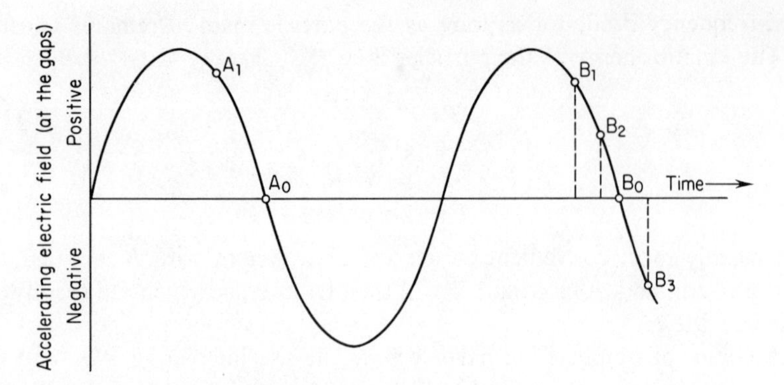

Figure 15-9. Phase stability diagram for particles in a resonance accelerator.

oscillator frequency decreases, a bunch of ions moving in a stable orbit at constant energy will find themselves emerging into the gap during an acceleration phase, as at B_1. Energy and mass will increase, angular velocity will decrease, but as the oscillator frequency continues to decrease there will be a continued gain in energy. Because of the tendency toward stable orbits, the programming of the rate of frequency decrease is not critical. Beam extraction can be timed to coincide with the minimum oscillator frequency.

The output beam of an FM cyclotron consists of a series of ion pulses at a repetition rate equal to the rate of frequency cycling. Figure 15-10 is a photograph of the 184-in. synchrocyclotron at Berkeley. A 4000-ton magnet provides a field of some 15,000 gauss, which permits proton energies up to 350 MeV.

Compensation for relativistic mass increase can also be made by using a magnetic field which increases with radius. The desired field variation is usually obtained by placing steel *shims* in the gap between the polefaces. In the *spiral-ridge* cyclotron, the shims are in the form of Archimedean spirals. This type of field provides both axial and radial beam focusing to keep the ions in the required orbits. The spiral-ridge construction permits a greater energy variation than is obtainable with conventional cyclotrons. In one design, proton energies of 3.5 to 15 MeV are available at an extraction radius of only 18.5 in.

15.08 The Betatron

The relativistic mass increase which limits ion acceleration by the cyclotron restricts even more severely its use in electron acceleration. A few MeV of electron energy can be obtained in this way, but, for higher energies, the *betatron* principle is much more effective.

Figure 15-10. The 184-inch synchrocyclotron. The yoke of the main magnet, the magnetic poles, and the housing for the dees are visible. The dee housing is connected to the large vacuum pump in the foreground. (Courtesy of Radiation Laboratory, University of California.)

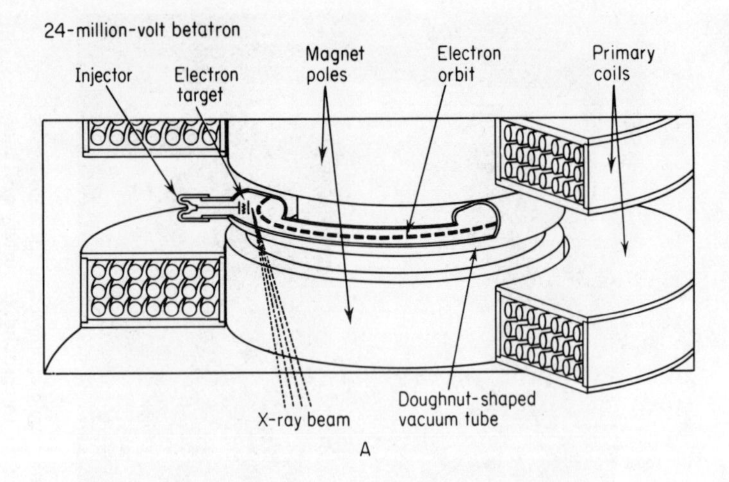

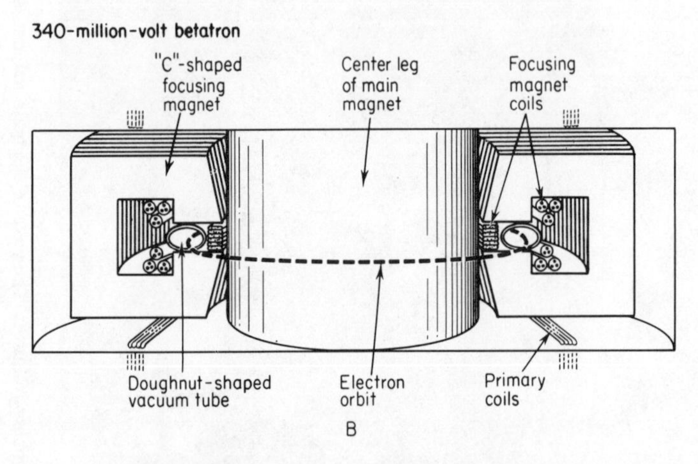

Figure 15-11. Schematic of betatron construction.

A betratron utilizes magnetic induction, much like a transformer, with the electrons taking the place of the usual secondary winding. We effectively have a transformer with an enormous number of secondary turns, for the electrons can make a large number of circuits around the orbit during one magnetic-field cycle.

As the schematic of Fig. 15-11 shows, the betatron orbit has a constant radius. A charge moving in this orbit must obey the basic centripetal force relation:

$$\frac{mv^2}{\rho} = Bev$$

$$mv = Be\rho \tag{15-7}$$

where B is the magnetic-flux density at orbit radius ρ. According to Faraday's law of magnetic induction, the electromotive force induced in a circuit linking a *total* flux ϕ will be

$$\mathscr{E} = \frac{d\phi}{dt} \tag{15-8}$$

Then in one circuit around the betatron orbit, the work done by the field in accelerating an electron will be

$$\mathscr{E}e = e\frac{d\phi}{dt} = 2\pi\rho F \tag{15-9}$$

where F is the accelerating force tangential to the orbit. By Newton's second law of motion, this force must be equal to the time rate of change of the momentum or

$$F = \frac{d(mv)}{dt}$$

whence

$$\frac{d(mv)}{dt} = \frac{e}{2\pi\rho}\frac{d\phi}{dt} \tag{15-10}$$

If we differentiate Eq. (15-7) with ρ constant,

$$\frac{d(mv)}{dt} = e\rho\frac{dB}{dt} \tag{15-11}$$

and

$$\frac{1}{2\pi\rho^2}\frac{d\phi}{dt} = \frac{dB}{dt} \tag{15-12}$$

If Eq. (15-12) is integrated and B_0 is set equal to zero,

$$\phi - \phi_0 = 2(\pi\rho^2)B \tag{15-13}$$

Now $\pi\rho^2 B$ is just the total flux that would be enclosed by the orbit if B were constant throughout the area. Equation (15-13) shows that the total flux enclosed must be twice this amount.

The required field relation can be obtained by shaping the polefaces to provide a strong central field with a lower field at the orbit. A detailed analysis shows that a stable orbit can be achieved if B does not decrease more rapidly than $1/\rho$. A decrease as $1/\rho^{0.75}$ provides a workable margin of stability. Since betatron acceleration requires changing fields, the magnet cores must be laminated to reduce eddy-current losses, as in a transformer. The coils can form part of a resonant circuit, tuned to the power frequency by a large condenser bank.

Electrons emitted from a hot filament are accelerated to perhaps 50 KV and injected into orbit in the betatron *doughnut* as an accurately timed electron pulse. At the moment of injection, the magnetic field is zero and increasing so that the electrons are accelerated. This acceleration continues until the field reaches its maximum value. An auxiliary deflecting coil is then pulsed to abruptly change the orbit so that the electrons either strike an X-ray target or emerge from the doughnut through a thin window. At betatron energies, efficiency of bremsstrahlung production is high, Eq. (8-4), so target cooling is not required. At usual betatron energies (20–100 MeV), photon production in the forward direction is highly favored. The beam half-angle at 100 MeV is only about 2°.

Energies of 100 MeV are readily obtained in a betatron because there is no relativistic limit imposed. An upper limit is set by radiative-energy losses from the accelerated charges. Radiative-energy loss is given by

$$L = \frac{1.2 \times 10^{13}\pi e}{\rho}\left(\frac{E}{E_0}\right)^4 \tag{15-14}$$

where L = loss in eV per revolution
e = charge in emu
E = electron energy
E_0 = rest energy of electron

Radiative-energy losses set a practical limit to betatron energies at about 500 MeV.

Each output pulse has a duration of the order of 10^{-6} sec with a repetition rate limited to 200 sec^{-1} because of the large inductance of the magnet. Thus the *duty cycle*, or the fraction of the time a useful beam is available, is low. Average beam currents will be in microamperes, even though the pulse currents are large.

15.09 The Electron Synchrotron

At high energies, the high central field required for betatron operation leads to a massive magnet, over 100 tons for a 100-MeV machine. Magnetic-flux capabilities also limit the energy that can be imparted to an electron per revolution. Many orbital circuits will be required to attain the higher energies, and during these circuits, radiative-energy losses will sharply reduce the net energy gain. Both of these limitations can be eased somewhat in the *synchrotron*, where orbital stability and acceleration are achieved by separate fields.

In the synchrotron, acceleration takes place at constant radius inside a ceramic doughnut. Constant radius is obtained by modulating the magnetic field as in the synchrocyclotron. The output beam will then be pulsed at the repetition rate of the field modulation. There is no central-field requirement as in the betatron, since the modulated, orbit-maintaining field provides no acceleration. Relatively small magnets produce the required field at the orbit radius.

A portion (perhaps one-eighth) of the synchrotron doughnut is made conducting by depositing silver or copper on the inner surface. This portion forms a resonant cavity which can be excited by pulses from an r-f oscillator. A bunch of electrons arriving at the cavity in proper phase will be accelerated by the electromagnetic field sweeping down the cavity. After leaving the cavity, the electrons will coast around the rest of the doughnut to receive another energy increment at the next revolution. Electron bunching will be maintained by the phase-orbit stability principle. At the end of the acceleration, the electron orbit may be abruptly changed, by a pulsed extraction field, to impinge upon a target. Alternatively, the r-f field may be cut off just before the peak of the orbit-maintaining field is reached. The final increase in this field, in the absence of a corresponding acceleration, will cause the orbit to spiral inward to strike a target near the inner radius of the doughnut. Pulses of the order of 10^9 electrons can be produced at repetition rates of perhaps 10 sec^{-1}.

Several variants of the basic synchrotron design have been constructed. In most machines, pulsed electron injection is followed by betatron acceleration up to 2–3 MeV. Synchrotron operation then takes over to bring the beam up to its final energy. In another design, the doughnut consists of four straight sections connected by four 90° bends where the magnetic deflection takes place. In this machine, all acceleration is obtained by synchrotron operation. Like the betatron, the electron synchrotron requires a large, heavy magnet, which makes the machine unwieldy and difficult to mount to provide for changing the output-beam orientation. For most applications, the linac has replaced magnetic-resonance machines as a source of high-energy electrons.

15.10 The Proton Synchrotron

Protons or even heavier particles appear to be the projectiles of choice for energies above 1 GeV. The radiative-loss barrier is much less restrictive for these particles, but magnet sizes for machines such as the synchrocyclotron become prohibitive. At the higher energies, it is imperative to accelerate at constant radius. Ions can be given an energy of several MeV by electrostatic generator or linac acceleration before injection into the main accelerator. Constant radius, there, is maintained by a modulated magnetic field as in the synchrotron; acceleration is obtained from several resonant cavities pulsed in sequence. In some respects, the proton synchrotron can be thought of as a synchrocyclotron with the central, low-energy part of the magnet eliminated.

The first proton synchrotron in the multi-GeV range was the *cosmotron* at Brookhaven National Laboratory, Fig. 15-12. A 2200-ton magnet maintained an orbit radius of 30 ft to provide protons up to 3 GeV. Pulses contained about 10^{10} protons at a repetition rate of 1 every 5 seconds.

Figure 15-12. The 3 GeV proton synchrotron. The four linear accelerating cavities can be seen between the 90-degree deflecting sections. (Courtesy of Brookhaven National Laboratory.)

Economics rather than technical considerations has limited the development of the proton synchrotron. Costs are strongly influenced by the magnet requirements and these in turn are determined by the beam size. Any improvement in radial or z-axis focusing will reduce beam size and magnet cost.

A notable advance in focusing came with the introduction of *strong*, or *alternating-gradient* (AG) *focusing*, which has been applied to several designs. The 33-GeV machine at Brookhaven is a representative example. In AG focusing, beam deflection is obtained from a series of magnets, 240 in the Brookhaven design, whose poles span the acceleration chamber. Each field has a strong radial gradient with alternating gradient directions in adjacent magnets. This focusing arrangement sharply reduces beam cross-section, so the total magnet weight is quite comparable to that in the older 3-GeV machine. Pulses of 3×10^{11} protons are readily obtained. Twelve accelerating sections are located around the one-half-mile circuit of the acceleration chamber.

The success of AG focusing has led to a design in which field pulsing is not required. These *fixed-field*, or FFAG, machines provide a substantial gain in beam current because of the increased duty cycle. Large beam currents can be further increased by *beam stacking*. In stacking, a bunch of ions brought up to full energy is removed from the acceleration region, either to a new orbit or to a second, nonaccelerating storage ring. Bunch after bunch can thus be stored until circulating currents of several amperes are attained. This intense beam can then be directed against a target.

Interest is increasing in collisions between two oppositely directed beams of high-energy particles. In this arrangement, the effective energy will be greatly increased over that obtained with a single-beam incident upon a stationary target. Very intense beams will be required to obtain detectable reactions, since reaction probabilities will vary as the square of the number of particles involved. One colliding-beam proposal envisages the use of two tangent storage rings, each supplied with stacked beams from FFAG accelerators.

Future accelerator developments will depend more upon available funds than upon scientific developments. There is no known barrier to obtaining particle energies up to at least 1000 GeV. Designs for proton accelerators in this range are now underway. Electron energies are lagging somewhat, but designs for 10 GeV are available.

15.11 Acceleration of Heavy Ions

The use of heavy ions, $A > 4$, as projectiles opens up new possibilities for studying nuclear structure. The increased number of nucleons involved leads to a wider variety of interactions with the target nuclei. Energetic heavy particles are also of biological interest, since they form a most important component of cosmic rays and space radiation.

Heavy particles are usually accelerated as multiply charged ions, the energy imparted to the particle being given by the product of the net charge and the accelerating potential. This energy is, of course, distributed equally among the A nucleons in the particle. To avoid confusion, it is customary to express heavy-ion energies as energy per nucleon, or E/A.

Under proper conditions, multiply charged ions can be produced in an arc discharge or by cathode sputtering of a solid. The energy required to remove the second, third, ... electrons from an atom rises rapidly with the degree of ionization. For example, O^{+1} requires 13.6 eV, while O^{+6} requires 140 eV. For effective ion production, the electrons in the discharge must have energies several times that needed for ionization.

Multiply charged ions can be effectively produced by the process of *stripping*. Energetic ions gain and lose electrons by interactions with the atoms of the material through which they pass until an equilibrium ionization level is established. Electron loss tends to exceed gain as long as the ion velocity exceeds the orbital velocity of the electron. Thus a high-speed beam of O^{+2} or O^{+3} ions becomes predominantly O^{+4}, O^{+5}, and O^{+6} after traversing some 2 μg cm^{-2} of argon. Fortunately equilibrium conditions are reached before beam scattering becomes serious. In addition to gas stripping, foils of nickel, carbon, and aluminum oxide in thicknesses of 5–100 μg cm^{-2} have been used successfully.

Any nonresonant accelerator can be used to accelerate multiply charged heavy ions. The tandem Van de Graaff is particularly suited to heavy-ion acceleration, since both ion source and target are at ground potential. Thus the cumbersome and short-lived arc and stripper are readily available for servicing.

The cyclotron requirement, Eq. (15-4), depends upon e/m, so a fixed-frequency machine adjusted to accelerate deuterons or alpha particles will also be suitable for heavy ions which have an e/m of $\frac{1}{2}$. Thus completely stripped ^{12}C, ^{14}N, ^{16}O, and ^{20}Ne have been accelerated in a deuteron-tuned cyclotron. Controlled stripping in the cyclotron has not been accomplished, however, so the machines become heavily loaded with unwanted ions. Cyclotron acceleration gives a low-intensity beam with a wide energy spread.

The proton linac is readily adapted to heavy-ion acceleration. Stripping presents no serious problems, and a variety of e/m values can be accelerated by merely changing the r-f driving voltage. Beam focusing is obtained by quadrupole magnets inside the drift tubes. Heavy-ion linacs up to 30 MeV A^{-1} appear feasible; beyond this, length becomes excessive. Higher energies will probably require a synchrotron with high-energy injection from a linac.

REFERENCES

Bohm, D. and L. Foldy, "Theory of the Synchrotron." *Phys. Rev.* **70,** 249, 1946.

Brobeck, W. M. et al., "Performance of the 184-inch Cyclotron." *Phys. Rev.* **71,** 449, 1947.

Kerst, D. W., "Acceleration of Electrons by Magnetic Induction." *Phys. Rev.* **60,** 47, 1941.

Livingston, M. S., "Standard Cyclotron." *Ann. Rev. Nuc. Sci.* **1,** 157, 1952.

Neal, R. B., ed., *The Stanford Two-Mile Accelerator*. W. A. Benjamin, New York, 1968.

Richardson, J. R. et al., "Development of the Frequency-Modulated Cyclotron." *Phys. Rev.* **73,** 424, 1948.

PROBLEMS

15-1. A cyclotron accelerating alpha particles to 16 MeV has an average beam current of 65 μA. To how many grams of equilibrium ^{226}Ra is this equivalent?

15-2. A beam of O^{-5} ions is accelerated to 1.5 MeV prior to injection into a tandem Van de Graaff generator operating at 6.3 MV. After one passage, the beam is stripped to O^{+2} and returned through the same tandem potential. What is the final energy of the ions, and the energy per nucleon?

15-3. The O^{-5} ion beam in Prob. 15-2 is contaminated with O^{-4} and O^{-6} ions, which must be removed before injection into the tandem accelerator. Calculate the three radii of curvature in an analyzing magnet with a field of 7100 gauss.

15-4. Calculate the relativistic mass of an electron accelerated to 600 MeV; to 2 GeV.

15-5. The accelerations in Prob. 15-4 take place in a length of 200 ft in laboratory coordinates, but to a moving system lengths undergo a relativistic contraction from 1 to $\sqrt{1 - v^2/c^2}$. Calculate the length of the accelerator as seen by the electrons of Prob. 15-4.

15-6. Calculate the relativistic mass of protons accelerated to energies of 1, 10, 100, and 1000 GeV. What will be the amount of length contraction at each of the final velocities?

15-7. A cyclotron magnet produces a field of 12,000 gauss over a usable poleface diameter of 36 in. What is the maximum energy that can be imparted to protons in this machine? What oscillator frequency will be required?

15-8. A synchrocyclotron is being designed to accelerate protons to 700 MeV. What percentage frequency change will be required in the r-f oscillator?

15-9. A 9.8-MHz oscillator supplies an effective voltage of 50 KV to the dees of a cyclotron accelerating alpha particles to a maximum energy of 28 MeV. Calculate the closest approach of adjacent turns of the spiral ion path, assuming the beam to lie between radii of 10 and 75 cm.

15-10. A 100-MeV betatron has an orbit radius of 35 cm in which the electrons acquire 480 volts per revolution. How far will the electrons travel in attaining full energy? How far will this distance appear to be to the electrons? The magnet is energized at 180 Hz and produces pulses of 2 μsec duration, each consisting of 3×10^9 electrons. What is the peak and the average beam current? What is the duty cycle?

15-11. A betatron with a doughnut radius of 45 cm operates at 180 Hz with a maximum field strength of 2000 gauss. How much energy will an electron gain per revolution?

15-12. A 350-MeV betatron has an orbital radius of 55 cm in which the electrons gain 2.5 KeV per revolution. What is the fractional loss of energy per revolution?

15-13. A linac operating at 500 MeV produces 60 pulses of 1.5 μsec duration every second, with 6×10^{11} electrons in each pulse. Assume a 95 per cent efficiency for energy conversion and calculate the average and peak rates of bremsstrahlung production. Assume the bremsstrahlung to be concentrated in a conical beam of 2° half-angle. What is the rad/min output of the machine in air at a distance of 1 meter from the target?

16

Neutron Physics

16.01 Discovery of the Neutron

Lord Rutherford anticipated the discovery of the neutron by a decade when he speculated in 1920 about the possibility of "an atom of mass 1 with zero charge." This speculation was temporarily forgotten in 1930 when Bothe and Becker observed a very penetrating radiation coming from a reaction assumed to be

$$^{9}_{4}\text{Be} + ^{4}_{2}\text{He} \longrightarrow ^{13}_{6}\text{C} + h\nu \tag{16-1}$$

Analogs to the reaction were observed with other light target elements. The penetrating radiations were assumed to be gamma rays, although the energies of 7–10 MeV estimated from the penetration seemed unusually high.

Irene Curie and Joliot obtained some even more remarkable results in a series of absorption experiments. Hydrogen-rich absorbers such as paraffin were found to enhance rather than to attenuate the radiation. Cloud-chamber studies showed the new radiation to be nonionizing, strengthening the conviction that it was an energetic photon. Long-range protons were ejected from the paraffin by what was assumed to be a Compton-type interaction. Photon energies of 40–50 MeV, unheard of as nuclear reaction products, were required to account for the observed proton ranges.

Chadwick extended the cloud-chamber studies using nitrogen as a chamber gas. He observed tracks of nitrogen ions recoiling from the point of collision with the new radiation. These recoil tracks could not be explained by a photon reaction, even of high energy. They were, however, consistent with a mechanical collision between a particle of about proton mass and the nitrogen. The lack of ionization in the original particle track indicated a neutral particle which Chadwick named the "neutron." These results showed that reaction (16-1) should be written

$$^{9}_{4}\text{Be} + ^{4}_{2}\text{He} \longrightarrow ^{12}_{6}\text{C} + ^{1}_{0}n \tag{16-2}$$

At the moderate energies used by Chadwick, nonrelativistic laws of momentum and energy conservation could be applied to the collisions. In the reaction,

$$^{11}_{5}\text{B} + ^{4}_{2}\text{He} \longrightarrow ^{14}_{7}\text{N} + ^{1}_{0}n \tag{16-3}$$

Chadwick knew the three atomic masses and the alpha-particle energy with fair precision. These data led to a mass of 1.0067 u for the neutron. A variety of later methods led to a presently accepted value slightly higher than the original value found by Chadwick.

When it became established that the neutron was heavier than the proton, Chadwick and Goldhaber predicted that the neutron was radioactive, decaying by beta emission to a proton. A 12.0-minute activity was detected and the products identified. A neutron beam from a nuclear reactor was led into an evacuated space, where some of the particles decayed. The effect of radiation arising in the reactor itself was reduced by coincidence counting of the proton and electron from the neutron decays. Figure 16-1 shows an energy spectrum of the beta particles from neutron decay. The maximum energy of 782 KeV is in excellent agreement with that expected from the n–^{1}H mass difference.

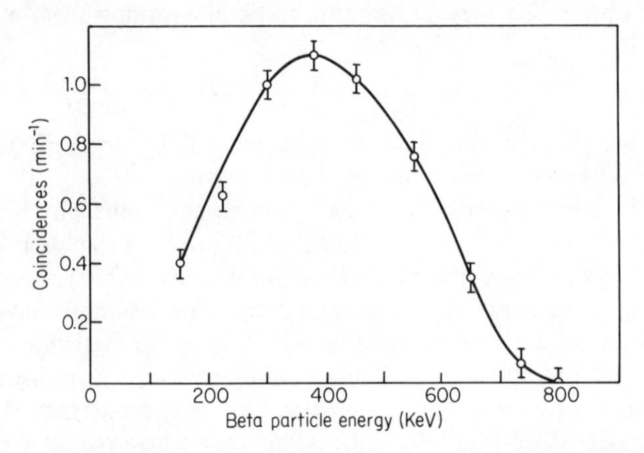

Figure 16-1. Energy spectrum of the beta particles from the radioactive decay of the neutron. (From J. M. Robson, *Phys. Rev.*, **83**, 349, 1951).

16.02 Energy Classification

Neutrons are usually classified according to either energy or velocity, although the boundaries between the various divisions are ill-defined. A common classification is:

High-energy	> 10 MeV
Fast	10 KeV–10 MeV
Intermediate	100 eV–10 KeV
Slow	0.03 eV–100 eV
Epithermal	~ 1 eV
Thermal	0.025 eV
Cadmium	> 1 eV

As neutrons traverse matter, they lose energy by a series of collisions, and like gas molecules eventually come into thermal equilibrium with the surroundings. The term *thermal neutron* refers to a neutron in equilibrium at ordinary room temperature. When in equilibrium, the neutron energies will have a Maxwellian distribution about the mean. This value is readily calculated from the Boltzmann equation:

$$E = \tfrac{1}{2}kT \tag{16-4}$$

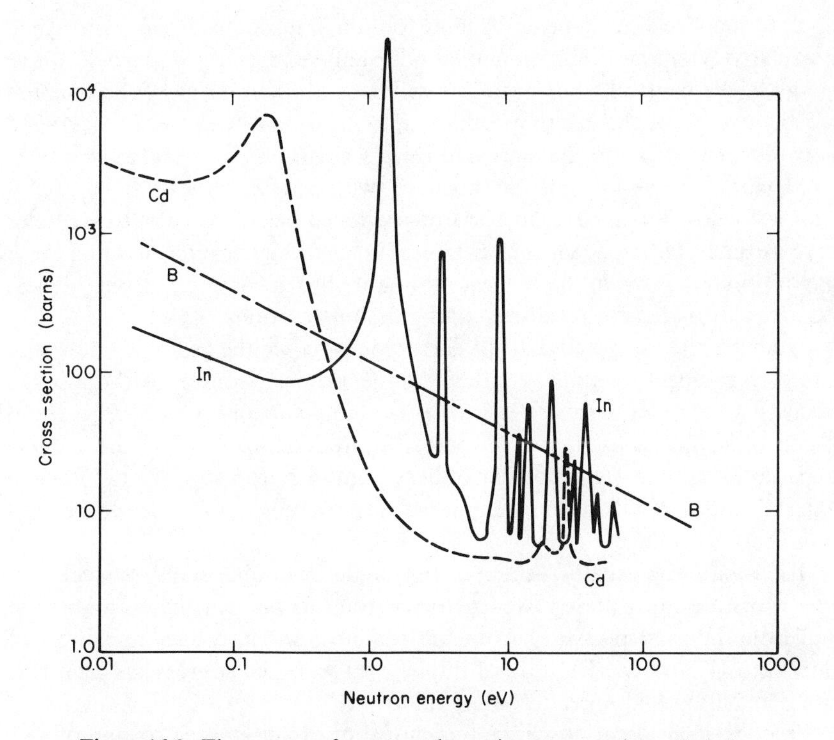

Figure 16-2. Three types of neutron absorption cross-sections. Cadmium shows one broad resonance with a sharp drop on the high-energy side, boron has a $1/v$ dependence, and indium shows many sharp resonances superposed on a $1/v$ absorption.

where E = ergs energy for each degree of freedom
k = Boltzmann's constant = 8.61×10^{-11} MeV °K^{-1}
T = absolute temperature

Convenient expressions relating temperature and velocity to the neutron energy are:

$$T = 1.159 \times 10^4 E \tag{16-5}$$

$$v = 13.83\sqrt{E} \tag{16-6}$$

where E is in eV, v in Km sec^{-1}, and T in °K.

The *cadmium neutron* is included in the classification because of the importance of this element as a neutron absorber. Cadmium has a very large absorption cross-section for neutrons up to about 0.4 eV, Fig. 16-2, but is relatively transparent at energies above 1 eV. Hence cadmium neutrons are those capable of passing through a cadmium absorber.

16.03 Neutron Sources

Nuclear reactions must be utilized to obtain free neutrons, in contrast to the relatively simple ionization methods that yield charged particles. The energy of the neutrons thus obtained will depend upon the Q of the reaction and the energy of the exciting radiation. Neutron energies can be degraded by collisions but cannot be increased by any sort of an accelerator.

Mixtures of alpha-emitting nuclides with low-Z targets make useful sources for low intensities. Accelerator-produced beams of charged particles serve as the initiating agent for neutron sources of higher intensity, and these may ultimately provide the highest attainable fluxes. At present, the highest fluxes are available from self-sustaining fission reactions.

We have already noted that the early workers used the (α, n) reaction with naturally radioactive alpha emitters as the primary source. Although the intensity from such a source is low, it is stable and provides a very useful source of neutrons for many purposes. Sources utilizing (α, n) are usually prepared by mixing a fine powder of beryllium or boron with the radioactive material. Intimate mixing is essential if the short-range alphas are to be used to the best advantage.

The mixture is usually welded into a single or double stainless steel capsule. Moisture must be removed before sealing to prevent the formation of dangerous internal pressures from the radiation-induced decomposition of water into H_2 and O_2. Rupture of improperly prepared sources has occurred, with a resulting widespread contamination.

Practical considerations limit the choice of alpha emitter to one of five nuclides. A ^{210}PoBe source is relatively free of gamma contamination (about 0.7 photon per neutron) but decays with a 138-d half-life. About 2.2×10^6

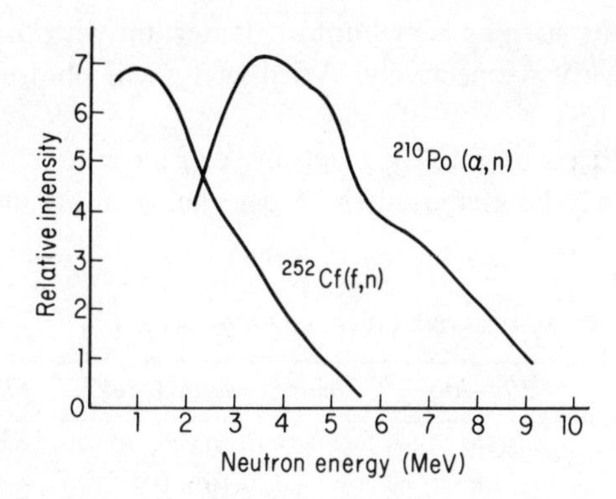

Figure 16-3. Energy spectrum of the neutrons emitted by ^{210}PoBe (α, n) and ^{252}Cf (f, n).

neutrons per second per curie are emitted from a well-prepared source. Neutron energies range up to about 11 MeV with an average value of 4.5 MeV, Fig. 16-3.

A ^{226}RaBe source has an almost constant output but is contaminated with all of the gamma radiation from the radium daughters. Neutron energies range from 13 MeV downward, with an average of 5 MeV. The yield may reach $10^7 n$ sec^{-1} Ci^{-1} of ^{226}Ra.

Radon–beryllium mixtures may be used for special purposes, but they have a relatively large gamma component and are severely limited by the short half-life of ^{222}Rn.

The long half-life of ^{239}Pu provides an almost constant emission from ^{239}PuBe with an even lower gamma contamination than can be obtained with ^{210}Po. The low specific activity of ^{239}Pu makes for a bulky source, about 16 grams being required to obtain 1 curie of alpha activity. Neutron yield is about $1.4 \times 10^6 n$ sec^{-1} Ci^{-1} with an average energy of 3.5 MeV and a maximum of 10.6 MeV.

A ^{238}PuBe source has several advantages over the other alpha-initiated reactions. The half-life of 86 years makes for a reasonably constant output without too much bulk. About $3 \times 10^6 n$ sec^{-1} Ci^{-1} can be obtained with an energy distribution about like that obtainable with ^{239}Pu.

Photoneutron sources. Neutrons produced from the photodisintegration of certain nuclei have some advantages over those produced by alpha bombardment. Neutrons produced by photodisintegration will be monoenergetic if the photon source is monoenergetic; these sources are easy to prepare, and can be made quite reproducible in terms of neutron output. A good photoneutron source requires the use of a target element in which a neutron is loosely bound, and a gamma emitter whose energy exceeds the photoneutron energy threshold. Practically, the first requirement restricts the target to the light elements such as beryllium or deuterium with threshold energies of 1.66 and 2.22 MeV, respectively. A list of typical photoneutron sources is given in Table 16-1.

Neutrons produced by photon reactions will have a small energy spread initially, but this may be increased to 15 per cent or more as some neutrons lose energy in escaping from the interior of the source. The National Bureau of Standards has developed neutron-flux standards utilizing the gamma rays in mixtures of ^{226}Ra and Be.

TABLE 16-1
REPRESENTATIVE PHOTONEUTRON SOURCES

Source	*Half-life*	*Neutron energy (MeV)*	*Yield* ($n\ sec^{-1}\ Ci^{-1}$)
^{124}SbBe	60 d	0.024	3.2×10^6
^{72}GaD$_2$O	14 h	0.16	0.64×10^6
^{24}NaD$_2$O	14.8 h	0.22	2.7×10^6
^{140}LaBe	40 h	0.62	0.04×10^6
^{24}NaBe	14.8 h	0.83	2.4×10^6

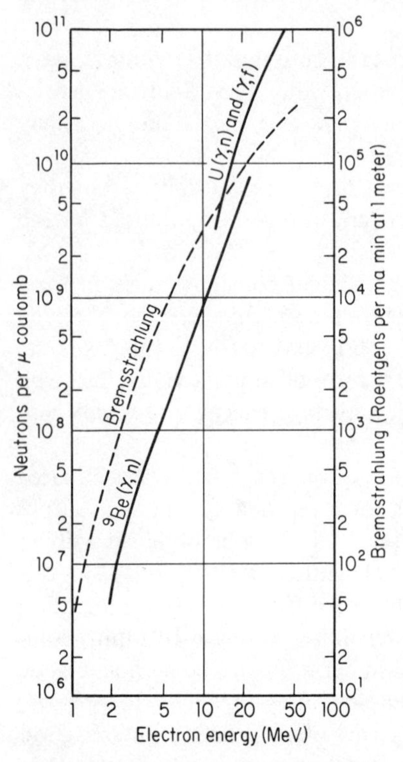

Figure 16-4. Neutron yields from high-energy bremsstrahlung on beryllium and uranium targets.

At high energies, the increased efficiency of bremsstrahlung production makes possible high-intensity neutron sources utilizing accelerator-produced photons. Neutrons will be produced by (γ, n), $(\gamma, 2n)$, or perhaps by (γ, f), Fig. 16-4. The neutrons from these sources will be heavily contaminated with unconverted photons, but some separation can be achieved by differential shielding and by the spatial arrangement for utilizing the particles. The residual photon beam will be strongly peaked in the forward direction, while the neutron flux will be more nearly isotropic. Because of the spectral distribution of the bremsstrahlung, the neutron output will have a wide energy distribution.

Positive-ion neutron sources. Intense neutron sources can be produced by bombarding low-Z targets with positive ions obtained from a cyclotron or some

TABLE 16-2
REACTION ENERGIES FOR NEUTRON PRODUCTION

Reaction	Energy (MeV)
$^{3}H(d, n)He$	17.6
$^{9}Be(\alpha, n)C$	5.6
$^{9}Be(d, n)B$	4.4
$^{2}H(d, n)H$	3.3
$^{3}H(p, n)He$	−0.76
$^{7}Li(p, n)Be$	−1.65

other type of particle accelerator. A good many of the reactions are exoergic, Table 16-2, so that energetic neutrons are produced at relatively low particle energies.

Neutron yields depend upon the bombarding energy and the target configuration. The $^{3}H(d, n)$ reaction on a target of $Zr^{3}H$ may produce $10^{7}n$ sec^{-1} per microampere of deuteron current. Be(d, n) may yield $10^{8}n$ sec^{-1} μA^{-1} at 1 MeV, 10^{10} at 8 MeV.

Spontaneous-fission sources. Most artificially produced transuranium elements undergo spontaneous fission with the release of several neutrons per fission. Of these nuclides, ^{252}Cf seems to have the greatest potential as a useful source, and large facilities have been built for its production. Californium-252 has a half-life of 2.65 years, which strikes a good balance between constancy of output over a reasonable time and source size. No voids need be left in the capsule to accomodate the accumulation of He, as with alpha emitters, and ^{252}Cf sources can be made significantly smaller than (α, n) sources of equal neutron output. ^{252}Cf (f, n) sources emit $2.3 \times 10^{12}n$ sec^{-1} per gram of Cf, or $4.4 \times 10^{9}n$ sec^{-1} Ci^{-1}. Neutron energies are somewhat lower than those from most (α, n) sources, Fig. 16-3, but are high enough to be generally useful.

16.04 Energy Moderation

Neglecting nuclear reactions for the moment, we may consider that fast neutrons are slowed by a series of elastic collisions which transfer kinetic energy to, but do not ionize or excite, the struck atom. Collision details, such as the angular distribution of the scattered particles, will depend upon the exact nature of the nuclear force interactions. For many purposes, a detailed description is not needed, so that average behavior taken over all possible collision modes may furnish adequate information.

Only in a vacuum will a neutron remain free long enough to undergo radioactive decay. In a condensed medium, a neutron will be captured long before decay. Before capture, the neutron will have been partially or completely *thermalized* by collisions.

On the average, neutron energies before and after collision with a nucleus of mass number A are related by

$$\frac{E_2}{E_1} = e^{-\xi} \tag{16-7}$$

where

$$\xi = 1 - \frac{(A-1)^2}{2A} \ln \frac{A+1}{A-1} \tag{16-8}$$

The average number of collisions required to reduce a neutron from energy E_0 to E will then be

$$\bar{\zeta} = \frac{1}{\xi} \ln \frac{E_0}{E} \tag{16-9}$$

An inspection of Eq. (16-8) shows that the fractional energy loss per collision increases with decreasing A. For hydrogen, $A = 1$, and, therefore, Eq. (16-8) becomes indeterminate but can be evaluated as $\xi = 1$. Materials effective in slowing neutrons without capture are called *moderators*. Pure carbon, usually in the form of graphite, and heavy water are regularly used as moderators.

16.05 Absorption Cross-section

Since all scattering angles from 0° to 180° are possible, an originally unidirectional neutron beam will cease to be so after a series of collisions. This requires a careful specification of the quantities used in referring to neutron effects. An unscattered, unidirectional neutron beam can be thought of as a neutron *current*, carrying energy in a given direction with *current density*, referring to the flow across a unit area. Specification may be in terms of *number* current density or *energy* current density, the two obviously being related by the amount of energy carried per particle. In a scattered, heterodirectional beam, current and current density must be determined by the net number or energy flow in the direction of interest.

Absorption processes are independent of particle direction and hence depend upon the total number crossing an area rather than upon the net number moving in a preferred direction. The total *rate* of particle flow across unit area, irrespective of direction, is known as *neutron-flux density*. Again number flux and energy flux must be distinguished. Particle-flux density will be, in general, a strong function of particle velocity.

When the particle-flux density F is integrated over an appropriate time interval, we obtain the particle *fluence* Φ, or the total number of particles that crossed unit area. Consider a beam of monoenergetic neutrons with velocity v incident upon a surface of 1 cm^2, Fig. 16-5. Let the neutron density in the beam be n cm^{-3}. From the figure it is evident that nv neutrons will cross the surface in 1 second, and in time t the total number crossing will be nvt. This is a factor frequently encountered in neutron calculations. From our definition of particle fluence,

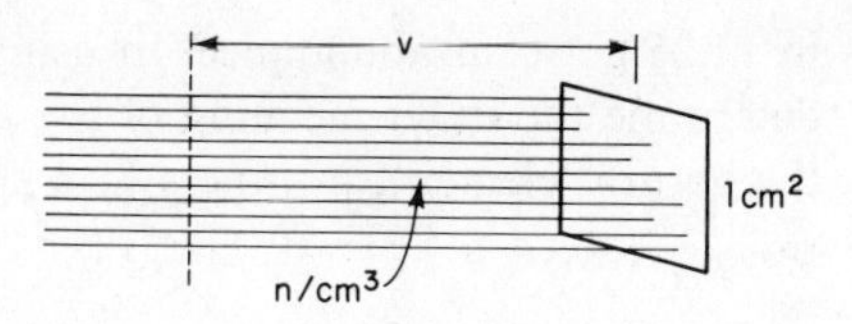

Figure 16-5. A monoenergetic neutron beam passing through an area of 1 cm².

$$\Phi = nvt \tag{16-10}$$

Equation (16-10) applies equally well if the particle-flux density is not unidirectional and is not monoenergetic.

If ϕ represents the number-flux density, we find the familiar form of the attenuation equation:

$$\phi = \phi_0 e^{\sigma_t n x} \tag{16-11}$$

where σ_t = total cross-section for absorption
n = atoms cm^{-3} in the absorber
x = absorber thickness

The total cross-section σ_t has several components:

$$\sigma_t = \sigma_{\text{elastic}} + \sigma_{\text{inelastic}} + \sigma_{\text{absorption}} \tag{16-12}$$

and $\sigma_{\text{absorption}}$ is in turn made up of several components to be discussed separately.

Neutron absorption cross-sections are conveniently expressed in units of 10^{-28} m^2, or barns. This is the order of magnitude of the "size" of a fast neutron in terms of its de Broglie wavelength. The de Broglie wavelength expression, Eq. (5-2), for a neutron is readily converted to

$$\lambda = \frac{2.87 \times 10^{-9}}{\sqrt{T}} \text{ cm} \tag{16-13}$$

where T is the kinetic energy of the neutron in eV. At thermal energies, the de Broglie wavelength has increased to about 10^{-8} cm, and we might expect that the cross-sections would increase accordingly, as some inverse function of the energy.

For low energies, well away from any resonances, theory leads to an absorption cross-section proportional to $1/v$, where v is the neutron velocity. The $1/v$ law is quite accurately obeyed in many cases, Fig. 16-2, and in others resonance peaks are superposed on a basic $1/v$ dependence.

Neutron cross-sections depend strongly upon the nature of the absorbing nucleus. There may be strong isotope effects. Natural cadmium, for example, has an absorption cross-section of 7000 b at 0.176 eV. This is due almost entirely to a cross-section of 20,000 b in ^{113}Cd, present in the natural material in 12.26 per cent abundance. In using tabular values, care must be taken to determine the exact meaning of the entries. In some cases they will refer to the natural element, in others to a specific isotope.

16.06 Resonance Capture

Many neutron-absorption curves show regions of large capture cross-section, usually at energies between thermal and a few KeV. Cadmium, Fig. 12-2, has such a *resonance* peak at 0.176 eV. Some elements show a number of such peaks within an energy range of only a few eV. Breit and Wigner in 1936 correctly interpreted these peaks as due to resonances between the energy of the incident neutron and nuclear-energy levels.

Breit and Wigner developed a theory of resonance absorption by analogy with the theory of optical dispersion in the vicinity of an absorption band. When a neutron is absorbed, a compound nucleus will be formed in a state of excitation. This excitation can be relieved by a variety of modes or *channels* of de-excitation, as by the emission of a photon, (n, γ), a neutron, (n, n), or a proton, (n, p). Each channel will have a resonant energy of transition and a corresponding width of the excited state, denoted by $\mathbf{\Gamma}$ with an appropriate subscript, as $\mathbf{\Gamma}_\gamma$, for the width of the excited state for photon emission. If an absorption curve were to be plotted as a function of energy, $\mathbf{\Gamma}$ would be the full width of the curve at the point where the absorption is one-half that at maximum.

Each de-excitation channel will have a characteristic width, the total width for a state being the sum of the components:

$$\mathbf{\Gamma} = \mathbf{\Gamma}_\gamma + \mathbf{\Gamma}_n + \cdots$$

If we consider the case of only two de-excitation channels, the Breit–Wigner relation for an (n, γ) reaction, $\sigma_{(n,\gamma)}$, at energy E close to a resonance energy E_r, is given as

$$\sigma_{(n,\gamma)} = \frac{\lambda^2 \mathbf{\Gamma}_n \mathbf{\Gamma}_\gamma}{4\pi[(E - E_r)^2 + \frac{1}{4}\mathbf{\Gamma}^2]} \tag{16-14}$$

where $\mathbf{\Gamma} = \mathbf{\Gamma}_n + \mathbf{\Gamma}_\gamma$ = width of the resonance peak at half its maximum

$\mathbf{\Gamma}_n$ = partial level width for the (n, n) reaction (neutron width)

$\mathbf{\Gamma}_\gamma$ = partial level width for the (n, γ) reaction (gamma width)

λ = de Broglie wavelength for the neutron of energy E

Experimentally, we may measure three quantities in this expression, namely, the neutron cross-section, the resonance energy, and $\mathbf{\Gamma}$, the latter being taken from the absorption curve. Thus we may be able to determine the partial level widths and, since these are proportional to the reaction probabilities, we can determine the relative probabilities of the two component reactions.

Illustrative Example

Estimate the relative probabilities of (n, n) and (n, γ) in indium, known to have a neutron resonance at 1.44 eV with a of Γ of 0.1 eV and a cross-section of 28,000 b.

At resonance, $E = E_r$, so that Eq. (16.4) simplifies to

$$\sigma_{(n,\gamma)} = \frac{\lambda^2}{\pi} \frac{\Gamma_n \Gamma_\gamma}{\Gamma^2}$$

The de Broglie λ for a neutron of 1.44 eV is $0.287/(1.44)^{1/2}$ or 0.24×10^{-8} cm. Thus

$$\Gamma_n \Gamma_\gamma = \frac{\pi(0.1)^2(2.8 \times 10^{-20})}{(0.24)^2 \times 10^{-16}} = 1.5 \times 10^{-4}$$

and since $\Gamma_\gamma \simeq \Gamma = 0.1$, we have

$$\frac{\Gamma_n}{\Gamma_\gamma} = 0.015$$

From the results of the illustrative example, we see that in the case of indium resonance, only about 15 neutrons are emitted per 1000 gammas. In most cases, the value of Γ_γ is greater than that of Γ_n by one or two orders of magnitude, so Eq. (16-14) assumes the form

$$\sigma_{\max} = \frac{\lambda^2 \Gamma_n}{\pi \Gamma_\gamma} \tag{16-15}$$

if we take $\Gamma_\gamma \simeq \Gamma$.

In general, artificial radioactivities are induced as a result of resonance capture. Since resonance peaks of various elements are distributed throughout the low-energy region, it is possible to measure neutron fluxes in specific energy ranges by using foils of appropriate elements. From the activities

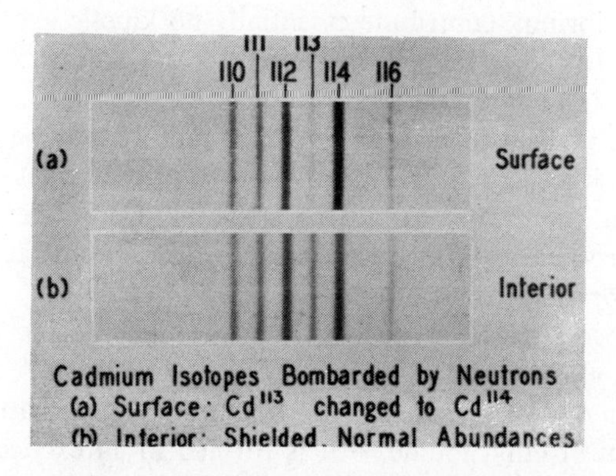

Figure 16-6. Surface depletion of ^{113}Cd from the natural metal by strong neutron absorption. (Courtesy of A. J. Dempster.)

induced in the foils and the known cross-section for the resonance level, one can calculate the neutron flux.

Resonance-absorption foils must be protected from the large contribution that would be made by thermal neutrons. This is usually done by covering the foils with a sheet of cadmium, about 0.08 mm being sufficient in most cases. The foils used for measuring resonance activation must be thin so that they do not shield themselves excessively from the neutron flux. With the very high cross-sections sometimes seen at resonance, it is relatively easy to seriously deplete the neutron flux in the outer layers of the foil, thus introducing a serious error in the flux determination. Figure 16-6 illustrates self-shielding in cadmium. The surface layer shows a large conversion of ^{113}Cd to ^{114}Cd with a resonance absorption so high that few neutrons penetrated to the interior to cause the same reaction. The relative line intensities demonstrate that the 113 isotope is responsible for the strong absorption.

16.07 Slow-neutron Reactions

Slow-neutron reactions have become increasingly important as high fluxes of moderated neutrons have become available in nuclear reactors.

a. *The* (n, γ) *reaction.* Nuclear capture of slow neutrons is favored by the absence of a potential barrier and the operation of the $1/v$ law. Some 8 MeV of binding energy will be brought to the the compound nucleus at neutron capture, and in general de-excitation will be prompt. Gamma-ray emission will be favored because again no potential barrier has to be penetrated. So many (n, γ) reactions, known as *radiative captures*, take place that it is almost impossible to obtain a pure neutron beam, uncontaminated with gamma radiation.

Thermal neutrons, produced by the complete moderation of fast neutrons to thermal equilibrium, contribute essentially no kinetic energy to the compound nucleus. In radiative capture, therefore, the gamma rays should have a total energy of about 8 MeV, as is observed.

The product nucleus formed by radiative capture may be either stable or radioactive. In either case, it will be an isotope of the target element, and in general chemical separation cannot be achieved. Consequently, (n, γ) reactions tend to produce radioactive products of low specific activity that may be unsuitable for some applications.

In a few special cases, a product with a high specific activity can be obtained by (n, γ) through the *Szilard–Chalmers reaction.* A classic example is the formation of ^{128}I by the neutron bombardment of naturally occurring ^{127}I in the form of ethyl iodide. ^{128}I is formed in an excited state which is relieved by gamma-ray emission. The recoil energy imparted to the ^{128}I

nucleus in order to achieve momentum conservation with the departing gamma ray is sufficient to liberate free iodine from the organic molecule. The reaction is

$$C_2H_5I + n \longrightarrow C_2H_5 + I^* \tag{16-16}$$

where the asterisk is used to denote a radioactive species. Chemical methods can be used to separate the elemental iodine from the organic-iodide target. The Szilard–Chalmers reaction is applicable in only a few cases.

The Szilard–Chalmers reaction emphasizes the fact that the recoil energy of an emitting nucleus will be sufficient to rupture the chemical bonds with adjacent atoms. Thus it is not possible to directly place a radioactive label on a particular molecular configuration. The radioactive product must be chemically separated from its target, and then synthesized into the desired structure.

b. *The (n, p) reaction.* In most cases, fast neutrons are required to initiate (n, p) reactions, because the escaping proton must acquire sufficient energy to penetrate the barrier, and so it must come from a relatively high state of excitation. In light nuclei, however, the barrier is low and (n, p) reactions are seen with ^{3}He, ^{14}N, and ^{35}Cl.

The slow-neutron (n, p) reaction in ^{14}N is of special interest because it is used to produce the very useful radioisotope ^{14}C according to

$$^{14}_{7}N + n \longrightarrow {}^{14}_{6}C + p + 0.63 \text{ MeV} \tag{16-17}$$

This reaction is also important in considering the effect of neutrons on living tissues. Part of the biological effect of neutron absorption will be due to the conversion of N to C by the reaction of Eq. (16-17), part to the ionizations produced by the recoiling nuclei, and part to the radioactivity of the beta-emitting ^{14}C.

c. *The (n, α) reaction.* As with (n, p), fast neutrons are usually required to produce an (n, α) reaction. Two exceptions are found in the light nuclei ^{6}Li and ^{10}B. Slow-neutron bombardment of lithium goes according to

$$^{6}_{3}Li + {}^{1}_{0}n \longrightarrow {}^{4}_{2}He + {}^{3}_{1}H + 4.78 \text{ MeV} \tag{16-18}$$

The thermal-neutron cross-section of 870 barns makes this reaction feasible for the large-scale production of tritium in nuclear reactors.

The ^{10}B reaction

$$^{10}_{5}B + {}^{1}_{0}n \longrightarrow {}^{4}_{2}He + {}^{7}_{3}Li + 2.81 \text{ MeV} \tag{16-19}$$

has a cross-section of 3900 barns for thermal neutrons.

All of the kinetic energies of the alpha particle and the recoiling lithium nucleus will go into producing a large pulse of ionization in an ion chamber. This large pulse can be detected in the presence of a large interfering gamma-ray intensity.

16.08 Fast-neutron Reactions

Fast neutrons contribute both kinetic energy and binding energy to the compound nucleus, and this high excitation state makes some new reactions possible, and increases the probability of some others. We would expect that charged-particle emission would be more probable and that this should compete with radiative capture which dominates at lower neutron energies.

a. *Inelastic scattering.* We have already considered the elastic scattering of neutrons and we have seen that the incident neutron upon striking the nucleus reappears with all of its original energy. Because of the identity of the incident neutron and any nucleonic neutron, we cannot distinguish between the two. Elastic scattering is usually the only process that takes place in elements of low Z below 1 MeV and in elements of high Z below about 100 KeV. At higher energies than these, inelastic scattering is also observed; that is, the scattered neutron comes away from the nucleus with less energy than that of the incident neutron, the difference in energy going into an excited state of the residual nucleus, or appearing as a prompt gamma ray. We may regard such scattering as (n, n) or $(n, n\gamma)$ reactions. The occurrence of inelastic scattering depends upon there being an excited state at a low enough energy so that the incident neutron can supply this energy of excitation. In the case of carbon, for example, the first excited state is very high (about 4 MeV) so that almost all fast-neutron scattering in carbon may be considered elastic. However, in most heavy elements, the first excited state is within reach of fast neutrons, being about 100 KeV for many elements; therefore, fast-neutron scattering in heavy elements is predominantly inelastic.

b. *The* $(n, 2n)$ *reaction.* In order for the $(n, 2n)$ reaction to proceed, the incident neutron must have a kinetic energy of greater than about 9 MeV. This kinetic energy coupled with the 8-MeV binding energy makes possible the emission of two neutrons. With increasing neutron energy, the $(n, 2n)$ process is observed with a higher yield until competition from the higher-energy $(n, 3n)$ reaction takes place. Over 100 $(n, 2n)$ reactions have been observed and in most of these the residual nucleus is a positron emitter. For example, the bombardment of ^{12}C with very fast neutrons (energy greater than 20 MeV) produces the following reaction:

$$^{12}C + n \longrightarrow {}^{11}C + 2n + Q \tag{16-20}$$

The residual nucleus ^{11}C exhibits a 20.5-min positron activity.

Positron emission or electron capture is the favored mode of decay in the majority of the residual nuclei produced in the $(n, 2n)$ reaction, but in heavier nuclei, negatron emission is observed. The latter might be expected, since for such nuclei the loss of a single neutron does not upset the neutron–proton balance as seriously as it does in lighter nuclei.

c. *Charged-particle emission.* As we have already mentioned, the (n, p) and (n, α) reactions are observed with slow neutrons for a number of exceptional cases, but as a general rule these reactions require fast neutrons. In the case of the (n, p) reaction in aluminum, we consider the reaction:

$$^{27}\text{Al} + n \longrightarrow {}^{27}\text{Mg} + {}^{1}\text{H} + Q \tag{16-21}$$

The mass balance requires the incoming neutron to supply about 2.1 MeV of kinetic energy before the reaction will go. There are individual differences, but in general (n, p) reactions require about 1–3 MeV, and $(n, 2n)$s about 10 MeV for initiation. If the reaction product is radioactive, it may be identified and used as a *threshold detector* that responds only to energies above the reaction threshold.

16.09 Neutron Activation

The high neutron fluxes available in nuclear reactors have permitted the production of artificial radioactive materials in fantastic quantities. With a few exceptions, these radioactive isotopes are produced or *activated* by thermal neutrons, utilizing the large capture cross-sections usually associated with low neutron velocities.

The production of these isotopes by a neutron flux density ϕ is just a special case of the general relations described in Eqs. (14-12) and (14-13). The number of radioactive nuclei formed is given by

$$R = \frac{\phi \sigma_{\text{act}} N}{\lambda}(1 - e^{-\lambda t})e^{-\lambda t'} \tag{16-22}$$

where σ_{act} = cross-section for the particular reaction
N = number of target atoms available
t = bombardment time
t' = time from bombardment end to measurement

The activity is given by

$$A = R\lambda = \phi \sigma_{\text{act}} N(1 - e^{-\lambda t})e^{-\lambda t'} \tag{16-23}$$

Illustrative Example

Calculate the maximum activity that can be induced in a 100-mg copper foil exposed to a thermal neutron-flux density of 10^{12} n cm^{-2} sec^{-1}.

For the very long irradiation time required to obtain an activity maximum, Eq. (16-23) becomes $A = \phi \sigma_{\text{act}} N$. Taking pertinent data from Table 16-3,

$$^{63}\text{Cu:} \quad N = \frac{m \times 0.691}{M} \times 6.2 \times 10^{23} = \frac{0.1 \times 0.691}{63.57} \times 6.2 \times 10^{23}$$
$$= 6.55 \times 10^{20}$$

where m = the mass of the sample and M = the molecular weight.

$$^{65}\text{Cu:} \quad N = \frac{0.1 \times 0.309}{63.57} \times 6.2 \times 10^{23}$$
$$= 2.92 \times 10^{20}$$
$$^{64}\text{Cu:} \quad A = 10^{12} \times 4.5 \times 10^{-24} \times 6.55 \times 10^{20} = 2.95 \times 10^{9} \text{ dps}$$
$$^{66}\text{Cu:} \quad A = 10^{12} \times 2.3 \times 10^{-24} \times 2.92 \times 10^{20} = 0.67 \times 10^{9}$$
$$3.62 \times 10^{9} \text{ dps}$$

Total activity:

$$\frac{3.62 \times 10^{9}}{3.7 \times 10^{7}} = 98 \text{ mCi}$$

Two qualifications must be placed on Eq. (16-22). First, the target must be so thin that all of the target atoms are exposed to essentially the same neutron flux, even though the element has a very large capture cross-section. This is important in a few elements at resonance energy, where the capture cross-section may be very large. Examples are gadolinium, samarium, and cadmium. Second, the fraction of the target that undergoes reaction should be small enough so that there is no significant decrease in the number of target atoms. The second qualification is not too stringent, since the number of atoms consumed by a nuclear reaction is usually small compared to the total atoms in the sample.

Neutron irradiation of two samples, one with an unknown amount of an element and the other with a known amount of the same element, may be used to determine the unknown quantity of the element. This method, employed in what is known as *activation analysis*, proves very useful in practice, for it permits a convenient determination of the amount of the unknown by comparison of the activities produced in it and in the standard sample, i.e., the one with a known amount of the element.

Activation analysis may be used to determine the presence of trace quantities of an element. For this to be useful, however, the neutron-activation cross-section must be fairly high; and the half-life of the induced activity must be somewhere in the range of from 1 minute to several days if high sensitivity is to be achieved.

Illustrative Example

A specimen is to be analyzed for trace quantities of manganese. It is irradiated to saturation in a nuclear reactor with an average thermal flux of $10^{11}\ n\ \text{cm}^{-2}\ \text{sec}^{-1}$. If the apparatus used for measuring the induced ^{56}Mn activity measures 20 per cent of the emitted radiation and has a lower limit of 80 counts per minute sensitivity, what is the least amount of Mn that can be measured?

From Table 16-3, the activation cross-section is 13.3 b, so that by substituting in rearranged form, we have

$$m = \frac{MA}{0.602 F \sigma_{\text{act}}} = \frac{55 \times 5(80/60)}{0.602 \times 10^{11} \times 13.3} = 4.6 \times 10^{-10} \text{ g}$$

Obviously, ability to detect this extremely small amount of manganese would depend upon the observer's ability to measure the induced activity against whatever masking activity was produced in the sample due to other induced activities.

TABLE 16-3
DATA FOR (n, γ) ACTIVATIONS

Target nucleus	*Percent abundance*	*Product half-life*	σ_{act}, *barns*
^{2}H	0.014	12.6 y	0.0006
^{23}Na	100	14.9 h	0.5
^{27}Al	100	2.3 m	0.24
^{30}Si	3.1	2.6 h	0.1
^{31}P	100	14.3 d	0.2
^{34}S	4.2	88 d	0.27
^{37}Cl	24.2	37.1 m	0.4
^{40}Ar	99.6	1.8 h	0.61
^{41}K	6.8	12.4 h	1.2
^{51}V	99.7	3.8 m	4.9
^{50}Cr	4.3	27.5 d	17
^{55}Mn	100	2.6 h	13.3
^{59}Co	100	5.3 y	20
^{64}Ni	1.16	2.5 h	1.5
^{63}Cu	69.1	12.9 h	4.5
^{65}Cu	30.9	5.3 m	2.3
^{64}Zn	48.9	250 d	0.46
^{84}Kr	56.9	4.5 h	0.10
^{107}Ag	51.3	2.4 m	35
^{127}I	100	25 m	6.4
^{191}Ir	38.5	74 d	700
^{197}Au	100	2.7 d	99
^{202}Hg	29.8	46.9	4
^{208}Pb	52.3	3.3 h	0.0005
^{209}Bi	100	5.0 d	0.015

16.10 Neutron Measurements

The (n, α) reaction in ^{10}B, Eq. (16-19), which follows a $1/v$ law over the low-energy range, is commonly used as a detector of slow neutrons. In its usual form, the boron chamber consists of an ionization chamber either filled with BF_3 gas or lined with boron metal, in each case preferably using boron enriched in ^{10}B. The associated electronic circuits can be adjusted to respond only to the massive ionizations produced by the alpha particles and to reject all smaller pulses. With this discrimination, a boron chamber will measure thermal or slow neutrons in the presence of the inevitable gamma rays.

The boron chamber measures the particle density n, since the probability of the reaction depends upon the neutron population inside the chamber rather than upon the number of neutrons that enter and leave. In a detector where absorption processes follow a $1/v$ law, a measurement of n can be used as a measure of the particle-flux density ϕ, provided that the velocity of the neutrons used in the chamber calibration is known. This presents no problem if the flux is truly thermalized, but under many conditions little is known about v and its distribution.

A recoil chamber suitable for measuring fast neutrons will be filled with hydrogen, or some hydrogen-containing gas such as methane. Collisions will transfer energy primarily to the hydrogen atoms which will recoil, leaving a trail of ionization that can be recorded as a series of pulses. The response of a recoil chamber will be proportional to ϕ rather than to n, since the chance of a recoil depends upon the rate at which fast neutrons enter the chamber.

Intermediate-energy neutrons present special problems which have not been completely solved. In this energy region, the $1/v$ law is breaking down, to be replaced by complicated patterns of resonance absorption. The probability of the B(n, α) reaction is decreasing with energy, yet the energies may not be high enough to create useful proton recoils.

Similarly, neutron-dose measurements have not attained the degree of precision that is customary with either charged particles or photons. The complex absorption patterns of each nuclide makes difficult the construction of tissue-equivalent ion chambers. Hydrogen recoils play a large role in fast-neutron absorption, and any chamber material must match closely the hydrogen content of tissues that it is simulating. Rossi and Failla developed suitable chambers using a special plastic and a special counting gas.

Neutron-equivalent plastic		*Counting gas*	
Hydrogen	10.1 per cent	Methane	64.4 per cent
Carbon	86.4	Carbon dioxide	32.4
Nitrogen	3.5	Nitrogen	3.2

In these formulations, the oxygen content of tissue is replaced by carbon, with little error. With these components, chambers can be built that will measure neutron doses within perhaps 5 per cent over a range of 0.5–15 MeV.

Fast neutrons may be detected by the proton recoil tracks produced in a photographic emulsion. Film badges containing a nuclear-track emulsion are commonly used for monitoring the neutron exposures of personnel potentially exposed. After processing, the emulsion is examined with a microscope, and the number of recoil tracks seen in a standard area is recorded, and compared with the number of tracks seen after exposure to a standard source.

Considerable training is required to insure proper track recognition, and track counting is tedious at best, but it does provide a useful measure of fast-neutron exposures.

16.11 Energy and Velocity Measurements

Neutron beams cannot be produced at a specific energy with the ease characteristic of charged-particle beam production, nor can neutron energies be measured by indirect means, as by an accelerator voltage or a magnetic stiffness. These facts have required the development of direct methods for measuring and selecting neutron velocities.

In early experiments, a *mechanical velocity selector* was used to sort and measure neutron velocities. Two rotating shutters, each consisting of alternating absorbing and transmitting sectors, permitted the transmission of particles, whose velocity was fixed by the spacing between the two shutters and by the common speed of rotation. Mechanical limitations restricted the use of the mechanical velocity selector to rather low neutron velocities. The mechanical arrangement has now been superseded by electronic techniques capable of operating at much higher velocities.

The development of electronic-pulse techniques has provided means for *time-of-flight* measurements over a wide range of neutron energies. In the time-of-flight method, an accelerator, used as the primary source, is pulsed to provide a short burst of neutrons. Modern pulse techniques applied to high-intensity sources permit the production of usable pulses of only a few nanoseconds duration.

The times required for the individual particles in a pulse of neutrons to reach a distant detector can be converted to voltages by well-known time–pulse-height converters. These pulses can be recorded and analyzed by standard pulse-height methods. Thus neutron-velocity distributions are determined from direct measurements of distance and time. As pulse durations have decreased and high-intensity sources have permitted greater source–detector separations, the upper energy limit amenable to analysis has increased into the MeV region. Figure 16-7 shows a portion of the spectrum emitted in the $^{10}B(d, n)C$ reaction initiated by 1-MeV deuterons.

Low-energy neutrons have de Broglie wavelengths comparable to the interatomic spacings in crystal lattices and hence produce diffraction patterns at measurable angles. Relatively high-intensity neutron beams are needed for neutron-diffraction studies, but these beams are available to provide a powerful analytical tool complementing X-ray diffraction analysis.

Figure 16-8 shows the geometrical arrangement of a crystal spectrometer for neutron diffraction. Constructive interference will occur, Sec. 5.02, when

$$n\lambda = 2d \sin \theta$$

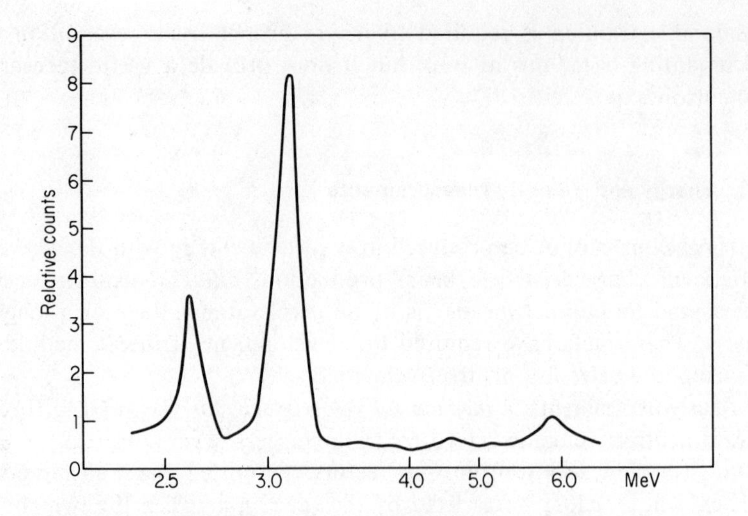

Figure 16-7. Portion of the neutron spectrum from $^{10}B(d, n)$ from time-of-flight measurements. (After G. C. Neilson, W. K. Dawson, and F. A. Johnson, *Rev. Sci. Instr.*, **30**, 963, 1959.)

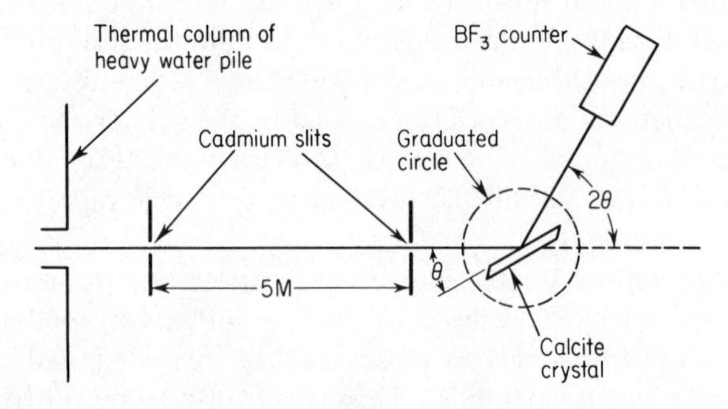

Figure 16-8. Schematic of crystal spectrometer for neutron diffraction.

whence

$$v = \frac{nh}{2dm \sin \theta} \tag{16-24}$$

where n = order of diffraction, 1, 2, . . .
d = distance between crystal planes
m = neutron mass
θ = angle of maximum intensity

Diffraction is a highly selective process, which permits the production of nearly monoenergetic neutron beams.

With beams of known velocity, neutron diffraction can be used to determine values of interatomic spacings in crystal structures. Since neutron-scattering cross-sections differ strikingly from the cross-sections for photon scattering, neutron diffraction provides structure analyses not obtainable with X-ray diffraction. Neutrons are most useful for in-depth analyses of organic compounds and for structures involving light atoms such as hydrogen in a matrix of heavy atoms. Hydrogen atoms may escape detection with X-ray diffraction because of the low scattering probability.

A crystal with a known lattice spacing can be used in an arrangement such as that shown in Fig. 16-8 to study the spectral distribution in low-energy neutron beams. High-intensity beams are needed, and the method is not suitable to the higher energies where the de Broglie wavelengths become very short.

Threshold detectors are used to obtain a rough idea of the energy distribution in a heterogeneous neutron beam. Figure 16-9 shows the reaction

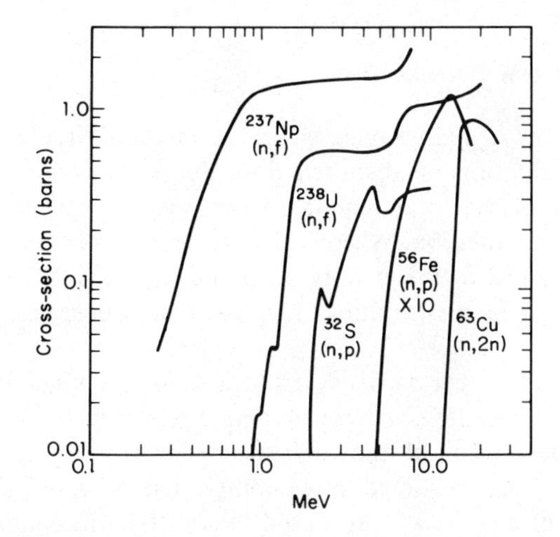

Figure 16-9. Activation cross-sections for commonly used threshold detectors.

cross-sections for a series of nuclides used as threshold detectors. Each reaction has a true threshold energy from which the cross-section rises more or less rapidly to more or less of a plateau. The effective threshold energy will be somewhat above the true threshold. Thus the true threshold for $^{238}U(n, f)$ is 0.7 MeV, but the effective threshold is more nearly 1.4 MeV.

Absolute counting methods serve to determine the activity produced in the sample foil. The neutron fluence can be calculated from a knowledge of the induced activity and the reaction cross-section.

Thermal neutrons are of special importance in many applications and it is usually desirable to know to what extent a thermal flux is contaminated with fast neutrons. This is usually done by determining the *cadmium ratio*. A boron chamber is first used to obtain a measure of the total particle density in the beam. A second measurement is then made with a thin cadmium shield placed over the chamber. This shield will effectively absorb all neutrons with energies below 1 eV, without seriously attenuating those that are more energetic, Fig. 16-2. The chamber response will now be due almost entirely to the fast neutrons, and the ratio of the two readings will serve to show the degree of contamination.

In the reverse technique, a heterogeneous beam is first measured with an unshielded boron chamber to obtain a reading characteristic of the entire spectral range. A second reading is then taken with a good moderator (such as pure carbon) placed over the chamber. Most of the fast neutrons will now be thermalized and will be detected more effectively. The ratio of the two readings gives some indication of the thermal: fast ratio in the original beam.

16.12 Fluence–Dose Calculations

The *kerma*, or *k*inetic *e*nergy *r*eleased in the *ma*terial, is a unit designed to facilitate the calculation of absorbed dose due to neutrons. The kerma is defined in terms of the total kinetic energy released by the absorption of indirectly ionizing radiation. Absorbed dose and kerma differ only by the amount of energy lost to the vicinity of the primary event from the production of penetrating bremsstrahlung. Except at high energies, the two units will be identical.

Calculations of the kerma delivered to a tissue are made by a stochastic method known as a *Monte Carlo calculation*. These calculations have become possible with the advent of the high-speed electronic computer, without which the method has restricted applicability. The Monte Carlo concept is, however, far from new. It was first used in statistical mechanics by Willard Gibbs under the name of the *random walk* of a particle in phase space.

In essence, the Monte Carlo technique applied to neutron absorption consists in determining in detail the case history of a neutron from the time it enters the absorber until it leaves, is absorbed, or is degraded in energy to a level of no interest.

All possible types of interactions and all possible values of the variables are coded to a series of random numbers. The first random number will determine the type and location of the first collision. Should this be a neutron absorption, the case history of this particular particle is ended, and it remains to trace in detail only the secondary products of the absorption. In a nonabsorptive collision, random numbers will determine the energy transfer, scatter angle, and all other pertinent details. Each of the secondary products involved will then be followed by random-number choices as long as they can contribute energy to the absorber. When the complete history of one neutron is obtained, the process is repeated until enough data are obtained to form a highly probable picture of the actual complex situation. Detailed Monte Carlo calculations require the use of large computers because of the many variables involved and the need for a large-capacity memory system.

Monte Carlo calculations of kerma have been made by Snyder and

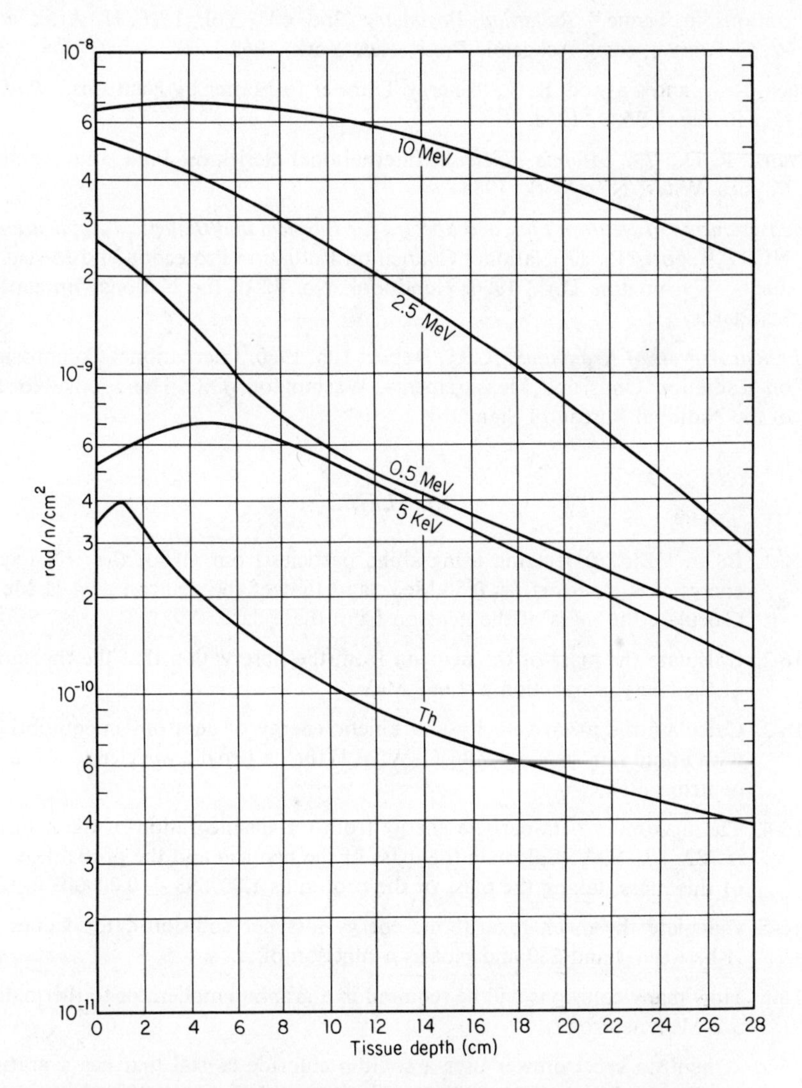

Figure 16-10. Energy deposition in an infinite slab of soft tissue. (After W. S. Snyder and J. Neufeld, *Brit, J. Radiol.*, **28**, 342, 1955.)

Neufeld. Their results for the dose in a slab of tissue of infinite extent are shown in Fig. 16-10. These calculations have been extended to apply to a variety of shapes approximating human bodies and organs.

REFERENCES

Auxier, J. A., Snyder, W. S., and Jones, T. D., "Neutron Interactions and Penetrations in Tissue." *Radiation Dosimetry*. 2nd ed., Vol. I (F. H. Attix and W. C. Roesch, eds.). Academic Press, New York, 1968.

Bach, R. L. and Caswell, R. S., "Energy Transfer to Matter by Neutrons." *Radiation Research* **35,** 1, 1968.

Evans, R. D., *The Atomic Nucleus*. International Series on Pure and Applied Physics. Wiley, New York, 1955.

Measurements of Neutron Flux and Spectra for Physical and Biological Applications. NCRP Report No. 23. National Council on Radiation Protection and Measurements, Washington, D.C., 1960. Handbook No. 72 of the National Bureau of Standards.

Physical Aspects of Irradiation. ICRU Report 10b, 1960. International Commission on Radiation Units and Measurements, Washington, D.C. Handbook No. 85 of the National Bureau of Standards.

PROBLEMS

16-1. In an $^{11}B(\alpha, n)$ reaction using alpha particles from ^{210}Po, the ^{14}N recoil energy was measured as 0.80 MeV, and that of the neutron as 4.31 MeV. Calculate the mass of the neutron from these data.

16-2. Calculate the mass of the neutron from the observation that the threshold of the $^{9}Be(\gamma, n)$ reaction is 1.668 MeV.

16-3. Calculate the mean velocity and kinetic energy of neutrons in equilibrium with liquid nitrogen at $-196°C$. What is the de Broglie wavelength of these neutrons?

16-4. The maximum beta-particle energy from the disintegration of the neutron is 782 ± 13 KeV. Calculate the mass of the neutron and the probable error of this mass, taking the mass of the proton as 1.007593 ± 0.000003 u.

16-5. Calculate the mean logarithmic energy loss per collision ξ for values of A between 1 and 250 and plot as a function of A.

16-6. How many collisions will be required in a graphite moderator to thermalize a 2-MeV neutron?

16-7. A neutron spectrometer uses a sodium chloride crystal that has a grating spacing of 2.81*A*. What will be the energy of the neutrons selected at a diffraction angle of 12° by first-order diffraction?

16-8. Cadmium-113, which has a natural abundance of 12.2 per cent, has an absorption cross-section of 20,000 b for thermal neutrons. What fraction of a thermal beam will be transmitted by an 0.3-mm foil of natural cadmium, density 8.6 g cm^{-3}?

16-9. Boron is frequently used in reactor control rods as an absorber of thermal neutrons. What is the relative absorption in boron of room-temperature neutrons and reactor neutrons in equilibrium at 260°C?

16-10. What is the kinetic energy of the neutron emitted in $^3H(d, n)$ by a deuteron of 200 KeV?

16-11. A 1-liter container is placed in the thermal-neutron beam from a reactor where the thermal flux density is $4 \times 10^{12} n$ cm^{-2} sec^{-1}. How much hydrogen will be produced by neutron decay in this volume during 1 year?

16-12. Silver-109 has a resonance absorption at 5.1 eV with a cross-section of 7600 b and a Γ of 0.19 eV. What fraction of the reactions at resonance will be (n, n)?

16-13. A sample of natural copper is exposed to a thermal-neutron flux density of $2 \times 10^{13} n$ cm^{-2} sec^{-1}. How long an irradiation will be required to convert 1 per cent of the available atoms to ^{64}Cu?

16-14. A $^{10}BF_3$-filled ion chamber was calibrated with thermal neutrons and was then used to measure a beam of 1 eV mean energy. What will be the relative fluxes for a given instrument reading?

16-15. An experiment requires 6 mCi of ^{32}P twelve days after it becomes available at the reactor, which has a thermal flux density of $6.5 \times 10^{12} n$ cm^{-2} sec^{-1}. How long must a 300-mg sample be irradiated to obtain the required activity? What will be the activity of the sample when it is removed from the reactor?

16-16. What maximum specific activity can be obtained in terms of ^{60}Co produced from natural cobalt in a reactor with a thermal neutron flux density of $4 \times 10^{12} n$ cm^{-2} sec^{-1}? How long an irradiation will be required to reach 25 per cent of the maximum activity?

16-17. A 10-g sample of natural cobalt is irradiated for 3 years in the reactor of Prob. 16-16. What will be the gamma-ray dose rate due to ^{60}Co at the end of the irradiation, 1 m from the source?

16-18. How many ion pairs will be formed in a $^{10}BF_3$-filled ion chamber when an epithermal neutron is captured?

17

Nuclear Fission

17.01 Discovery and Early Experiments

Shortly after the discovery of the neutron, Enrico Fermi and his collaborators in Italy began neutron irradiation of many elements, one of them being uranium. Using a Ra–Be neutron source, they found that the emitted neutrons produced several artificial radioactivities which could not be identified with any of the elements in the range of $Z = 86$ to 92. The new activities were believed to be due to transuranium elements produced as a result of neutron capture and subsequent beta decay. Actually, the first fission reaction had been induced, but it required some five more years for a complete explanation of the new nuclear reaction.

Fermi's neutron experiments gave considerable stimulus to neutron–uranium research in other countries, for the prospect of discovering elements beyond atomic number 92 proved very enticing. In 1938, Curie and Savitch chemically separated from neutron-irradiated uranium an element that precipitated out with lanthanum and had a 3.5-hr activity. Lanthanum was not known to have any such characteristic activity and the investigators concluded that it must be assigned to one of the elements beyond uranium.

Hahn and Strassman in Germany repeated the Curie–Savitch experiment and proved that the 3.5-hr activity must be assigned to barium. They were reluctant, however, to specify that the neutron-induced process was anything so drastic as splitting the uranium atom in half and it remained for L. Meitner and O. Frisch to conclude: "It seems possible that the uranium nucleus has only a small stability of form, and may, after neutron capture, divide itself into two nuclei of roughly equal size." To this radically new nuclear reaction they gave the name *fission*. They pointed out the analogy between the fission process and the division of a small liquid sphere into smaller drops as a result of physical deformation of the droplet. Furthermore, they pointed out that in very heavy nuclei the mutual repulsion of protons becomes of the same order as the cohesive forces within the nucleus. Thus, a small amount of energy added to a uranium nucleus might be able to produce such a deformation that it would be split asunder. The two heavy nuclei formed in fission (called *fission fragments* or *fission products*) would each carry off very high kinetic energy as a result of the breakup of the uranium nucleus. Meitner and Frisch showed that the barium, lanthanum, and other elements formed in fission should have an excess of neutrons and hence should be highly unstable. It was clear that the activities previously held due to transuranium elements must be attributed to new activities among the fission products.

A basis for theoretical understanding of fission was laid by Bohr and Wheeler in the fall of 1939. Their calculations predicted that the fission of uranium under slow-neutron bombardment should be attributed to the ^{235}U isotope of uranium rather than to the heavier most-abundant isotope ^{238}U. Within a year, this prediction was confirmed by the experiments of Dunning and his colleagues at Columbia University working in association with A. O. Nier. A very small sample of ^{235}U was separated from ^{238}U by magnetic-deflection methods, and thermal-neutron fission of the light isotope established. It was found that ^{235}U was responsible for thermal fission in uranium and a cross-section of between 400 and 500 barns was estimated for the process. At the time, it was recognized that more than one neutron was produced per uranium fission. Measurements ranged from one to three neutrons per fission, but the results of Zinn and Szilard, indicating 2.3 neutrons per fission, seemed most reliable.

In the summer of 1940, the possibility of using atomic energy for military purposes imposed restrictions, first voluntary and later mandatory, on fission research, and a cloak of secrecy enshrouded nuclear physics laboratories. Already the basic phenomena of fission had been investigated. Thorium, protactinium, and uranium had been fissioned with fast neutrons, but only ^{235}U fissioned with slow neutrons. It was known that the fission fragments ranged from $Z = 34$–74 and that these fragments carried off most of the approximately 200 MeV released in fission. Many of the activities of the fission products had been studied and some decay chains identified. The cross-section for thermal fission in ^{238}U was known to be much smaller than that for thermal capture. ^{238}U was known to have a large resonance absorption at 7 eV, which did not lead to fission.

In 1940, the large-scale release of nuclear energy through fission appeared feasible. Many technical problems required solution, but no obstacles seemed insurmountable. With plutonium discovered and its fissionability confirmed, an alternative approach was opened up. Late in 1940, the United States government authorized a large-scale effort to develop a nuclear explosive. Within $3\frac{1}{2}$ years, the first nuclear detonation occurred at Alamogordo, New Mexico.

17.02 Types of Fission

a. *Thermal fission.* Fission of ^{235}U and of ^{239}Pu by thermal neutrons are the most important reactions leading to the release of nuclear energy. Since a thermal neutron adds negligible kinetic energy to a fissionable nucleus, it is of importance to determine why thermal fission takes place readily in ^{235}U and not in ^{238}U. The explanation lies in the δ term of the binding-energy equation, Eq. (7-17). In applying this expression to neutron-induced fission, we must remember that the fissioning structures are the compound nuclei, ^{236}U and ^{239}U in the present case. Both ^{236}U and ^{240}Pu are even–even nuclei in which the neutron is more tightly bound than in the even Z–odd N structure of ^{239}U. A neutron captured in ^{235}U makes available more energy than will a capture by ^{238}U. With a critical energy threshold for the fission process, the nucleon pairing terms can be decisive.

Figure 17-1 shows the fission cross-sections for the two common uranium isotopes, together with the energy spectrum of the neutrons emitted in fission.

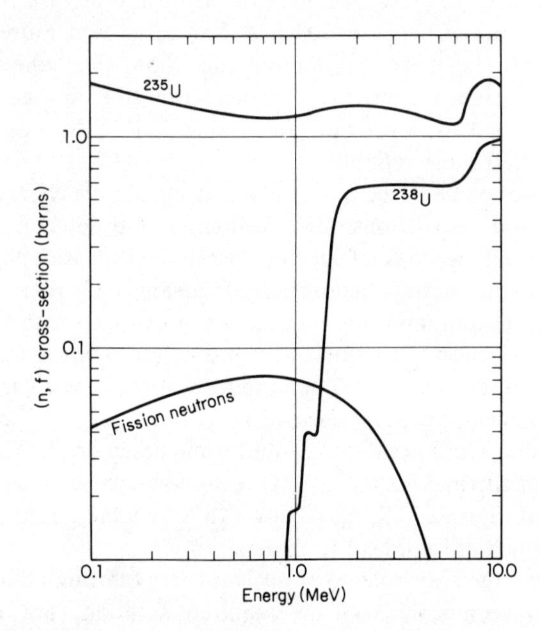

Figure 17-1. Fission cross-sections for ^{235}U and ^{238}U with a typical spectrum of neutrons emitted in fission.

It is evident that the fission probability is greater in ^{235}U than in ^{238}U at all energies. When the two fission cross-sections approach equality above 1.5 MeV, the number of fission neutrons is decreasing rapidly. The chance of a thermal fission or a self-sustaining chain reaction in pure ^{238}U is zero. The relative cross-sections for some other reactions can be seen from Table 17-1.

TABLE 17-1
THERMAL NEUTRON CROSS-SECTIONS (BARNS)

	^{235}U	^{238}U	*Natural U*	^{239}Pu
Fission	580	0.0	4.2	750
Activation	110	2.8	3.5	310
Scattering	10	8.2	–	10
Absorption	690	2.8	7.7	1060

b. *Fast Fission.* ^{238}U can be fissioned by fast neutrons, and there is a small probability of some of these fissions in a chain reaction sustained by some other species. A nuclear reactor containing some ^{238}U in its fuel will derive some of its energy from ^{238}U fissions, by what is known as the fast effect. From Fig. 17-1, it is evident that this contribution will amount at most to a few per cent of the total fissions.

Other isotopes of uranium and other elements enter into (n, f) reactions with fast neutrons. These nuclei are said to be *fissionable,* in contrast to ^{235}U and the other nuclei that can be fissioned with a neutron capture at zero energy. The latter nuclei are *fissile.* Uranium fission is the most widely exploited and will receive the most attention here.

c. *Charged-particle fission.* Fission in elements above atomic number 90 has been produced by bombardment with protons, deuterons, and alpha particles. Protons at about 7 MeV will excite the (p, f) reaction in uranium, and deuterons of slightly more energy suffice for the (d, f) reaction. The latter can be considered to be neutron-induced because of the polarization of the deuteron as it approaches the nucleus. High-energy charged particles will induce fission in elements even in the middle of the periodic table. The reactions here are highly endoergic. About 50 MeV is required in the case of ^{63}Cu fissioning to ^{38}Cl and ^{25}Al.

d. *Photofission.* High-energy photons will induce fission in the heavier elements. In the case of ^{238}U, there is a fairly sharp threshold for (γ, f) at 5.1 MeV. Cross-sections are small, of the order of 10^{-3} barns.

17.03 Spontaneous Fission

In 1940, Flerov and Petrazhak discovered that natural uranium undergoes spontaneous fission. Libby had attempted to measure the effect in 1939, but the detection equipment then available was not able to demonstrate it. Subsequent measurements have revealed spontaneous fission in over 30 nuclides, some occurring naturally and some artificially produced. In almost every case, spontaneous fission is in competition with alpha emission, which is the predominant mode of decay. In natural uranium there are only about 25 spontaneous fissions per hour per gram of the element.

Spontaneous fission is predicted by the empirical nuclear-mass equation. Consider specifically the case of symmetrical fission, where the original nucleus splits into two equal parts. Neglecting the pairing term δ, we have, for the differential mass,

$$\Delta M = M(Z, A) - 2M(Z/2, A/2) \tag{17-1}$$

$$M(Z, A) = a_1(A - Z) + a_2 Z - a_3 A + a_4 A^{2/3} + a_5 Z^2 A^{-1/3} + a_6(A - 2Z)^2 A^{-1}$$

where the coefficients are to be identified with the numerical values in either Eq. (7-16) or Eq. (7-17).

$$2M\,(Z/2, A/2) = 2\left[a_1 \frac{A - Z}{2} + a_2 \frac{Z}{2} - a_3 \frac{A}{2} + a_4 \frac{A^{2/3}}{2} + a_5 \frac{Z^2 2^{1/3}}{4A^{1/3}} a_6 + \frac{(A - 2Z)^2}{2A}\right]$$

The terms in a_1, a_2, a_3, and a_6 cancel in the subtraction and

$$\Delta M = a_4\left(A^{2/3} - \frac{2A^{2/3}}{2^{2/3}}\right) + a_5\left(\frac{Z^2}{A^{1/3}} - \frac{2Z^2 2^{1/3}}{4A^{1/3}}\right)$$

$$= a_4 A^{2/3}(1 - 2^{1/3}) + a_5 \frac{Z^2}{A^{1/3}}\left(1 - \frac{1}{2^{2/3}}\right)$$

Evaluating the constants in the MeV system,

$$\Delta M = -3.58A^{2/3} + 0.22\frac{Z^2}{A^{1/3}} \text{ MeV} \tag{17-2}$$

Fission will take place if the two fission fragments, each a sphere of radius $r = r_0(A/2)$, are separated to the point of being just in contact. The electrostatic energy of this configuration is

$$E = \frac{Z^2 e^2}{4} \times \frac{1}{2r} = \frac{Z^2 e^2 2^{1/3}}{8r_0 A^{1/3}} \text{ ergs}$$

Evaluating the constants and converting to MeV,

$$E = 0.15\frac{Z^2}{A^{1/3}} \text{ MeV} \tag{17-3}$$

If spontaneous fission is to occur, the energy requirement of Eq. (17-3) must come from the energy loss of Eq. (17-2). This requires that

$$-3.58A^{2/3} + 0.22\frac{Z^2}{A^{1/3}} > 0.15\frac{Z^2}{A^{1/3}} \tag{17-4}$$

For spontaneous fission, therefore, we have the requirement

$$\frac{Z^2}{A} > 51 \tag{17-5}$$

The binding-energy equation, based on the liquid-drop model, predicts the spontaneous fission of heavy nuclei, for only in them can the inequality of Eq. (17-5) be satisfied. Fission at somewhat lower than the predicted values of Z^2/A are to be expected, because no account was taken in the derivation of the possibility of barrier tunneling. Spontaneous fission has been detected for Z^2/A as low as 35.

About two-thirds of the known examples of spontaneous fission occur in even-even nuclei. In this group the logarithm of the half-life for spontaneous fission ($\log T_{SF}$) is nearly a linear function of Z^2/A, Fig. 17-2. When similar data are plotted for other nucleon combinations, the points tend to fall at much longer half-lives, as if some hindrance were involved. In all cases spontaneous fission is much less probable than decay by alpha emission. Because of its important relation to the fission process, the factor Z^2/A is known as the *fissionability parameter*.

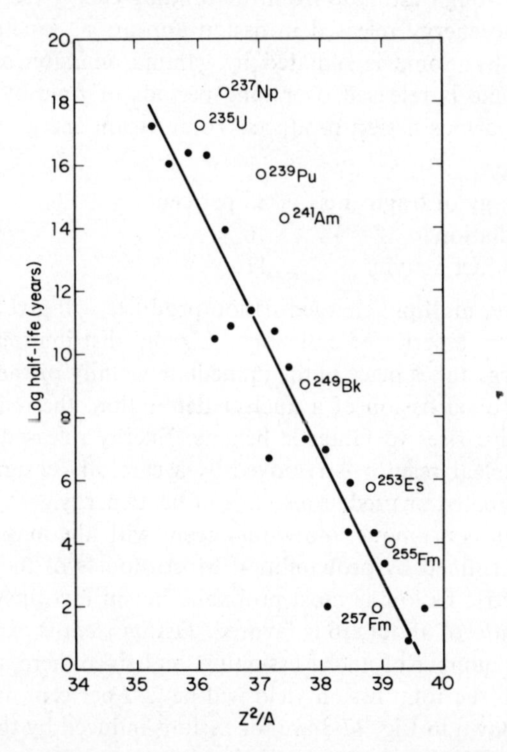

Figure 17-2. Dependence of the half-life for spontaneous fission on the parameter Z^2/A. Even–even nuclei are denoted by solid circles.

17.04 Energy Release in Fission

An examination of Eq. (17-3) shows that the symmetric fission of a uranium nucleus should increase the binding energy by slightly less than 1 MeV per nucleon. A total energy release of about 200 MeV would be expected from the fission of a uranium nucleus, or one of its neighbors.

A more exact value of the fission energy can be obtained from the mass balance of a fission reaction. A typical nonsymmetric fission reaction is given by Eq. (17-6), which shows the final, stable end products of the reaction rather than the primary fission products:

$$^{235}_{92}U + ^{1}_{0}n \longrightarrow ^{139}_{57}La + ^{95}_{42}Mo + 2^{1}_{0}n \tag{17-6}$$

Mass numbers are balanced in reaction (17-6) in accordance with the requirements of nucleon conservation. Atomic numbers are not balanced because each stable end product is the last member of a *fission chain* formed by several negatron decays from the primary fission product. From tables of mass values, we find that reaction (17-6) is exoergic by 208 MeV, in good agreement with our rough estimate from the binding-energy curve.

Most of the energy released in fission appears as kinetic energy of the fission fragments. Some is radiated by gamma emission at the instant of fission, and some is released over long periods of time by the radioactive decays of the various fission products. Total fission energy is divided about as follows:

Kinetic energy of fragments	83 per cent
Prompt radiation	6
Fission-product activity	11

The massive, multiply charged fission products will give up their kinetic energy locally in a series of collisions. A rapid distribution of the bulk of the fission energy takes place in the immediate vicinity of the initial reaction. In the uncontrolled fission of a nuclear detonation, the temperature of the exploding device rises to fantastic heights. Energy released by steady-state fission in a nuclear reactor is removed by a carefully designed cooling system, to be dissipated or used as a source of heat energy.

Each fission is a unique, individual event with the masses of the fission fragments determined by probabilities. Most models of the fission reaction predict symmetric fission as most probable, but in fact unsymmetric fission with a mass ratio of about 1.6 is favored. *Fission yield* is expressed as a percentage of the number of nuclei fissioning, and since there are two products for each event, the total fission yield will be 200 per cent and not 100. The yield curves shown in Fig. 17-3 are for fissions induced by thermal neutrons. Symmetrical fission becomes more favored as the exciting energies increase, and at 400 MeV it is the most probable reaction.

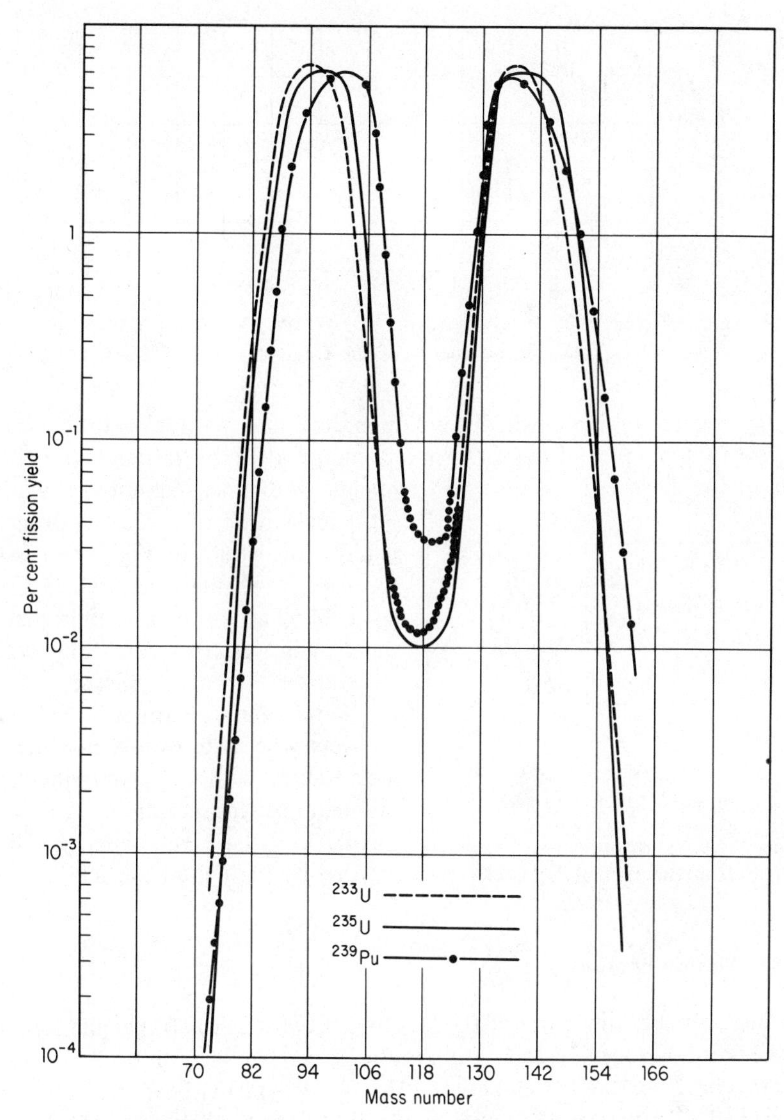

Figure 17-3. Fission-product yields showing the minimum at symmetry that is typical of thermal fission.

Fission-product mass numbers range from 72 to 158, including some 200 radioactive nuclides of 34 elements. About 60 nuclides have been identified as primary products, the rest originating from the radioactive decays of these primaries.

It is evident from the shape of the binding-energy curve that the total

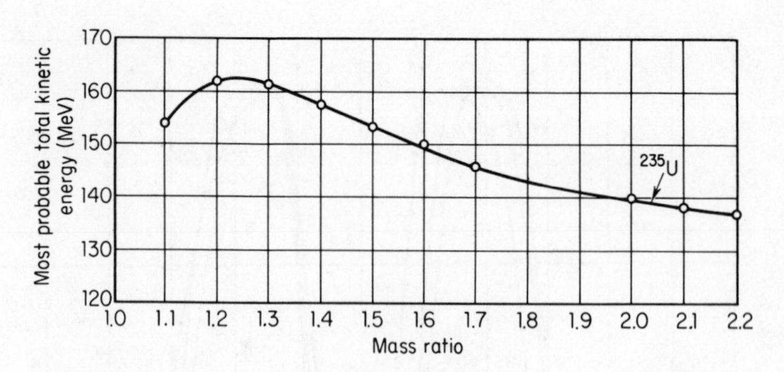

Figure 17-4. Total kinetic energy of the fragments from ^{235}U fission.

energy released in fission will depend upon the Z and A values of the products. Figure 17-4 is a plot of the total kinetic energy available in fission as a function of the mass ratio of the two fragments. A distinct maximum is evident at a mass ratio of 1.25. The observed energy distribution, Fig. 17-5, has a most probable mass ratio of $93.1/61.4 = 1.52$. The most probable fission process is not, therefore, that which maximizes the energy release.

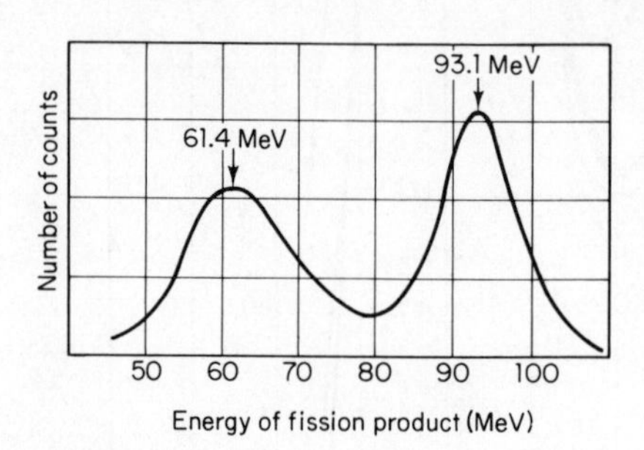

Figure 17-5. Distribution of fragment energies from the thermal fission of ^{235}U.

Total kinetic energy will be divided between the two fission products in accordance with the requirements of momentum conservation. About 60 per cent of the kinetic energy will be carried by the lighter particle.

17.05 Fission Energetics

We have already made use of the fact that if two nuclear fragments separate to the point of tangency, separation is almost certain to continue. At the point of tangency, nuclear forces between the two fragments have just decreased to zero. This position represents the height of the potential barrier, beyond which all mutual forces will be coulomb repulsion. The height of the potential barrier is easily calculated. In the general case,

$$E_B = \frac{Z_1 Z_2 e^2}{r_1 + r_2} \tag{17-7}$$

where the two nuclear radii are calculated from $r = r_0(A)^{1/3}$.

The energy thus calculated is shown as E_B in Fig. 17-6. Before any separation started, $r = 0$, and the system was in an energy state E_0. Except for the few that tunnel through the barrier, a *critical energy* $E_B - E_0$ must be added to each nucleus to induce fission. Thus the probability of fission depends strongly upon E_0, and hence upon the shape of the potential-energy curve below $r = r_B$. Three possible potential-energy functions must be considered, as depicted in Fig. 17-6.

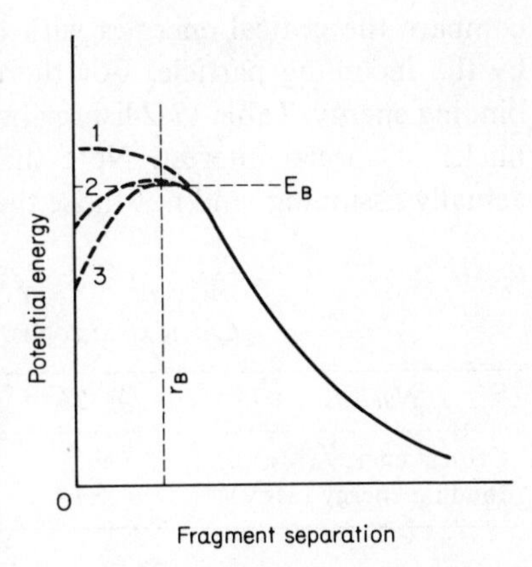

Figure 17-6. Potential energy functions inside and outside the breakaway radius.

1. $E_0 > E_B$. Here no energy need be supplied to surmount the barrier, and fission proceeds spontaneously. This situation, represented by curve 1, Fig. 17-6, is characteristic of the transuranium elements that are missing in nature because of their instability.
2. $E_0 < E_B$. When the energy difference is small, some tunneling can occur with relatively long half-life. These are the spontaneous fissions seen in some of the heavy, naturally occurring nuclei. We are particularly interested in those cases where $E_B - E_0 \simeq 6$ MeV, which is about the energy supplied to a heavy nucleus by a thermal neutron.
3. $E_0 \ll E_B$. This is the case of medium-mass nuclei, where a large activation energy is required to induce fission. In the limiting case at about $A = 80$, no energy is released. At this point, curve 3, Fig. 17-6, would start from the origin.

It is of interest to consider the variations of E_0 and E_B with mass number. The latter can be calculated from Eq. (17-7), and the former can be obtained either from measured mass values or from the nuclear-mass equation. Plots of the two energies are shown in Fig. 17-7. Each vertical distance between the two curves represents the activation energy that must be supplied to insure fission. Note that the activation energy decreases steadily as A increases. The curves cross at $A = 250$, the transition point from case 2 to case 1.

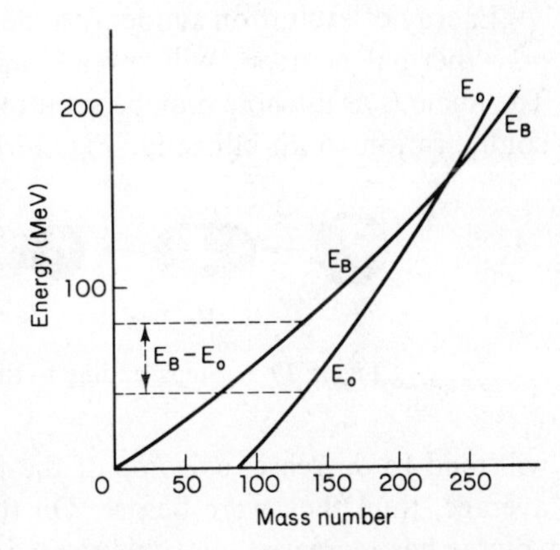

Figure 17-7. Variations of barrier energy and energy inside the well with mass number.

To determine the energetic possibilities for fission, it is necessary to compare the critical energies with those brought to the compound nucleus by the incoming particle. For thermal neutrons, the latter will be just the binding energy. Table 17-2 lists critical and neutron binding energies for some nuclei of special interest. Note that the nuclei are listed as the structures actually fissioning, and not those that initially captured the neutron.

TABLE 17-2
CRITICAL AND NEUTRON BINDING ENERGIES

Nucleus	^{233}Th	^{234}U	^{236}U	^{239}U	^{240}Pu
Critical energy (MeV)	6.5	4.6	5.3	5.5	4.0
Binding energy (MeV)	5.1	6.6	6.4	4.9	6.4

An inspection of Table 17-2 reveals that ^{235}U can enter into fission by capturing a thermal neutron, while the more abundant ^{238}U cannot.

17.06 The Fission Mechanism

Meitner and Fritsch were the first to suggest that fission might be understood in terms of a liquid-drop model. Bohr and Wheeler developed a theory of fission on this model, studying the energy changes which take place when a spherical nucleus is deformed into an ellipsoid.

The model assumes the nucleus to consist of an aggregation of incompressible nucleons, held together by a surface tension-like force against the coulomb repulsion of the protons. The latter are assumed to be rather uniformly distributed throughout the volume.

Energy of excitation suddenly added to the structure, as upon the capture of a thermal neutron, will spread rapidly throughout the nuclear volume. The nucleus as a whole may be set into oscillation, changing from a spherical configuration to an ellipsoid, Fig. 17-8. The electrostatic force of repulsion

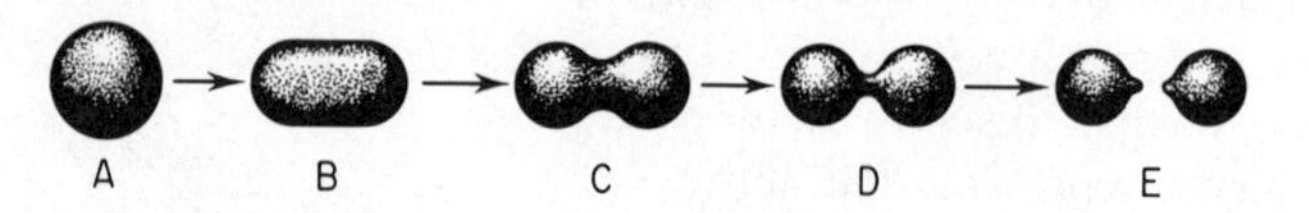

Figure 17-8. Steps leading to the fission of a liquid-drop model.

will tend to decrease, as some of the protons are now farther apart, on the average, than they were before. On the other hand, the ratio of surface to volume has increased, and this reduces the effectiveness of the nuclear force. As these force variations come into play, the nucleus executes a series of rapid oscillations. If the amplitude of the oscillation exceeds some critical value, a central constriction will appear. The nuclear force across this reduced area will no longer be able to hold together the two parts of the dumbell-shaped nucleus. The constriction then narrows rapidly and pinches off, releasing the two fission fragments. The whole process requires perhaps 10^{-12} sec.

Many of the predictions of the liquid-drop model agree with experience. When the elliptical nucleus is considered in detail, a value of $Z^2/A = 45$ is obtained as the limit of spontaneous fission. This value is in closer accord with observation than is the higher value that we obtained by a simpler treatment.

The liquid-drop model predicts symmetrical fission as most probable, a result at sharp variance with the facts. Attempts have been made to modify the liquid-drop model by introducing some features of the shell model, but these have not been helpful in explaining the phenomena of fission. Magic numbers have been invoked, but again the results are not convincing. A complete description of the fission mechanism is not yet available.

Three-body fission is predicted by the liquid-drop model and has been observed in about 1 fission out of 10^4. It is surprising that *tripartition* occurs so rarely, for it is energetically favored by at least 10 MeV. Low probabilities of forming proper nucleon groupings appear to overwhelm energy considerations.

Ternary fission is a term sometimes applied to a division into two fission products of normal mass and a high-energy alpha particle. In ^{235}U this occurs in about 1 out of 3000 events. Evidence indicates that these alpha particles are released as a part of the fission process and are not nuclear evaporations from one of the products of a binary fission.

17.07 Prompt and Delayed Neutrons

In each fission, one or more neutrons will be emitted along with the fission fragments. It is, of course, the average number emitted per fission, Table 17-3, that determines the probability of producing a self-sustaining chain reaction.

TABLE 17-3
AVERAGE NEUTRON EMISSION PER THERMAL FISSION

Capturing nucleus	^{232}Th	^{233}U	^{235}U	^{238}U	^{239}Pu
Neutrons per fission	1.88	2.48	2.40	2.30	2.90

There is a strong correlation between the directions of the neutron emissions and the directions of the fission fragments, suggesting that the neutrons come from the fragments after separation, rather than originating as a part of the primary process. Emission from the fragments is not surprising, because they will have an excess of neutrons for maximum stability.

More than 99 per cent, the *prompt* neutrons, will be emitted within 10^{-16} sec of the initial event. Prompt-neutron emission increases somewhat with the energy of the original neutron. Prompt neutrons are emitted with a continuous energy distribution ranging from almost zero to several MeV. Figure 17-9 shows the energy distribution for the thermal fission of ^{235}U. Several mathematical expressions have been used to describe the distribution, $e^{-E} \sinh \sqrt{2E}$ being quite satisfactory. The most probable prompt-neutron energy is 0.85 MeV, the average is about 2.0 MeV.

Delayed neutrons are also observed after fission. About 0.65 per cent of all ^{235}U fissions emit delayed neutrons, whose number decreases with time according to several distinct half-lives. Six important groups of delayed neutrons are observed from ^{235}U fission, with half-lives ranging from 0.23 to 55.7 sec. Only 0.21 per cent of ^{239}Pu fissions emit delayed neutrons. Six distinct groups have been detected here also, with half-lives ranging from 0.26 to 55 sec. Figure 17-10 shows the decay of neutron activity with time after ^{239}Pu fission.

Although their number is small, delayed neutrons are very important to the positive control of nuclear reactors. Without delayed neutrons, the flux buildup would be too rapid for control by mechanical means. The fraction of delayed neutrons permits mechanical devices to act rapidly enough to maintain the reaction rate at the desired level. A reactor is said to be "delayed

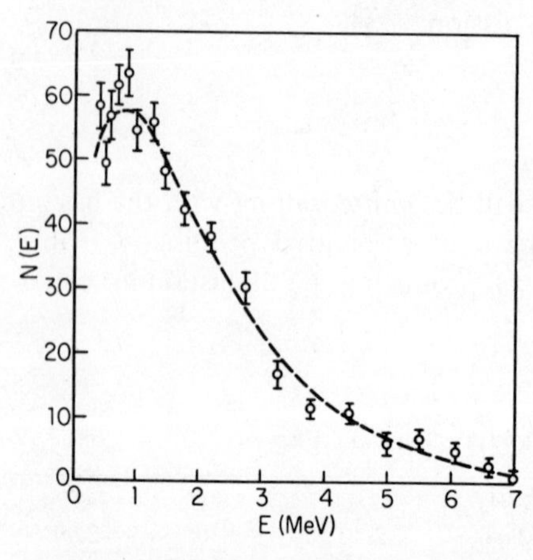

Figure 17-9. Energy distribution of the neutrons emitted in the thermal fission of ^{235}U. Observed points plotted. The curve is a plot of $e^{-E} \sinh 2E$. (From N. Nereson, *Phys. Rev.*, **85**, 601, 1952.)

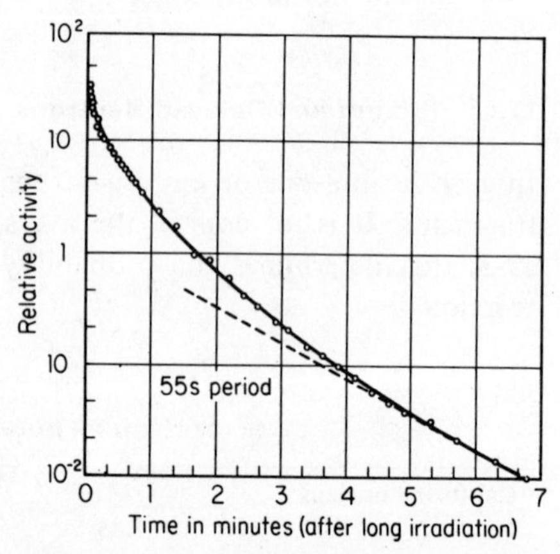

Figure 17-10. Delayed neutron activity after the fission of ^{239}Pu, showing the transition into a 55s half-life.

critical" or "prompt critical," the latter meaning that the reactor is operable without the contribution of the delayed neutrons. Under this condition, the reactor is potentially out of control.

17.08 Fission Product Radioactivity

A symmetrical binary fission of the compound nucleus $^{236}_{92}U$ with two free neutrons emitted leads to two product nuclei, $^{117}_{46}Pd$. These nuclei contain seven neutrons more than the heaviest stable isotope of Pd and would, therefore, be expected to be unstable to beta decay. Negatron emission would be expected, for this will reduce the neutron excess to produce daughters nearer to the stable neutron–proton ratio.

A neutron excess is also found in the products of asymmetric fission, Fig. 17-11. Except for the loss of a few neutrons at fission, the products will have the same N/Z ratio as the compound nucleus ^{236}U. The products will lie then along the straight line joining the compound nucleus and the origin rather than along the curved stability line.

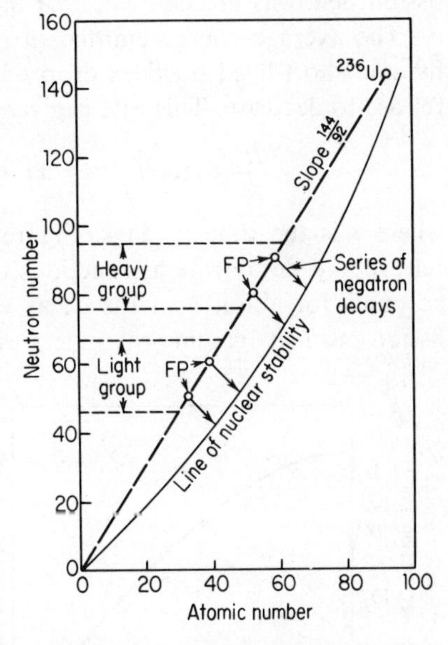

Figure 17-11. Neutron–proton plot of the fission products from ^{235}U. The curvature of the stability line has been exaggerated.

In general, the fission products have such a neutron excess that several negatron emissions are required to attain stability. These successive decays give rise to *fission chains*. Three decays are observed on the average, but chains of six are known.

Experimental identification of the primary-fission products is difficult because the great degree of initial instability leads to very short half-lives. One or more decays may have taken place in the earliest samples practically obtainable. A few cases are known where the product is formed between unusually neutron-rich stable isobars. Such a product must then be a primary product or be formed from one by prompt-neutron emission. An example is ^{82}Br, which is isobaric with stable ^{82}Se and ^{82}Kr.

A plot of fission-product yield against atomic number gives a double-maximum curve much like that for the mass-number distribution. This indicates that nuclear material is not very polarizable, tending to maintain a rather constant specific charge Z/A. The average specific charges of light and heavy fission fragments are not identical, which leads to some uncertainties as to the most probable nucleon division in fission. Three possibilities have been advanced:

1. *Equal specific charge*, in which each fragment has the same Z/A as the original compound nucleus. This is the assumption on which Fig. 17-11 was drawn.
2. *Minimum decay energy*, in which the product Z's divide in such a way as to make the total radioactive-decay energy of the two chains a minimum.
3. *Equal chain lengths*, where the most probable division of Z is that making for equal chain lengths in the two products.

A clearcut choice between the three possibilities is difficult because their predictions lie close together. Present evidence tends to favor the third possibility, at least for low neutron energies. At high energies, more prompt-fission neutrons are emitted, Z/A increases, and the fission chains shorten.

The average energy emitted in radioactive decay is about 20 MeV per fission. Short-lived nuclides drop out with time, causing the rate of energy release to decrease. This rate is given by

$$\frac{dE}{dt} = (3.9t^{-1.2} + 11.7t^{-1.4}) \times 10^{-6} \text{ MeV sec}^{-1} \tag{17-8}$$

where t is the time in days. Figure 17-12 shows the variations of energy release and decay rate as functions of time. As a rough rule, beta emission accounts for about $\frac{2}{3}$ of the decay energy; gamma rays, about $\frac{1}{3}$. Average fission-product gamma-ray energies are 0.7 MeV; average beta-particle

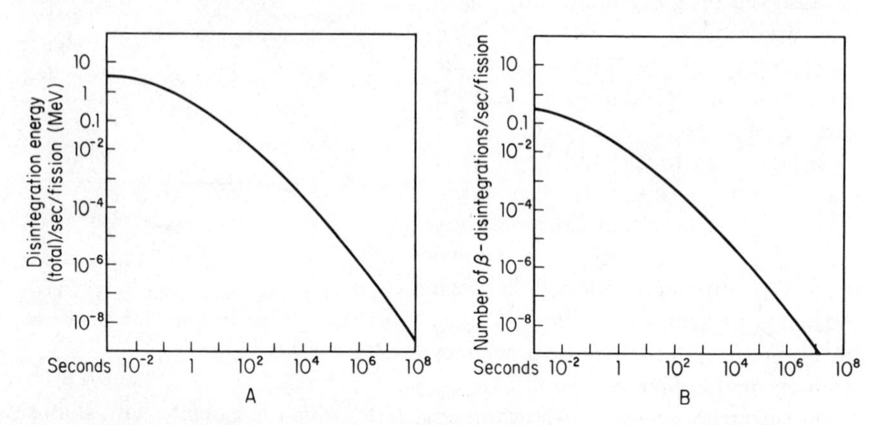

Figure 17-12. (A) Rate of energy release from fission product decay. (B) Rate of beta decay of fission products. (From K. Way and E. Wigner, *Phys. Rev.*, **73**, 1318, 1948.)

energies are 0.4 MeV; average values of maximum beta energies, 1.4 MeV.

In a nuclear reactor operating at a constant flux level, the shorter-lived fission-product activities will successively reach equilibrium values, while the longer-lived activities continue to increase, Eq. (16-23). The gross decay rate for mixed fission products will, therefore, depend upon the length of time the sample was irradiated. Figure 17-13 shows the relative activity of fission

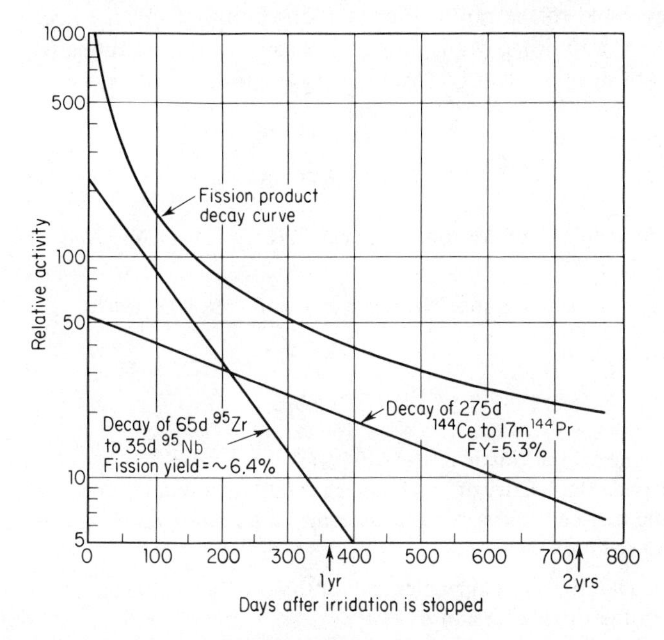

Figure 17-13. Decay of fission products after 100 days' operation of a nuclear reactor.

products accumulated during 100 days operation of a reactor. The contributions of two high-yield long-lived fission products, 65-day ^{95}Zr and 275-day ^{144}Ce, are also shown. The former contributes significantly to fission products which have cooled for several weeks, while the latter assumes greater importance for pile products several months after removal from the nuclear reactor. It is common practice in dealing with the high activities of pile products to allow for a cooling period of several weeks or even months so that the hazard of dealing with intense radioactivity from short-lived products is minimized.

Equation (17-8) obviously will not fit for very old fission products, since these will consist of only a few radioactive nuclides of long half-life, each decaying at its characteristic rate. ^{90}Sr (28 y) and ^{137}Cs (30 y) are examples of long-lived products formed in good yield. These nuclides, present in the nuclear debris of a nuclear detonation, or fallout, have given concern to many people as an undesirable hazard of nuclear-weapons testing.

Large-scale controlled fission, as in power reactors, has provided large quantities of previously rare elements in the middle of the periodic table. Hitherto unknown properties have been measured and some of the elements have been put to practical use. On the whole, reactor products constitute dangerous and expensive waste material. Millions of gallons of liquid reactor wastes are now stored in stainless steel tanks, buried underground. Solutions are concentrated to the point of continuous boiling from the heat generated by their own radiations. Management of wastes will constitute a real problem in the future development of fission energy on a large scale.

REFERENCES

Fraser, J. S. and J. C. D. Milton, "Nuclear Fission." *Ann. Rev. Nuc. Sci.* **16,** 379, 1966.

Halpern, I., "Nuclear Fission." *Ann. Rev. Nuc. Sci.* **9,** 245, 1959.

PROBLEMS

17-1. Assume that a kinetic energy of 8 MeV per charge is required to penetrate the potential barrier of ^{235}U, and calculate the excitation produced in it by: a thermal neutron, a proton, a deuteron, an alpha particle, and an annihilation photon.

17-2. Calculate the total kinetic energy of the products ^{98}Zr and ^{138}Xe resulting from the thermal fission of ^{239}Pu.

17-3. How much ^{235}U and ^{238}U will be lost in 1 year from spontaneous fission in a stockpile of 1000 Kg of natural uranium? How much will be lost from alpha decay?

17-4. Reaction equations such as Eq. (16-22) are valid only if there is a negligible depletion of the target, a situation which may not obtain in heavily irradiated reactor-fuel elements. Show that with depletion the total number of fissions produced in a constant thermal flux F will be given by $N = fN_0(1 - e^{-F\sigma t})$, where f is the fraction of the absorptive processes that leads to fission, and σ is the cross-section for all absorptive processes.

17-5. Show that the activity of a particular fission product with decay constant λ, produced with a fractional yield k in a constant thermal-neutron flux F, is given by

$$A = \frac{fkN_0F\lambda\sigma}{\lambda - F\sigma}(e^{-F\sigma t} - e^{-\lambda t})$$

17-6. What fraction of a sample of ^{235}U, exposed in a nuclear reactor to a thermal flux density of $2 \times 10^{13} n\ \text{cm}^{-2}\ \text{sec}^{-1}$, will be fissioned in 1 year?

17-7. What thermal-neutron fluence will be required to reduce the natural abundance of ^{235}U in a sample of natural uranium to 90 per cent of its original value?

17-8. What are the energy relations involved in the symmetrical fission of ^{64}Zn to $2 \times {}^{32}P$?

17-9. The energy released in a nuclear detonation is usually expressed in kilotons (KT) of TNT, which releases 10^3 calories per gram. How many fissions are involved in a 50-KT explosion? How many grams of U are fissioned? How much mass is converted?

17-10. How many grams of ^{90}Sr will be produced in the detonation of a 100-KT fission weapon? What is the activity of this amount of ^{90}Sr? Repeat the calculation for ^{137}Cs.

17-11. Under most field conditions, the $t^{-1.2}$ term in Eq. (17-8) is sufficient to describe the fission-product decay. Assume this and show that the total radiation dose from time t to infinity is given by $D = 5t$(dose rate at time t).

17-12. Five hours after a nuclear detonation, a survey party encounters a fission-product field with an exposure rate of 3.8 R hr^{-1}. What will be the total exposure at this point for the complete decay of the activity? What exposure would be received in the first 24 hours after entry?

17-13. Old reactor wastes consist almost entirely of ^{90}Sr and ^{137}Cs. Assume that all of the energy radiated in a 10^5-liter holding tank is absorbed in the liquid and goes into evaporating water up to a limit of 10^3 liters per day. How much activity can be stored in the tank?

18
Nuclear Reactors

18.01 Production of Nuclear Fuel

The possibility of producing a self-sustaining chain reaction became evident when it was established that more than one neutron was emitted in each fission. In principle, it is only necessary to arrange things so that at least one neutron from each fission initiates another fission. Any neutrons in excess of those needed to sustain the reaction can be used for other purposes.

It soon became apparent that the large-scale utilization of nuclear energy would, for some time to come, be based on uranium. Other elements could be made to fission and were available in low concentrations, but any large-scale exploitation of them seemed out of the question.

Natural uranium will not support a chain reaction because of the small fast-fission cross-section and large capture cross-section of ^{238}U. At 2 MeV, the fission cross-section is only 0.5 barn, the scattering cross-section is 10 times as large. Consequently, neutrons emitted in fission will be quickly and preferentially reduced in energy below the threshold energy for ^{238}U fission. These neutrons might still induce fission in the ^{235}U component, but in natural uranium an insufficient number of neutrons can escape the strong resonance capture in ^{238}U to sustain a chain reaction.

As soon as these facts were recognized, a large-scale effort was launched to develop methods of separating ^{235}U from the more abundant ^{238}U. Many processes such as centrifugation, thermal diffusion, pressure or gaseous diffusion, electrolysis, and electromagnetic separation were available, and several were tried during the development of the nuclear-energy program. *Gaseous diffusion* has proved to be the most practical method for large-scale production. This can be followed by electromagnetic separation when the highest purity is desired.

Gases forced through an orifice by pressure flow at a rate given approximately by Graham's law:

$$R = \frac{\text{constant}}{\sqrt{M}} \tag{18-1}$$

where R = rate of diffusion
M = molecular weight of the gas

Uranium hexafluoride, UF_6, a corrosive gas with a good many undesirable properties, is the only gaseous compound available for uranium separation. When UF_6 is forced through a barrier consisting of many fine openings, the lighter $^{235}UF_6$ molecules, having slightly greater velocities, will flow through preferentially. An isotopic enrichment in accordance with Eq. (18-1) will be achieved. Enrichment in a single diffusion stage is small, but many stages can be connected in series to obtain any desired degree of enrichment.

Isotopic separation is basic to all processes utilizing the energy released in nuclear fission. Several gaseous-diffusion plants have been built in the United States, each an enormous installation capable of a yearly production of ^{235}U measured in tons.

18.02 Neutron Utilization

Even when uranium enriched in ^{235}U is available, it is necessary to design a system with some care in order to achieve a self-sustaining reaction. The essential problem is to prevent the loss of neutrons to nonfission reactions.

Consider the fates of neutrons produced in a sphere of uranium, Fig. 18-1, enriched in ^{235}U but containing a considerable concentration of the heavier isotope. Several possibilities face a neutron released in a fission within the sphere:

a. Initiation of another ^{235}U fission after being thermalized by collisions
b. Initiation of a fast fission in ^{238}U before any significant amount of energy is lost
c. Escape from the fissionable volume
d. Reflection back into the fissionable volume by a collision outside
e. Absorption by a nonfissionable nucleus
f. Nonfission capture by ^{238}U
g. Capture by ^{235}U without fission

Processes (a) and (b) generate the neutrons needed to perpetuate the reaction. Of the two, (a) is the more important, although (b) is not negligible. The relative importance of the two will depend upon the degree of enrichment and the amount and disposition of any moderator that may have been introduced to thermalize the neutrons.

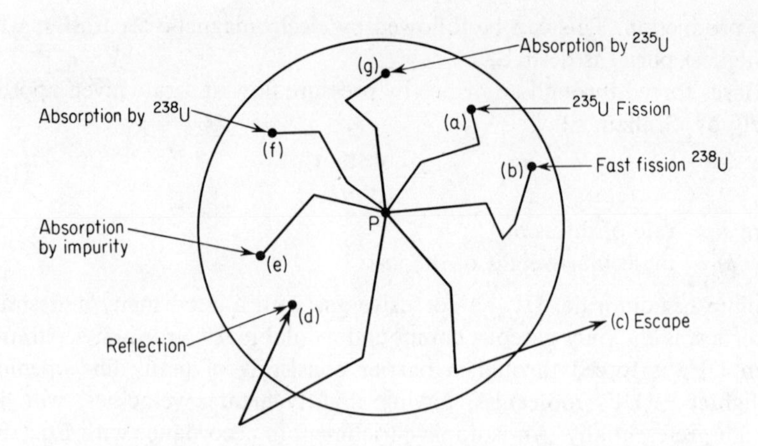

Figure 18-1. Schematic representation of the various fates of neutrons in a chain-reacting system.

Process (c) depends primarily upon the amount of fissionable material in the system. The rate of fission within the sphere will be proportional to its volume and hence to the cube of its radius r. Neutron escape rate will be proportional to the surface area of the sphere or to r^2. The ratio neutron production/neutron escape is then $r^3/r^2 = r$, which becomes more favorable as r increases. This leads to the concept of a *critical size* or *critical mass.* With a given nuclear fuel and geometrical configuration, there will be a critical size below which no self-sustaining reaction is possible because losses will exceed production. A system smaller than this is said to be *subcritical*; one larger is *overcritical.*

The sphere has the most favorable volume/surface ratio and hence requires the least mass to become critical. Larger masses will be required to reach criticality in the shapes more adapted to the usual methods of construction, but the general argument used for the sphere will still apply.

Critical mass depends upon many factors, one of which is process (d), Fig. 18-1, in which neutrons are reflected back into the system. An isolated mass of uranium may be subcritical but may go critical when it is surrounded by an effective reflector. Such a reflector will probably consist of relatively heavy nuclei, with very small cross-sections for all neutron-capture processes.

Some of the neutrons in our uranium sphere will be lost by capture in nonfissionable materials. Even the most efficient moderator will have a small capture cross-section. Products formed from previous fissions will be present in concentrations determined by the previous fission history of the assembly. Some of the fission products have extremely large capture cross-sections and hence are effective absorbers even in low concentrations.

Resonance absorption in ^{238}U, process (f), plays a most important role

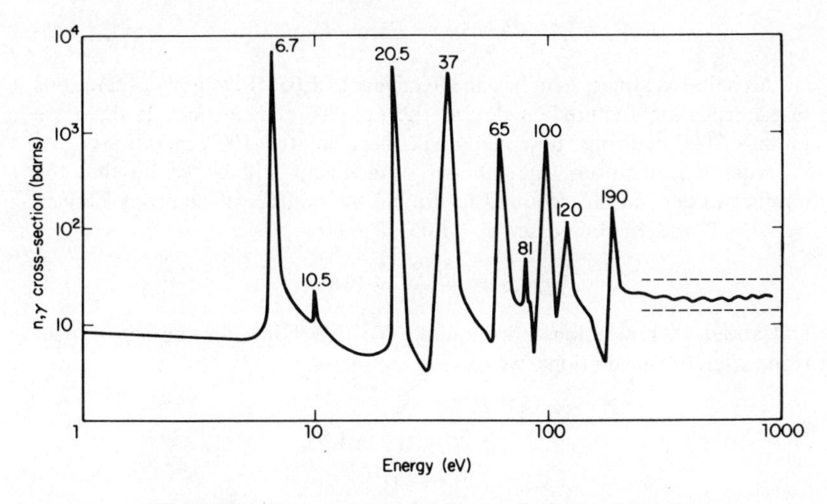

Figure 18-2. Cross-sections for the (n, γ) reaction in ^{238}U.

in all chain-reacting systems. The capture cross-section has several strong resonances in the 7–200 eV region, Fig 18-2. A moderator may be designed to reduce as many neutrons as possible to energies below the 7-eV resonance before they encounter another fuel element. Not all neutrons will be thermalized, however, and hence capture in ^{238}U and the subsequent formation of ^{239}Pu enter into the operation of almost all chain reactions.

An occasional neutron capture by ^{235}U will initiate an (n, γ) reaction rather than a fission. Process (g), Fig. 18-1,

$$^{235}_{92}U + ^{1}_{0}n \longrightarrow ^{236}_{92}U + \gamma \tag{18-2}$$

has a cross-section of about 100 barns, or about one-fifth that for fission. The product nucleus is an alpha emitter with a half-life of 2.4×10^7 y.

All of these seven processes, and some others of lesser importance, are active in any chain reaction. All cross-sections are complicated functions of neutron energy; some are truly enormous.

18.03 The Reproduction Factor

Once criticality is exceeded, the rate of buildup in a chain-reacting system is extremely rapid. A quantity k, the *reproduction* or *multiplication factor*, is defined as the ratio of the number of neutrons in the nth *generation* to the number in the $(n - 1)$th generation. If k is exactly unity, fission will go on at a constant rate, with a constant neutron-flux density in the system. If k exceeds unity by even a small amount, the increase in the number of free neutrons in the system will increase very rapidly.

Illustrative Example

A chain-reacting system has an effective k of 1.05. How many generations of neutrons are required to double the number of neutrons? If there are initially 1000 neutrons, how many will there be after 100 generations?

After n generations, the number of neutrons will be k^n, so that the number of generations required to double the number of neutrons is given by $k^n = 2$ and for our case, $k = 1.05$. Therefore,

$$n = \frac{\ln 2}{\ln 1.05} = 14.2$$

and about 14 generations are required. To determine the number of neutrons after 100 generations, we solve

$$(1.05)^{100} = N$$

$$\ln N = 100 \ln 1.05$$

$$N = 131.5$$

Since the initial number of neutrons was 1000, we multiply this by 131.5, yielding a result of 131,500 neutrons.

Figure 18-3 shows semilogarithmic plots of N versus n. Note that N is extremely sensitive to small changes in the value of k. If the average time between collisions of the neutron is of the order of a millisecond, the buildup of neutrons within a system will take place very rapidly. Consider, for example, a system with $k = 1.010$; in 0.1 sec, the neutron increase will be a factor of e and this will jump to e^{10} in 1 sec. It is easy to see that a runaway nuclear chain reaction might result in an explosive release of energy.

A nuclear reactor—or *pile*, as it was originally called—is a device for releasing nuclear energy through fission at a controlled rate. A reactor must have control devices fast enough and of ample range to counteract any possible rise in the reproduction factor. No controls will be used in a nuclear weapon, where the most rapid and largest possible energy release is desired.

Any explosion of a nuclear reactor will be of a very low order compared to that obtainable from a weapon. Power reactors are now operated in *containment vessels*, designed to withstand any pressure produced by a runaway reaction.

Three broad categories of nuclear reactors can be distinguished, depending upon whether the fissions are induced by thermal, intermediate, or fast neutrons. Each category will include a number of designs, depending upon the principle use of the reactor. Thermal reactors are the most common, and only these will be considered at this point.

In a thermal reactor, the reproduction factor k is the product of four components:

η is the average number of neutrons released in the fuel from the absorption of one thermal neutron. Note that is not quite the same as ν. The latter

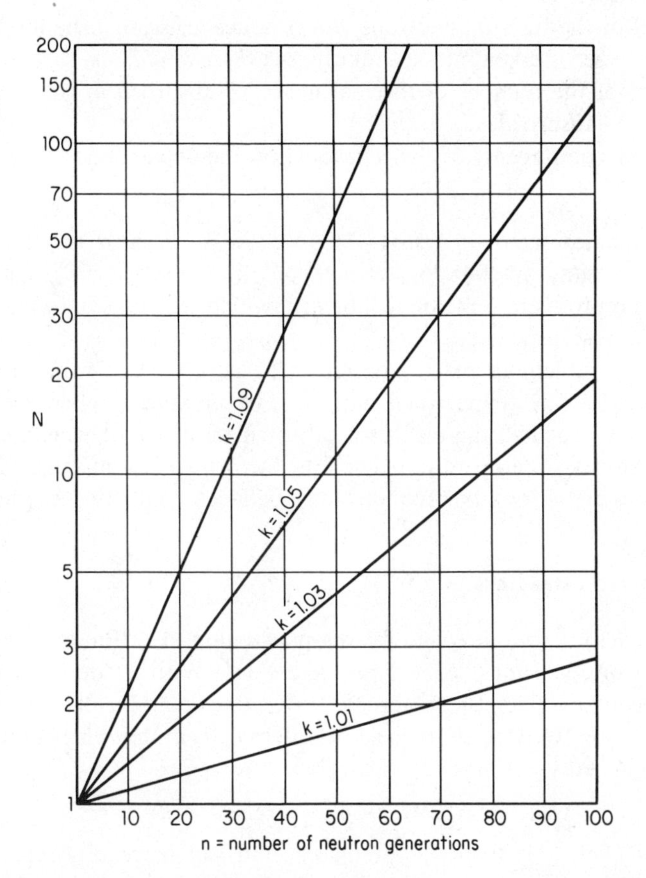

Figure 18-3. Increase in the neutron population as a function of the multiplication factor k.

is the number of neutrons emitted per fission, but since some neutrons are absorbed without fission, $\nu > \eta$.

Fast fission is taken into account by the *fast-fission factor* ϵ. This factor is the ratio of all of the fission neutrons, produced by all fissions, to the number produced by thermal fissions alone. ϵ will depend upon the relative amounts of ^{235}U and ^{238}U, and upon the design parameters of the reactors. Typically, ϵ will have a value of 1.03–1.08.

Some of the fast neutrons produced by fission will be lost to intermediate-energy resonances before they can be thermalized by collisions. Note the series of strong resonances in the (n, γ) cross-section for ^{238}U, shown in Fig. 18-1. The *resonance-escape probability* p is just the average probability that a neutron will reach thermal energy before capture. This factor will obviously depend upon many of the characteristics of each particular reactor.

Not all of the thermal neutrons will produce a fission. The loss to non-fission processes is taken into account by the *thermal utilizion factor f*, which is the ratio of the number of thermal neutrons absorbed in the fuel to the total number absorbed.

The four components of the reproduction factor combine:

$$k_\infty = \eta \epsilon p f \tag{18-3}$$

The reproduction factor in Eq. (18-3) is written k_∞ because no account has been taken of any possible loss of neutrons from the system. In an actual device, the reproduction factor will be k_{eff}, which will be somewhat smaller than k_∞ because of leakage.

A basic problem in reactor design is to adjust the parameters in Eq. (18-3) so that a k_{eff} greater than unity can be obtained. When the desired power level is reached, the reactor is adjusted to $k = 1$, the necessary condition for steady-state operation. The four factors are not mutually independent, so any design will be a compromise between competing requirements.

18.04 The First Reactors

The first reactors were designed for the production of ^{239}Pu, predicted from even–odd considerations to be fissile like ^{235}U. With a copious supply of neutrons available from the fission of ^{235}U, some could be allowed to undergo capture by ^{238}U. Two short-lived beta decays from the compound nucleus produced 24,000-y ^{239}Pu:

$$^{238}_{92}\text{U} + ^{1}_{0}n = ^{239}_{92}\text{U} \longrightarrow \beta^- + ^{239}_{93}\text{Np} \longrightarrow \beta^- + ^{239}_{94}\text{Pu} \tag{18-2}$$

The ^{239}Pu nucleus is fissile, as predicted, and can be readily separated by chemical means from the parent uranium.

Fermi and his associates at Columbia University put together a lattice of graphite and uranium in the summer of 1941. The structure was a cube about 8 feet on edge and contained some 7 tons of uranium oxide lumped throughout the graphite. This *exponential pile* was a subcritical assembly made with the same internal geometry as planned for the actual full-scale pile. Using a Ra–Be neutron source, the neutron density was measured at a number of points in the structure and k was determined experimentally. A second exponential pile built at Columbia University gave a value for k_∞ of 0.87. This value was soon increased to 0.98 by using purer pile materials. In July, 1942, a year after the first exponential-pile construction, improvements in materials and design led to a value of $k_\infty = 1.007$, and it became clear that a self-sustaining nuclear chain reaction was within sight.

Based upon data gained from exponential-pile experiments, a pile was designed to sustain the nuclear chain reaction. This pile was a graphite–uranium oxide assembly built in a squash court at the athletic field of the

Figure 18-4. The first operational chain reactor under construction. (Courtesy of U. S. Atomic Energy Commission.)

University of Chicago. Its successful operation was first demonstrated on December 2, 1942. The only photograph made of this pile during its construction is shown in Fig. 18-4. In November, 1942, when the photograph was taken, the pile had been built up to its nineteenth layer. A total of 56 layers were required to attain criticality. About 6.2 tons of uranium metal and oxide were imbedded in the graphite lattice in alternate layers of the carbon blocks, as shown in the illustration. In its final form, the pile took on an oblate spheroid form rather than a spherical geometry which had been planned initially. Control of the pile, as well as insurance that there would be no "startup" while the assembly was being made, was accomplished by inserting long cadmium control rods. Because of its high neutron absorption, the cadmium served to keep the pile subcritical until startup. Operation of the pile was accomplished by slowly withdrawing the control rods, thus allowing more and more neutrons to escape cadmium absorption and contribute to thermal fission and to the chain reaction. The buildup of neutrons within the pile as measured by a BF_3 neutron counter was used to gauge the approach to criticality. An actual k of 1.0006 was realized with all neutron-absorbing rods in a full-out position. In its early operation, the pile was operated at a 0.5-watt power level and later raised to 200 watts. Further increase in the power level was inadvisable, since radiation from the pile would have required substantial shielding to protect personnel nearby.

A high-power pile, designed primarily to serve as a pilot plant for plutonium production, was built in 1943 at Oak Ridge, Tennessee. Originally plans were made for a uranium–graphite pile that would produce significant quantities of plutonium; for this purpose, a power level of 1000 KW was specified. A sketch of the basic components of a reactor of the first Oak Ridge type is shown in Fig. 18-5. Although a spherical design would be more

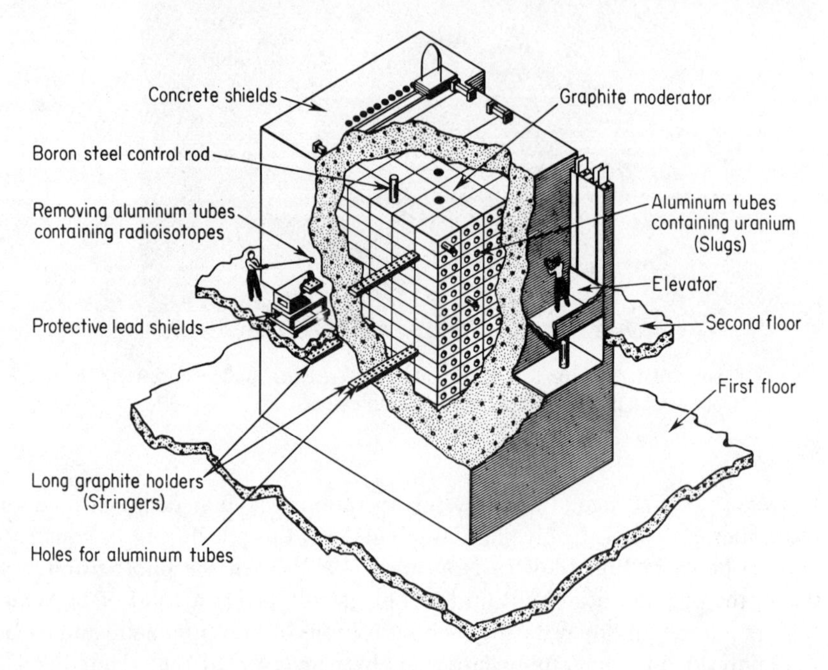

Figure 18-5. Schematic of an early production reactor at Oak Ridge. (Courtesy of U. S. Atomic Energy Commission.)

desirable in principle, a cubical array of graphite blocks was more practical. As finally constructed, the Oak Ridge reactor was a cube 24 feet on edge, including in this dimension a graphite reflector. The requirement for heat removal and for ease of replacement was met by placing the uranium in the form of rods positioned in holes bored through the graphite blocks.

The uranium used was in the form of aluminum-jacketed slugs 1.1 in. in diameter by 4 in. in length; these were placed in 1248 fuel channels so arranged that the distance between the uranium cylinders was 8 in. Air pumped through the fuel channels kept the slugs below 245°C and the moderator at an average temperature of 130°C. Control of the chain reaction was effected by means of boron-surfaced control rods whose position was adjustable by servomechanisms. Additional safety rods were provided in the event of accident or control-rod failure, so that the pile would automatically be shut down before a runaway chain reaction would lead to thermal disruption of the pile components. To protect personnel, a 7-ft concrete shield surrounded the reactor core. Shielding was sufficient so that normal operations could be conducted around the shield surface without exceeding industrial standards for radiation exposure.

18.05 Modern Reactors

Since the early days, much has been learned about reactor design, and many new materials have become available in industrial quantities. As a result, sophisticated reactor designs for specific applications are made routinely.

Light elements appear to be the most desirable moderators, Eqs. (16-8) and (16-9), because they can achieve thermalization with the fewest number of collisions and hence with the least chance of loss by capture. Table 18-1 lists the collision parameters for some of the more promising moderators. Practical considerations such as cost and availability limit the choice of moderators to two: graphite and heavy water.

TABLE 18-1
PROPERTIES OF VARIOUS MODERATORS

Material	ξ	ζ
Hydrogen	1.0	18
H_2O	0.93	19
D_2O	0.51	32
Beryllium	0.20	87
Graphite	0.16	110
Nitrogen	0.14	130
Oxygen	0.12	150

Separated ^{235}U and ^{239}Pu are available in quantity, and reactor cores can now be constructed with any desired degree of enrichment with fissile material. Metals that were formerly only laboratory curiosities are now available for encasing or *cladding* the fuel elements. New alloys with exotic properties are now in routine use in many reactors.

A few criticality excursions, some planned and some accidental, have led to improved control systems and safety devices. One of the most basic safety features is a built-in negative temperature coefficient. If the reactivity decreases as the temperature rises, the device can be made self-limiting, so that the chain reaction will cease as the power level rises. Every reactor design accepted for licensing in the United States must be shown to have a negative temperature coefficient.

Reactors are being designed and constructed in ever-increasing sizes for the production of electric power. In a power reactor, some coolant removes heat from the reactor core and in turn converts water to steam at high pressure. The steam is then used to drive turbines and produce electric power in the usual way. Future installations, particularly in the large sizes, will probably combine power production with water desalinization, utilizing heat normally wasted. In common with all thermal power systems, a reactor–turbine generating plant has a relatively low thermodynamic efficiency. Thus a plant might be rated as 200 megawatts thermal and 70 megawatts electrical (200 MW_{th} and 70 MW_e).

Production reactors are designed for the irradiation of samples, usually with the object of making them radioactive through (n, γ) reactions. These reactors have provided the bulk of the radioactive materials used so extensively in medicine and industry.

High-flux reactors are available for the accelerated testing of the effects of high neutron fluences on materials. These reactors are also useful for the production of high specific-activity samples, since the ultimate activity attainable depends upon the flux, Eq. (16-23).

Nuclear reactors have proved very useful in some types of ship propulsion. Steam is generated in the reactor and used in conventional turbine drives. Tentative designs for a reactor-propelled aircraft were developed, but the project was abandoned, primarily because of the enormous weight of the radiation shielding required. Prototype reactors for deep-space travel have been tested, and this program can be expected to continue.

18.06 Reactivity and Reactor Control

In a reactor that is supercritical, the neutron density at any time t can be shown to be given by

$$\phi = \phi_0 e^{(k-1)t/l} \tag{18-5}$$

where k is the effective multiplication constant and l is the mean lifetime of the neutrons in the reactor. The term $(k - 1) = \Delta k$ is known simply as *excess k*. The ratio $\Delta k/k$ is the *reactivity* ρ. A term, *pile period T*, is used to denote the time required for the flux density to increase by a factor of e. From Eq. (18-5),

$$T = \frac{l}{k-1} \tag{18-6}$$

The mean lifetime of neutrons in a reactor is about 10^{-3} sec and, by Eq. (18-6), this leads to very short pile periods for any reasonable value of k. On this basis, almost any value of excess k would lead to such rapid increases in flux, and hence energy release, that they could not be checked by mechanical control devices. Disaster is averted by the presence of delayed neutrons, Sec. 17.07. The number of delayed neutrons is small, but it is sufficient to permit mechanical control of the reactivity.

Total k consists of two components, one with the short mean life of the prompt neutrons, the other with a mean life of about 10 sec applying to the delayed emissions. The marginal condition for safe reactor operation is

$$(1 - \beta)k = 1 \tag{18-7}$$

where β is the fraction of the emitted neutrons that is delayed. If k exceeds the value given by Eq. (18-7), the reactor is said to be *prompt critical* and is able to rapidly build up the power level with no help from the delayed neutrons. When k is less than the marginal value given by Eq. (18-7), the reactor is *delayed critical* and is subject to mechanical control.

The reactivity ρ is related to k by

$$k = \frac{1}{(1 - \rho)} \tag{18-8}$$

From this expression and Eq. (18-7), the marginal condition for prompt criticality becomes

$$\rho = \frac{k - 1}{k} = \beta \tag{18-9}$$

Although β is small (0.0065 for ^{235}U and 0.0021 for ^{239}Pu), any reactivity approaching β will have a profound effect on reactor operation and will be approaching the dangerous condition of prompt criticality. Obviously, a reactor operating on ^{239}Pu is more sensitive than a uranium system.

The amount of reactivity given by Eq. (18-9) is known in the United States as *one dollar*. An obvious subdivision is 1 per cent of this or *one cent*, which in the case of ^{235}U amounts to a reactivity of 0.000065. Materials inserted in a reactor for irradiation should have a negative reactivity (absorbance) of much less than 1 dollar, lest their inadvertant withdrawal cause the reactor to go prompt critical.

18.07 Fission-product Poisons

Reactor control is achieved by the deliberate introduction into the reactor core of neutron absorbers in the form of control rods to compensate for any excess reactivity. Reactor startup is accomplished cautiously, making small changes in reactivity by the withdrawal of the rods. A reactor operated for the first time will differ from one that has been run for some time. The latter will contain an accumulation of fission products, formed in its core during previous operation.

Two fission products, ^{135}Xe and ^{149}Sm, have such large capture cross-sections and are formed in such quantities that they have a profound effect on reactor operation. Of the two, ^{135}Xe, with its cross-section of 2.7×10^6 b is the more important. About 5 per cent of the ^{135}Xe is formed directly in fission; the rest is formed in a fission chain headed by ^{135}Te:

$$^{135}\text{Te} \xrightarrow{2\text{m}} {}^{135}\text{I} \xrightarrow{6.7\text{h}} {}^{135}\text{Xe} \xrightarrow{9.2\text{h}} {}^{135}\text{Cs} \xrightarrow{2\times10^4 y} {}^{135}\text{Ba} \tag{18-10}$$

Cross-sections for all members of the chain except ^{135}Xe are small enough to be neglected.

Following reactor startup, ^{135}Xe will build up until an equilibrium concentration is reached after a few days' operation. Equilibrium is established when the loss of ^{135}Xe by decay to ^{135}Cs and by neutron capture, $^{135}\text{Xe}(n, \gamma)^{136}\text{Xe}$, equals the rate of production. ^{136}Xe has a small capture cross-section and does not enter appreciably into the neutron economy. As the ^{135}Xe increases, the reactivity of the system must be restored by withdrawing the control rods.

When a reactor is shut down, the burnup of ^{135}Xe by neutron absorption ceases, but its concentration builds up from the decay of the ^{135}I inventory, until a maximum is reached in about 12 hours. ^{149}Sm behaves in a similar fashion to a lesser extent. If one wishes to restart the reactor during this period, there must be enough excess reactivity available to overcome the absorption by the poisons. The reactivity requirements for a large power reactor would be about like:

Requirement	$\Delta k/k$
Depletion of nuclear fuel:	0.110
Xenon poisoning (equilibrium):	0.031
Samarium poisoning (equilibrium):	0.007
Temperature range 450°F:	0.026
Total:	0.174

As nuclear fuel is consumed, the reactivity available decreases, and eventually the reactor core must be replaced. To avoid frequent refueling, which is a major operation, it is desirable to start with a large initial fuel load. This in turn requires a large capacity in the control rods in order to insure control at all times. Some of the control-rod requirement can be removed by introducing a burnable poison such as ^{10}B into the reactor core. As reactor poisoning by fission products increases, the amount of burnable poison decreases, and the reactor can be maintained with a somewhat constant reactivity over the life of the core.

18.08 Radiation Damage

All materials within the reactor core are subject to irradiation with neutrons and gamma rays. Fuel elements also experience the short-range effects of fission products and alpha particles. Burnup of fissionable material removes uranium atoms from the material, and for each split atom a pair of fission products is formed. As a result of prolonged burnup, uranium "grows," i.e., it undergoes changes in dimensions and also exhibits decreased thermal conductivity. Radiation damage to fuel rods is minimized by using uranium alloys which have been carefully heat-treated.

In control rods, the use of boron as a poison involves an (n, α) reaction, which produces two high-speed particles and a concomitant radiation damage, which limits the usefulness of boron in reactor control.

Moderating materials are also subject to radiation damage. Graphite, for example, is subject to such damage, principally from the action of fast neutrons. When a neutron impinges upon an atom in a solid, it may knock it from its normal position in the crystal lattice. Neutron bombardment of graphite adds to the lattice energy so that energy is stored up. Should this stored energy continue to build up over a period of time, it is possible for a precipitous energy release to occur. Such a happening could have catastrophic consequences if the temperature of the graphite soared high enough to destroy part of the reactor structure. This effect was predicted by Wigner and named after him, prior to the construction of the large graphite-moderated piles at Hanford, Washington. Operation of these reactors did indeed produce a swelling of the moderator, but it was found that the stored energy could be released harmlessly by annealing the moderator at relatively low temperatures before it released spontaneously. The unexpected release of Wigner energy led to the fuel meltdown in the Windscale reactor in Great Britain.

Although the theory of radiation damage has received great attention, the mechanisms involved in this process are complex and do not lend themselves to ready solution. For this reason, new reactor materials are treated experimentally by exposing them to prolonged irradiation in high-flux piles. Most of these are located at the AEC's test site at Arco, Idaho.

18.09 Reactor Shielding and Safety

High-power reactors build up large inventories of fission products and induced activities, so that even after shutdown they form an intense source of radiation. Reactor designers must therefore build in the means of coping with this radiation hazard and in addition must take precautions against the possibilities of an accident which could release radioactivity to the atmosphere.

In the case of stationary reactors, an external shield of concrete or some concrete aggregate is usually the cheapest way to provide protection against the penetration of neutrons and gamma rays. Mobile or special reactors, where weight is a consideration, require a more compact shield, and this is built up of composites involving iron, lead, and hydrogenous substances. It is necessary to shield not only the reactor core but also pumps, valves, and heat exchangers where a radiation hazard might exist. For example, where water is circulated through a reactor, one must provide not only for the possibility of fission-product contamination but also for absorption of the 6.3-MeV gamma radiation emitted by ^{16}N formed in the coolant. Various coolants will involve specific induced activities depending upon their composition. They may also pick up corrosion products from the reactor core.

It is of course absolutely essential that no effluents be allowed to escape from the reactor core. Power reactors have their cores retained within massive pressure shells; these usually take the shape of a cylinder with a hemispheric head. The Atomic Energy Commission has set up rigorous procedures to insure that reactors built in the United States comply with adequate regulations for reactor safety. The control systems for all proposed reactors together with provisions for containment of any possible accident are analyzed carefully by the Atomic Energy Commission before the reactor is finally approved for operation.

Altogether, the reactor safety record of the United States, viewed on an industrial accident scale, has been remarkable. Some fatalities have occurred, but no major accident involving a populated region has taken place. The most notable reactor accident involved a meltdown of some fuel elements in a British Windscale reactor on October 10, 1957. Some 20,000 curies of radioiodine-131 were released to the atmosphere and some 200 square miles of milkshed were considered contaminated to an excessive level. Temporary interdiction of the milk supply was deemed sufficient to deal with the short-lived toxic hazard of iodine-131.

18.10 Breeder Reactors

If one could so design a reactor that the neutron economy allowed for an excess of neutrons to where more ^{239}Pu or ^{233}U is produced than is burned up in fission, then all of the uranium would be available as fuel. A reactor in which this neutron economy obtains is known as a *breeder*, since it breeds more fuel than it consumes. Values of η for ^{235}U, ^{233}U, and ^{239}Pu are 2.07, 2.30, and 2.10, respectively. Since one neutron is required to maintain the chain reaction, and another one to just maintain breeding, it is evident that the restrictions on losses in ^{235}U and ^{239}Pu are severe. A thorium breeder is complicated by absorption in the intermediate nuclei ^{233}Th and ^{233}Pa, which makes this breeder a marginal proposition. All things considered, ^{239}Pu appears to be the most promising breeder fuel.

The design of breeder reactors is restricted because of the requirements for a very high neutron flux. This is necessary if one is to produce large power output because the fission cross-sections at high energy are quite low. A breeder reactor has to have a high fuel inventory compared with a thermal reactor of the same power rating. Because the neutron energy must not be severely degraded in the reactor core, there are only a limited number of choices for coolants and structural elements. The high neutron fluxes required for high power output introduce a radiation-damage prospect for any materials in the core. For these reasons, and because the short time constants of a fast reactor make the control problem very difficult, only a few breeder designs have been constructed.

A breeder design is not allowed the option of a light element for a coolant, since the neutrons would be too quickly degraded in energy by collisions with light nuclei. On the other hand, because of the condensed size of the fast reactor core, the coolant must have the ability to remove large amounts of heat from the core. Experience has shown that liquid Na–K alloy serves as an excellent coolant even though this highly reactive mixture poses severe engineering difficulties in practice. It is obviously desirable to develop a successful breeder, and much effort has been expended in this direction. Although some progress has been made, a completely satisfactory design has yet to appear.

REFERENCES

Kramer, A. W., *Boiling Water Reactors.* Addison-Wesley, Reading, Mass., 1958.

Lane, J. A., "Economics of Nuclear Power." *Ann. Rev. Nuc. Sci.* **16,** 345, 1966.

Murray, R. L., *Introduction to Nuclear Engineering.* Prentice-Hall, Englewood Cliffs, N. J., 1954.

Starr, C. and R. W. Dickinson, *Sodium Graphite Reactors.* Addison-Wesley, Reading, Mass., 1958.

Weinberg, A. M. and E. P. Wigner, *The Physical Theory of Neutron Chain Reactors.* Univ. of Chicago Press, Chicago, 1958.

PROBLEMS

18-1. A nuclear reactor has an effective k of 1.03. How many generations will be required to double the number of neutrons? What time will be required for doubling, assuming an initial energy of 1.5 MeV, and a mean-free-path for fission of 8.0 cm?

18-2. A single neutron initiates a chain reaction that results in the fission of 1 Kg of ^{239}Pu. How many neutron generations will be required, if the multiplication factor is 2.11?

18-3. At shutdown, the fuel elements in a reactor contain 2.0 per cent of the $^{135}I \longrightarrow {}^{135}Xe$ equilibrium concentration of ^{135}Xe. Calculate the time at which ^{135}Xe activity will be maximal, and the amount of this activity relative to that at shutdown.

18-4. Liquid sodium used as a reactor coolant has a cross-section for thermal activation of 600 mb. The circulating sodium spends about $\frac{1}{4}$ of its time in the reactor core, where the thermal flux is $8.6 \times 10^{12} n \text{ cm}^{-2} \text{ sec}^{-1}$. What is the specific activity of the coolant at equilibrium? How long will be required for the coolant to reach $\frac{3}{4}$ of equilibrium activity?

18-5. A nuclear-powered aircraft carrier requires 300,000 horsepower at the propeller shafts. Assume an overall efficiency of 22 per cent and calculate the rate of ^{235}U burnup if all of the power is derived from fission.

18-6. Assume that 0.6 per cent of the fission neutrons are delayed, and that they have an effective life of 8 sec. What is the pile period when it is operating on maximal reactivity on delayed neutrons? While being irradiated in a steady state, a sample worth 30 cents slips out of position. How long will it be before the alarm, set for a 10 per cent flux increase, will sound?

18-7. A power reactor operates at a power level of 300 KW_{th} for 3 months. How much ^{90}Sr will be in the reactor core at the end of this period? What is its activity?

18-8. Assume a nuclear detonation with a multiplying factor of 2.14 to be started by a single neutron. What will be the equivalent KT of a reaction that terminates in the seventieth generation?

18-9. A power reactor operating at a power level of 10^5 watts goes prompt critical due to a failure of the control-rod system. Assume a multiplying factor of 1.09, and that the negative temperature coefficient stopped the reaction in 5 msec. Estimate the thermal energy generated during the runaway.

18-10. Calculate the activity of 53-d ^{89}Sr in a 1-g sample of ^{235}U that has been in a reactor operating at a flux of 7×10^{11} thermal neutrons $\text{cm}^{-2} \text{ sec}^{-1}$ for $\frac{1}{2}$ yr.

19

Nuclear Fusion

19.01 Energy from Fusion

We must not forget that fusion reactions, as well as fission, may be exoergic. In fact, it is evident from the shape of the binding-energy curve, Fig. 7-3, that per nucleon, the release of energy by fusion may be several times that available from fission.

Nature makes use of controlled fusion to release nuclear energy on a grand scale in the stars, and in particular, in our sun. Man has succeeded in initiating uncontrolled fusion in the thermonuclear or so-called hydrogen bomb. Controlled fusion reactions, however, are proving to be very elusive. In principle, controlled fusion appears to be as feasible as did fission in the late 1930s. Exoergic fusions reactions are obviously possible, but the conditions needed to achieve them are very difficult to attain.

Atoms to be used in fusion are brought to a high temperature, which strips them of all of their orbital electrons and forms a *plasma* of electrons and positive ions. If the ions have enough energy to breach the potential barrier, they will coalesce with the release of binding energy. The requirements of the potential barrier, among other things, limit the choice of reactants to low-Z nuclei. It is almost essential that at least one of the reactants be hydrogen. Even in hydrogen the potential barrier is high when it must be surmounted with purely thermal energies. The potential barrier between two protons, separated by twice the nuclear radius, is about 500 KeV. This is equivalent, by the Boltzmann relation, to the most probable thermal energy at 5×10^{9}°K. Because of the polarization, or orientation effect, deuterons may require thermal energies of only 10^{9}°K, which is still a formidable temperature. Tunneling will permit barrier penetration at lower temperatures, but with sharply reduced cross-sections, Fig. 19-1.

Several exoergic reactions are of interest:

$$^2H + {}^2H \longrightarrow {}^3He + {}^1n + 3.26 \text{ MeV} \tag{19-1a}$$

$$^2H + {}^2H \longrightarrow {}^3H + {}^1p + 4.04 \tag{19-1b}$$

$$^2H + {}^3H \longrightarrow {}^4He + {}^1n + 17.6 \tag{19-1c}$$

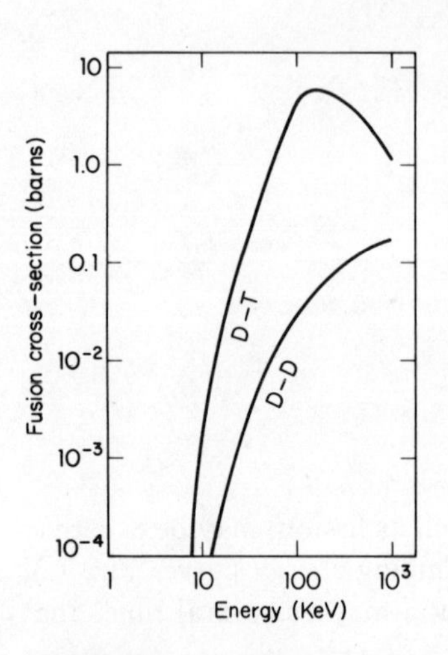

Figure 19-1. Cross-sections for *d, d* and *d, t* fusions.

The $^3H(d, n)$ reaction yields the most energy, and has the highest cross-section, about 5 barns at a deuteron energy of 100 KeV. The $^2H(d, n)$ reaction has a cross-section of only 16 mb at the same energy, but it has the advantage of using only stable and cheap components that are readily available in large quantities.

Fission-type detonations will provide the temperature needed to initiate an uncontrolled fusion reaction. The total energy released from such a composite device may be measured in megatons of TNT equivalent, instead of the kilotons associated with simple fission devices. If very small fission initiators were available, a controlled fusion reaction might be possible. Unfortunately, the demands of critical mass are such that a fusion reactor would be destroyed by the fission trigger before the release of any fusion energy.

19.02 Sources of Solar Energy

For centuries, man has pondered the origin of the energy released from our sun. The role played by gravitational energy has been recognized for many years, but the contributions from nuclear fusion have only recently become understood.

Gravitational forces were probably responsible for the original assembly of the solar material from the rarefied, more uniformly distributed matter. As the size grew, the gravitational energy of assembly became enormous, until, finally, 1 gram of matter pulled in to the surface of the sun would release almost 2 ergs, or more than 10^6 MeV. The energy of assembly alone would provide for the solar radiation for some millions of years, but it cannot account for it over the known solar lifetime of more than 4 billion years.

At solar assembly, the gravitational energy released raised the temperature to the point where fusion reactions were initiated. Bethe has shown that two series of fusion reactions can account for the energy released by our sun, and by other stars, over the long period of their existence. Slow-reacting steps in each reaction sequence keep the rate of energy release under control, so that up to this point our sun has not undergone the explosive energy release that leads to the formation of a supernova.

19.03 The Solar Energy Cycles

There are three steps in the *p–p* or *proton cycle* by which thermonuclear reactions can release energy:

Reaction	Reaction time	
$^{1}H + {}^{1}H \longrightarrow {}^{2}H + e^{+} + \nu$	7×10^{9} y	(19-2a)
$^{2}H + {}^{1}H \longrightarrow {}^{3}He + \gamma$	10s	(19-2b)
$^{3}He + {}^{3}He \longrightarrow {}^{4}He + 2{}^{1}H$	3×10^{5} y	(19-2c)

Reaction (19-2a) involves a weak interaction and as a consequence is so slow that it controls the reaction rate of the entire sequence. Six protons are required to complete the entire cycle. Two protons are recovered as products of reaction (19-2c), so the net result is the loss of four protons and the production of one ^{4}He nucleus. The net energy release of the entire cycle is 27 MeV. A portion of this energy may be carried away by the neutrino, but in a body as large as the sun all of the kinetic energy of the particles, and essentially all of the gamma-ray energy, will be absorbed in the sun itself.

The proton cycle (19-2) accounts reasonably well for most of the energy production by our sun, but it does not suffice to explain the mass–temperture–energy relations that exist in many of the other stars. Another set of reactions, with a different temperature coefficient, is needed. Lithium, beryllium, and boron are relatively scarce and cannot be enlisted as nuclear fuels. Their scarcity is probably a reflection of the relative speeds with which they reacted long ago. Carbon is abundant in our sun and in many of the stars, and Bethe proposed the *carbon* or *C–N* cycle, which satisfactorily complements the proton cycle as a source of thermonuclear energy:

Reaction	Reaction time	
$^{12}C + {}^{1}H \longrightarrow {}^{13}N + \gamma$	10^{6} y	(19-3a)
$^{13}N \longrightarrow {}^{13}C + e^{+} + \nu$	10 m	(19-3b)
$^{13}C + {}^{1}H \longrightarrow {}^{14}N + \gamma$	2×10^{5} y	(19-3c)
$^{14}N + {}^{1}H \longrightarrow {}^{15}O + \gamma$	3×10^{7} y	(19-3d)
$^{15}O \longrightarrow {}^{15}N + e^{+} + \nu$	2 m	(19-3e)
$^{15}N + {}^{1}H \longrightarrow {}^{12}C + {}^{4}He$	10^{4} y	(19-3f)

The overall rate of the carbon cycle is controlled by the slow reaction (19-3d). Carbon is recovered at the end of the cycle, so the net result is the conversion of four protons into ^{4}He. About 2 MeV will be lost from the system through the two neutrinos. The remaining 25 MeV will be absorbed locally to maintain the high temparature and produce the thermal radiation.

Because of the higher Z values involved, the carbon cycle is more temperature-sensitive than is the proton cycle. The two cycles release about equal amounts of energy at 1.7×10^7°K. Above this, the carbon cycle predominates; below, the *p–p* cycle is the main contributor. The combination of the two cycles provides a very satisfactory explanation of the energy-release rates in a large number of stars, covering a wide range of masses and temperatures.

19.04 The Solar Future

In our sun, both of the energy-producing cycles will continue at relatively constant rates for millions of years to come. Both cycles take place most rapidly in the interior, where temperatures are highest. Radiation pressure prevents a collapse of the outer layers, even though the gravitational forces are enormous. Eventually, however, the hydrogen fuel will be exhausted and the reactions will slow. Gravitational collapse will quickly follow the reduction in the radiation pressure, and the resulting energy release will produce a substantial temperature increase. At the new, higher temperature, ^{4}He can enter into energy-producing cycles and will become a nuclear fuel.

At this point in the solar history, ^{4}He will be abundant, because of the previous hydrogen burning. Several helium-fusion reactions are possible. One of the most important is

$$3\,{}^4\text{He} \longrightarrow {}^{12}\text{C} + \gamma \qquad (19\text{-}4)$$

This reaction is favored over the seemingly more probable

$$2\,{}^4\text{He} \longrightarrow {}^8_4\text{B} \qquad (19\text{-}5)$$

No stable eight-nucleon structure is known, because of the saturation nature of the nuclear force, so in fact reaction (19-5) does not occur. The three-particle fusion, seemingly very improbable, is adequately favored by the presence of some strong resonances.

The fact that reaction (19-4) can go with a reasonable probability eliminates one of the stumbling blocks to one of the otherwise plausible theories of the development of the elements. Proponents of the *big-bang theory* of the creation assume that at the start the universe consisted of a mixture of elementary particles and radiation. Neutrons quickly decayed into protons and electrons. Element buildup proceeded smoothly until it was halted by the nonexistence of nuclei $A = 5$ or 8. Reaction (19-4) permits a leap over the missing structures. From ^{12}C, nuclear synthesis can proceed upward to the heaviest stable structures.

Still farther in the future, solar helium will be depleted, another gravitational collapse will ensue, and still heavier nuclei will take up the role of providing energy through nuclear fusions. At this point, the fuel will consist of nuclei well up on the binding-energy curve, where the curve is beginning to flatten out, corresponding to a smaller energy release per nucleon. Eventually, there will be a gravitational collapse which will not stimulate new nuclear reactions to the point where radiation pressure can withstand the gravitational forces. At this point, gravitational collapse will proceed unchecked with the rapid release of an enormous quantity of energy. Very probably, this energy release will be so rapid that a disruptive explosion will result. A supernova will be formed, and the solar material will be blown out into interstellar space.

19.05 Fusion-reaction Parameters

Three conditions must be satisfied in order to initiate a self-sustaining fusion reaction. The plasma must be raised to the required temperature, an adequate plasma density must be attained, and this density must be sustained until there have been enough barrier penetrations to insure a positive energy balance. These three requirements are not independent, and are, indeed, mutually incompatible.

There can be little compromise on plasma temperatures, which must be high enough to insure favorable reaction cross-sections. At this temperature, the plasma will behave much like a perfect gas, exerting a pressure given by

$$P = NkT \tag{19-6}$$

where N is the number density of the particles, k the Boltzmann constant, and T the Kelvin temperature.

Reaction rates will be proportional to N^2, so a high plasma density is desirable. Magnetic fields appear to offer the only practical means for confining the plasma and maintaining the necessary density. Material walls are out of the question, for they would rapidly lower the temperature below the reaction point. Charged particles cannot penetrate into a magnetic field, but move at right angles to it. A magnetic field of B gauss can exert a constraining pressure of $B^2/8\pi$ dynes cm^{-2}. If a plasma is to be confined,

$$\beta = \frac{NkT}{B^2/8\pi} < 1.0 \tag{19-7}$$

Practical considerations limit the field strength B, and the total magnetic flux $\phi = B \times$ area. Then from Eq. (19-7), any restriction on B results in a limitation on N. This in turn requires an increase in the length of time the plasma must be held together, in order that a sufficient number of reactions may take place.

The product $N\tau$, where τ is the confinement time, turns out to be a critical parameter. According to this, self-sustaining fusion can be initiated by holding a relatively dense plasma for a short time, or a rarer plasma for a longer time. Temperature requirements lead to a critical value of $N\tau = 10^{14}$ particles cm^{-3} sec^{-1}. Failure to approach this requirement has prevented the realization of controlled fusion reactions.

Plasma densities on the order of 10^{15} cm^{-3} appear to be most useful, and these require confinement times of 0.1 sec or more. Confinements on the order of 10^{-6} sec were obtained rather early in the fusion research program, and hopes were high that improved designs and highly purified materials would lead to self-sustaining reactions. Improvements have been slow and expensive, and the required conditions are still almost two orders of magnitude away from realization. Success seems certain eventually, but only after continued slow progress along present lines, or until an entirely new approach is utilized.

19.06 Experimental Machines

Much effort has gone into designing and constructing devices intended to produce the conditions necessary for nuclear fusion. The many design variants can be considered in three broad categories.

a. *Pinch-effect devices.* The plasma is usually produced by an arc discharge at low pressure in the appropriate gas. The magnetic field associated with the arc current will exert a force of compression on the plasma particles, tending to force them into a small, dense column or *pinch* along the axis of the discharge tube. Calculations show that arc currents of the order of 10^6 amperes are required to generate effective fields. Steady currents of this magnitude seem out of the question, but a series of condenser discharges is quite feasible.

Devices based on the pinch effect are relatively simple, and several were constructed early in the thermonuclear research program. Photographs showed that the originally diffuse plasma was indeed compressed into a small pinched volume. The pinch could not be maintained, however. Plasma instabilities invariably appeared, and the plasma escaped from the confining field and dissipated rapidly at the wall of the discharge tube.

Several generations of pinch devices have been built, and each has improved over earlier models. Instabilities of several different kinds remain, and the values of $N\tau$ needed for an overall exothermic reaction have not been achieved. Pinch devices have, however, been most useful in providing basic information on the nature of plasmas and their interactions with fields.

b. *Stellerator.* Magnetic confinement in the stellerator is achieved by winding an external coil around the walls of a toroidal discharge tube. The resulting axial field will cause the charged particles in the plasma to move in tight helices around the lines of magnetic force. Unfortunately, there is also a spatial separation of positive and negative charges which will set up strong electric fields in the plasma. The direction of the charge separation can be reversed in one half of the tube if the plane of the toroid is twisted through 180° to form a distorted figure-8. In later designs, the need for the twist was eliminated by adding auxiliary field windings outside the discharge tube.

Several devices based on the stellerator principle have been constructed. Plasma confinement has been obtained, but instabilities and escape to the walls intervened before stable fusion conditions were obtained. At each stage of their evolution, stellerators have provided important information on plasma behavior, but instabilities remain as a barrier to successful fusion reactions.

c. *Magnetic bottles.* A wide variety of magnetic-field configurations have been proposed or used. Typical are the so-called mirror machines. In a mirror machine, the plasma is generated in a linear discharge tube. An axial field confines the plasma into the central portion of the tube. Near the ends of the tube, the field strengths are greatly increased. A particle approaching this strong field is turned back toward the midportion of the tube, thus effectively increasing the confinement time. The strong magnetic field acts much like a magnetic mirror.

Some of the early magnetic-bottle devices gave confinement times of several milliseconds, but the plasma pressures were low. At higher plasma densities, instabilities again entered to prevent the hoped-for improvements.

Although no absolutely interdicting principle has been discovered, it is evident that attaining the conditions required for self-sustaining fusion is a difficult task. As magnetic devices have become more and more complex in attempts to overcome the instabilities, it has become increasingly doubtful that this approach will lead to practical fusion reactors. Some drastically different approach is needed.

19.07 Nonmagnetic Devices

The stakes in obtaining a fusion reactor are great, and the scientific challenge alone has led to a large effort over the past years. Many devices have been considered, and not a few of them tried in small-scale models.

Preliminary heating by a shock wave, followed by a magnetic compression, has been incorporated in a device known as Scylla. An intense electric discharge creates a shock wave in the plasma contained in an auxiliary discharge tube. Some of the hot plasma, now at about 2×10^5°K, is injected into the main discharge tube, where it undergoes magnetic compression and adiabatic heating. The Scylla series of devices have produced some encouraging results, but ultimate success is yet to be achieved.

An entirely different approach involves the use of high-powered laser beams to develop the required high temperature in a small pellet of fusionable material. In this case, the fuel might consist of tiny liquid pellets of some fusionable material such as deuterium-tritium molecules cooled to the liquid state. The required particle density is then initially present without the need for elaborate confining fields. If a high-powered laser beam could be concentrated on the pellet, perhaps 200 microns in diameter, it might raise the temperature to the point where fusion would be initiated. Expansion associated with the energy release would act to rapidly reduce the particle density, but it may be possible to obtain the necessary confinement times. Presently available lasers do not have the required power capability, but they fall short by only about two orders of magnitude. The possibility of obtaining fusion without the complications associated with plasma instabilities is very enticing.

In the pellet-type reactor, the energy would be released in a series of relatively small uncontrolled bursts as each pellet is injected into the laser beam. Energy could be extracted from the reactor by some sort of coolant and heat exchanger.

One intriguing but apparently impossible reaction involves the production of μ-mesic atoms, Sec. 14.14. A negative muon captured by a proton or a deuteron will have an orbital radius only $\frac{1}{206}$ that of the usual orbital electron, Eq. (6-18). At the smaller distance, the muon will partially shield the nuclear charge from an approaching particle. If the latter were also μ-mesic, the two could approach nearly to the range of the nuclear force before the coulomb barrier became appreciable. With these structures, fusion should be possible at much lower temperatures than are required in the usual plasmas. Muons would be released to act as unconsumed catalysts, free to repeat the process with other nuclei. Unfortunately, the free life of the muon seems to be too short to make this type of reaction feasible, and no other particle now known appears to have the required properties.

19.08 The Lure of Fusion Energy

Controlled fusion energy offers a good many advantages over energy sources operating on fission. It is because of this that several large research programs in the fusion field are being supported, even though progress has been painfully slow and the projected reactors appear to be necessarily complex and expensive.

Fusion fuel will be plentiful and cheap. For several reasons, deuterium will probably be the favored reactant. Although this nuclide exists with a natural terrestrial abundance of only 0.013 per cent, the total content of the oceans will support any forseeable rate of energy consumption for centuries to come. Uranium also is plentiful, but much of it is in the form of very low-grade ores. When these must be utilized, the costs of fission energy are bound to rise, perhaps even going out of competition with fossil fuels. The cost of separating deuterium is low, and this should remain relatively stable for a long period of time.

Fusion reactors will be relatively safe. The safety record of fission reactors has been outstanding, but they are complicated nuclear-mechanical devices and some failures are bound to occur. Any fission reactor that has been in operation for some time will have a large inventory of radioactive fission products which could create a serious radiation hazard if they escaped confinement. In addition a fission reactor contains a large amount of unfissioned, long-lived nuclides whose presence will aggravate any destructive accident.

All trends indicate that a fusion reactor will contain only small quantities of the reactants at any time, so any accident should be strictly limited. The large-scale, destructive release of energy in a fusion reactor appears extremely remote. The end products of all proposed fusion reactions are stable nuclides. Thus the use of fusion will avoid the problems associated with the storage of vast quantities of long-lived radioactive nuclides such as the products of fission reactions. The radiation hazard from a fusion reactor will not necessarily be zero. Neutrons will be released from most of the fusion reactions that appear to have promise, and these will require extensive shielding in order to protect operating personnel. In the event of a destructive accident, there could be a release of a large amount of radioactivity induced in the reactor components by the neutrons. Normal operation may result in the release of substantial quantities of tritium. However, fusion-reactor hazards appear to be orders of magnitude less than those associated with sources of fission energy.

Practically all of the energy released from fission fuels is converted to electrical energy for general distribution. Steam at high pressure is generated at the high reactor temperature. Steam then drives a conventional turbine, and the latter an electric generator. The overall energy conversion is so poor that scarcely one-third of the thermal energy is eventually realized as electrical energy. The unavailable energy, thrown away in rivers, lakes, or estuaries, may constitute an unacceptable source of thermal pollution.

Energy from a fusion reactor may be converted in the same fashion. It is too early to say whether the efficiency of the fusion conversions will be better or worse than those of the fission cycle. In any case, the differences will probably not be very great if one is forced to go through the steam–turbine–generator sequence. Fusion reactors, on the other hand, offer the real possibility of the direct conversion of plasma energy to electrical energy. Moving charges set up magnetic fields, and it may be possible to couple the fields in a fusion reactor through transformers directly to power-distribution systems, thus eliminating all steam-generating equipment and rotating machinery. A development in this direction would give a tremendous advantage to the fusion cycle over fission.

REFERENCES

Bethe, H. A., "Energy Production in Stars." *Phys. Today* **21,** 36, 1968.

Glasstone, S. and R. H. Loveberg, *Controlled Thermonuclear Reactions.* Van Nostrand, New York, 1960.

Salpeter, E. E., "Energy Production in the Stars." *Ann. Rev. Nuc. Sci.* **2,** 41, 1953.

Thorne, K. S., "Gravitational Collapse and Death of a Star." *Science* **150,** 1671, 1965.

PROBLEMS

19-1. The rate of energy delivery to the earth from the sun, 1.5×10^8 Km distant, is known as the solar constant, with a value of 2.0 cal cm^{-1} min^{-1}. Calculate the rate at which the sun is losing mass to supply this radiated energy. What fraction of its mass of 2×10^{33} g does the sun lose per year?

19-2. Calculate the energy radiated during the assembly of the sun mass into its present diameter of 1.4×10^6 Km. What energy would be released if the diameter were to suddenly shrink by 10 per cent?

19-3. The total volume of water in the oceans is 1.34×10^9 Km^3. What is the total mass of deuterium contained in the water? What will be the total energy available from all of this deuterium through reaction (19-1b)? How much mass will be lost in this energy release?

19-4. Assume a laser-induced lithium fusion reaction to be carried out with 0.4-ng pellets of ^6Li and ^2H. Use the most exoergic reaction available and calculate the energy that will be released at each pellet fusion. At what rate must pellets be fused to obtain a thermal-energy release rate of 1 megawatt?

19-5. Consider the possibilities of obtaining fusion energy from ^6Li and ^1H.

19-6. What will be the height of the potential barrier seen by the incoming deuteron in Eq. (19-8)? To what temperature does this correspond?

20

Cosmic and Terrestrial Radiation

20.01 Discovery of Cosmic Rays

By the early part of this century, ion chambers and electrometers had been developed into highly refined and sensitive instruments. Even with the use of the finest insulating materials, and with the utmost care in construction, each system showed an unaccountable leakage current. V. F. Hess correctly suspected that this leakage was due to a previously unrecognized source of radiation.

Previously, physicists had attributed this leakage to ionization produced by radioactive materials in the earth. In 1911, Hess sent instruments away from the surface of the earth, in balloons, and discovered that the leakage increased with altitude, rather than decreased, as would be the case with a terrestrial source. Hess concluded that the ionization was produced by "a very penetrating radiation coming mainly from above and being most probably of extra-terrestrial origin."

In the early 1920s, R. A. Millikan and his collaborators at the California Institute of Technology directed a series of balloon flights which carried instruments to a height of almost 10 miles and measured the radiation increase with altitude. These experiments were followed by absorption measurements made with ionization-chamber electroscopes sunk to great depths in lakes at different altitudes. These investigations proved conclusively that the mysterious radiation was far more penetrating than any gamma rays emitted by naturally radioactive substances; at the same time, the extraterrestrial origin of the radiation was established, leading Millikan to introduce the name "cosmic rays" for the penetrating particles. Millikan's work marked the beginning of an era of intensive cosmic-ray exploration, which has added an immense store of knowledge to man's comprehension of fundamental particles and nuclear processes.

Ionization intensities were measured at many points on the earth's surface, and at altitudes as high as could be reached by balloons. It soon became evident that the character of the ionizing agent changed with altitude and latitude, and that it was necessary to distinguish between the primary radiation and the secondaries produced by it. Early work demonstrated the effect of the earth's magnetic field on the cosmic radiation, and showed that geomagnetic latitude was a more direct parameter of the radiation than was geographical latitude, Fig. 20-1.

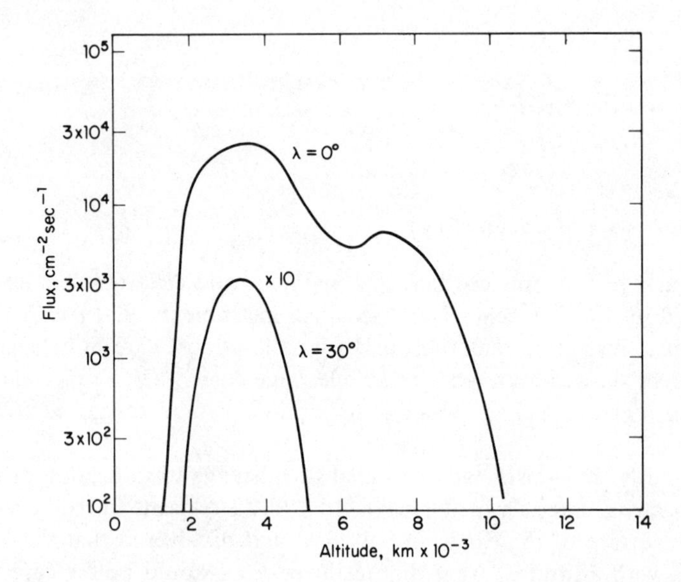

Figure 20-1. Variation of cosmic ray intensity with altitude and geomagnetic latitude.

Although the earth's field is weak, it extends out for large distances, and so acts on incoming charged particles for an appreciable length of time. Even this weak field is sufficient to deflect some of the less-energetic particles away from the earth. More-energetic particles will be deflected less and may penetrate through the field to the earth. The shape of the field is such that the magnetic effect is greatest at the geomagnetic equator, least at the poles.

In dealing with magnetic deflections, momentum is a more convenient parameter than kinetic energy, because the former is simply related to the magnetic stiffness per electronic charge, Eqs. (4-14) and (4-15). The minimum momentum required for vertical penetration at geomagnetic latitude λ is given by

$$p_{\min} = 15 \cos^4 \lambda \text{ GeV}/c \tag{20-1}$$

Thus about 15 GeV/c is required for penetration at the equator, while only 4 GeV/c is needed at 45° geomagnetic latitude. Extraterrestrial fields are also present, making Eq. (20-1) of doubtful validity above latitudes of 50–60°.

Cosmic-ray energy distributions measured near the earth's surface are related to the type and number of secondaries produced by the primaries through a variety of absorptive processes. As very high altitude measurements became possible, the true nature of the primaries was revealed. At the top of our atmosphere, the primary flux consists almost entirely of positive ions, completely stripped of their orbital electrons. Some electrons are also present, but their numbers are small compared to the flux of the incoming nuclei.

Primary-particle energies as high as 10^9 GeV have been detected, although the bulk of the particles have more nearly 6 GeV. The discovery of these enormous energies settled one of the controversies that had developed during the early days of cosmic-ray research. One school interpreted the cosmic rays as the "birth cries" of atoms, as heavy structures were formed by the fusion of light components. Another school visualized them as the "death rattle" of atoms that were being annihilated and converted to radiation. Neither of these processes can account for the observed energies. The complete annihilation of a uranium nucleus would yield only some 200 GeV, and the fusion process would yield even less.

Even today, after years of intense research, the origin of the cosmic radiations is not established with certainty. It seems clear, however, that the energies must come from accelerations in intergalactic fields, either at the point of origin or during the long, long journey through space.

20.02 Galactic Cosmic Rays

As measurements became more sophisticated and covered greater areas of space, it became possible to distinguish several components in the cosmic-ray flux. *Galactic cosmic rays* make a relatively constant contribution to the total radiation flux incident upon the top of our atmosphere. Ions ranging from protons, $Z = 1$, to iron, $Z = 26$, have been identified. The number distribution is compared with the "universal abundance" of the corresponding elements in Table 20-1.

The predominance of protons apparently reflects the universal abundance of hydrogen. Most surprising is the large number of C, N, and O ions seen in the cosmic radiation, compared to the amount of these elements now existing. Perhaps they were more abundant at the time the cosmic rays were being produced.

TABLE 20-1
COMPOSITION OF THE PRIMARY COSMIC-RAY FLUX

	Element					
	H	He	C,N,O	Mg	Ca	Fe
Atomic no.	1	2	7	12	20	26
Flux cm^{-2} min^{-1}	75	10	5	1.6	0.5	0.2
Abundance	100	8	10^{-6}	0.2	0.003	10^{-6}

Two-dimensional arrays of G–M-counter tubes connected for coincidence recording form a *cosmic-ray telescope* with which the directions of the arriving particles can be determined. Galactic cosmic radiation is found to be essentially isotropic in space, with no preferred direction of arrival. Thus no clue is provided as to the place or places at which the rays originated. This result is not surprising. After centuries of travel in space, and encountering innumerable magnetic fields, a particle trajectory is not apt to bear any discernable relation to its point of origin or to its original direction of propagation.

The observed isotropy does indicate that the galactic component originates well outside our solar system. Distances inside the system are too small for any of the known fields to produce an appreciable deflection of the high-energy particles.

Little is known about the long-time variations of cosmic-ray intensities. Constancy is assumed in dating calculations involving neutron-induced products, and the results obtained seem to be self-consistent. However, these are very short times on a geological scale. Presumably, some of the cosmic radiation now reaching the earth was emitted with modest energies many centuries ago.

At least some of the galactic cosmic radiation is probably produced in *neutron stars*, whose intense fields are quite capable of ejecting particles with the observed energies. One example of such a source is the pulsating star or *pulsar* in the Crab nebula. This star, formed in the creation of a supernova in AD 1054, is probably typical of many older sources, no longer visible, which are capable of emitting the energetic particles with little need for further accelerations by intergalactic fields.

20.03 The Solar Wind

Space probes have shown that our earth is moving through a stream of charged particles that are being emitted continuously by the sun. This plasma, known as the *solar wind*, is a sort of supersonic expansion of material from the solar corona. The plasma, consisting almost entirely of protons and electrons, is called "collisionless" because of the low density, 3 ions cm^{-3}, and the correspondingly long mean-free-path of 10^7 cm.

The charged particles in the solar wind cannot penetrate the magnetic field of the earth, and instead are deflected by it. The earth–plasma encounter takes place at a relative velocity of 5×10^{-7} cm sec^{-1}, and this creates a sort of shock front ahead of the earth. A few particles will be extracted from the plasma and will be trapped in the earth's field, to create more densely ionized regions. A long tail of ionization extends behind the earth in the antisolar direction.

Figure 20-2 is a simulation of the plasma–earth encounter. A hydrogen plasma was projected from the hydrogen arc at the right of the photograph,

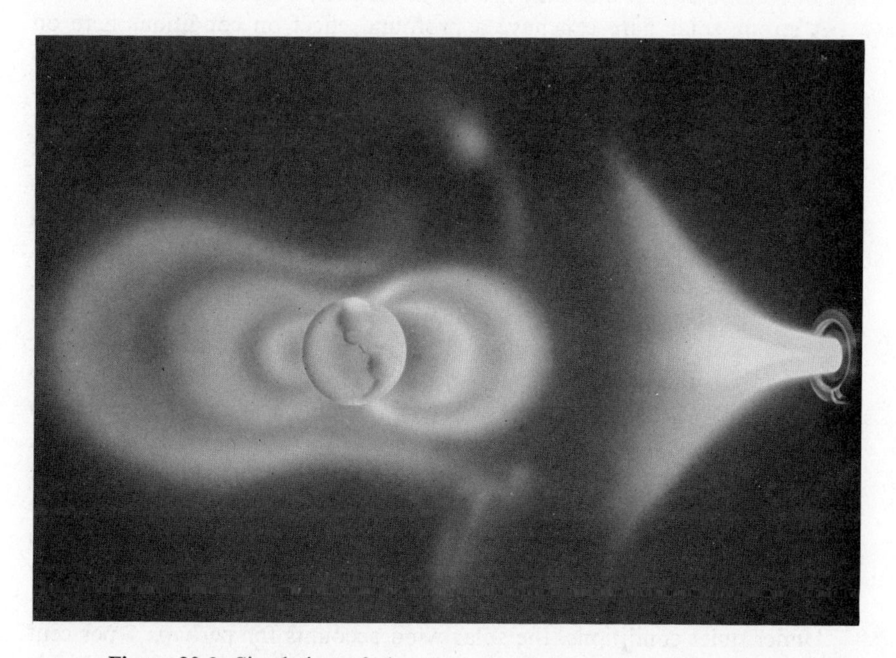

Figure 20-2. Simulation of the encounter between the earth and the solar wind. (Courtesy of NASA Lewis Research Center, Cleveland, Ohio.)

with a velocity comparable to that of the solar wind. The earth model contains a magnetic dipole strong enough to produce radii of curvature comparable with the dimensions of the equipment. The deflection of the incoming particles can be plainly seen. Two ionized regions extend from the magnetic poles of the simulated earth in a manner strongly suggestive of the actual Van Allen belts. A long tail of ionization, shaped by the magnetic field, streams out behind. Quantitative measurements on simulated solar winds of this type will be most useful in our understanding of high-speed plasma–field interactions.

During the time of a "quiet" sun, the plasma density will be of the order of 10 ions cm^{-3}. Solar activity, particularly in localized areas, can change dramatically in a matter of minutes. A region on the solar surface may rise from its normal temperature of 6000°K to perhaps 10^6°K, to form a *solar flare*. Flares may vary widely in magnitude, with an extremely large event occurring on the average only once in several years. A solar flare may be visible for 30–50 minutes. During this time, there is an increased emission of photons of all energies from the low radio frequencies up through the X-ray region. Plasma ejection is also enhanced, which results in an increased rate of particle arrival that may last for as long as 48 hours.

A strong solar flare can have a profound effect on conditions here on earth. The normal geomagnetic field will be distorted and this may lead to changes of an order of magnitude or more in the cosmic-ray intensities. Radio communications may be partially or totally disrupted for an hour or more. Although the local fields may be distorted, terrestrial inhabitants will still be protected from the lethal effects of the high radiation intensities by the absorbing atmosphere, and by the fact that the magnetic field still prevents many of the particles from reaching the surface of the earth.

Solar flares present the greatest radiation hazard to space travel. At the present time, there is no sure way of predicting the advent of a large solar flare. Flares develop out of sunspots, but most sunspots subside without intensifying to the flare stage, and so they have little prognostic value. At best, sunspots will give an advance warning of only a few days. Astronauts on an extended mission in space are vulnerable to the large radiation doses that will result from a strong unanticipated flare. Effective shielding against the flare radiation is impractical. Partial shielding may only enhance the radiation dose through the increased production of low-energy radiations that are more effectively absorbed in the body than are the high-energy primaries. Complete shielding requires impossible masses of material.

Under quiet conditions, the solar wind accounts for perhaps 1 per cent of the total cosmic-ray ionization. The solar component is largely responsible for the rapid fluctuations, both from the enhanced primary flux and from the indirect effect on the earth's field.

20.04 Geomagnetically Trapped Radiation

The shock front in the solar windstream partially isolates a region around our earth known as the *magnetosphere*. Charged particles that enter this area become trapped by the earth's field. Near the geomagnetic equator, the particles will move along the magnetic lines of force toward one of the magnetic poles. As the particles approach the poles, they will be reflected back by the magnetic effect of the stronger field. The effect is similar to the magnetic-mirror principle being used in some of the thermonuclear devices to confine a plasma. Magnetic reflections take place near each magnetic pole, so the trapped particles oscillate back and forth along magnetic field lines. Magnetic inhomogeneities and the earth's rotation distribute the charges to all longitudes.

The trapped-radiation regions, first discovered by the Explorer X space vehicle in 1958, are known as the Van Allen belts, after the man who studied them most intensively.

The inner Van Allen belt, extending from about 500 to 10,000 Km above the earth, is a plasma composed primarily of protons and electrons. Proton fluxes here may reach 10^4 cm^{-2} sec^{-1}, with a minimum energy of 40 MeV. Intensities in the inner belt fall off as altitude increases, to gradually merge with the outer belt, which extends from perhaps 15,000 to 50,000 Km. Electrons predominate in the outer belt. Most of the electrons have energies measured in tens of KeV, but there are small numbers up to 2 MeV. The outer flux is more variable than the inner, as might be expected, since at the large distances the strength of the earth's field is low and more subject to modulation by fields from outside sources.

Presumably, the particles that are trapped in the belts were extracted from the solar wind and the galactic radiation. The existence of the trapped radiation was suspected before it was detected, but its persistence was not anticipated. Charged particles have been injected into the trapping region by high-altitude nuclear detonations. Many of the injected particles are still in the belts, and will probably remain there for as long as 25 years.

Radiation doses within the belts are so high that any prolonged human stay is precluded. The dose rate within a typical space vehicle will be 20–30 rads hr^{-1} in the inner belt, 100–1000 rads hr^{-1} in the outer. Orbital missions that remain at distances well below the inner belt will experience only modest radiation-dose rates. Missions outside the belts are accomplished by a rapid transit and re-entry through the ionized regions, and the total exposure is acceptable.

20.05 Absorption of Cosmic-ray Primaries

Many of the cosmic-ray primaries enter the earth's atmosphere with energies greater than the total binding energy of any nucleus with which they may collide. The tremendous excitation energy brought into a struck nucleus usually leads to the complete disintegration of the nucleus by *nuclear evaporation* or *star formation*. Figure 20-3 shows an example of a star initiated by

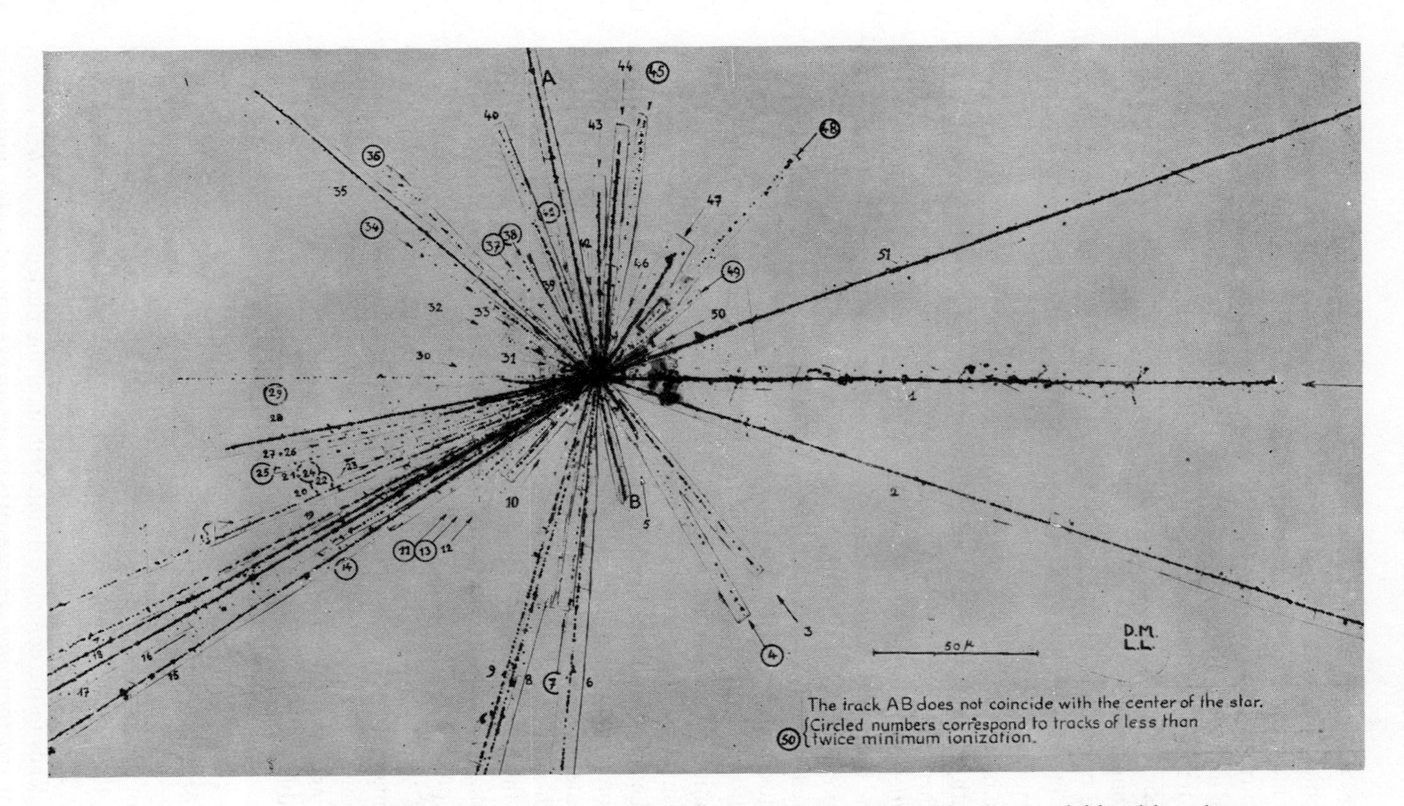

Figure 20-3. A 51-prong star resulting from the evaporation of a silver nucleus. The star was initiated by a heavy primary, $Z = 17$, striking a silver nucleus in a photographic emulsion. (From A. L. Leprince-Ringuet, *Phys. Rev.* **76**, 1273, 1949.)

the impact of a heavy primary on a silver nucleus in a photographic emulsion. The products of the initial nuclear evaporation contain so much energy that they initiate other nuclear disintegrations. A massive, densely ionized area or star is formed at the initial site, and a number of penetrating secondaries leave the area.

Figure 20-4 depicts a meson shower triggered by a high-energy alpha particle. Some densely ionizing tracks may be seen, along with those of the sparsely ionizing mesons.

If the heavy primary does not enter into a cataclysmic nuclear reaction, it will lose its energy gradually in a closely spaced series of collisions. Many of the delta rays will have energies of several hundreds of electron volts, and these will produce a dense column of ionization along the track of the primary. As the latter slows and acquires electrons, its ionizing capabilities are reduced, and a typical *thindown track* is seen, Fig. 13-1.

As the primary flux enters into reactions with terrestrial nuclei, many new components appear. Neutrons, pions, and muons are among the debris created by the high-energy interactions. Some of the neutrons will be captured to form radioactive products. The $^{14}N(n, p)$ reaction has maintained a natural abundance of ^{14}C in spite of its short half-life on a geological scale.

Many of the radiations have an extremely high LET. Absorbed doses in living organisms will be highly nonuniform. A single massive thindown might conceivably destroy the function of a vital center in the central nervous system, while the vast majority of the brain tissue was untouched. Little is known of the probability of such an event. It seems clear, however, that our classical ideas of radiation dose do not apply here, and that some new criteria are needed.

20.06 Dose from Background Radiation

Cosmic rays are not the only source of ionizing radiation to which human beings and instruments are exposed. Inescapable radiation exposure of people living on the earth comes from three sources: cosmic rays, local sources of gamma radiation, and radioactive materials incorporated in the body. The first two components vary widely from place to place. Representative values at sea level might be

Cosmic rays:	35 millirads per year
Local gamma emitters:	50
Internal emitters:	20
Total:	105

At low altitudes, the cosmic-ray contribution can be properly evaluated, because most of the extremely high-energy, high-LET components have

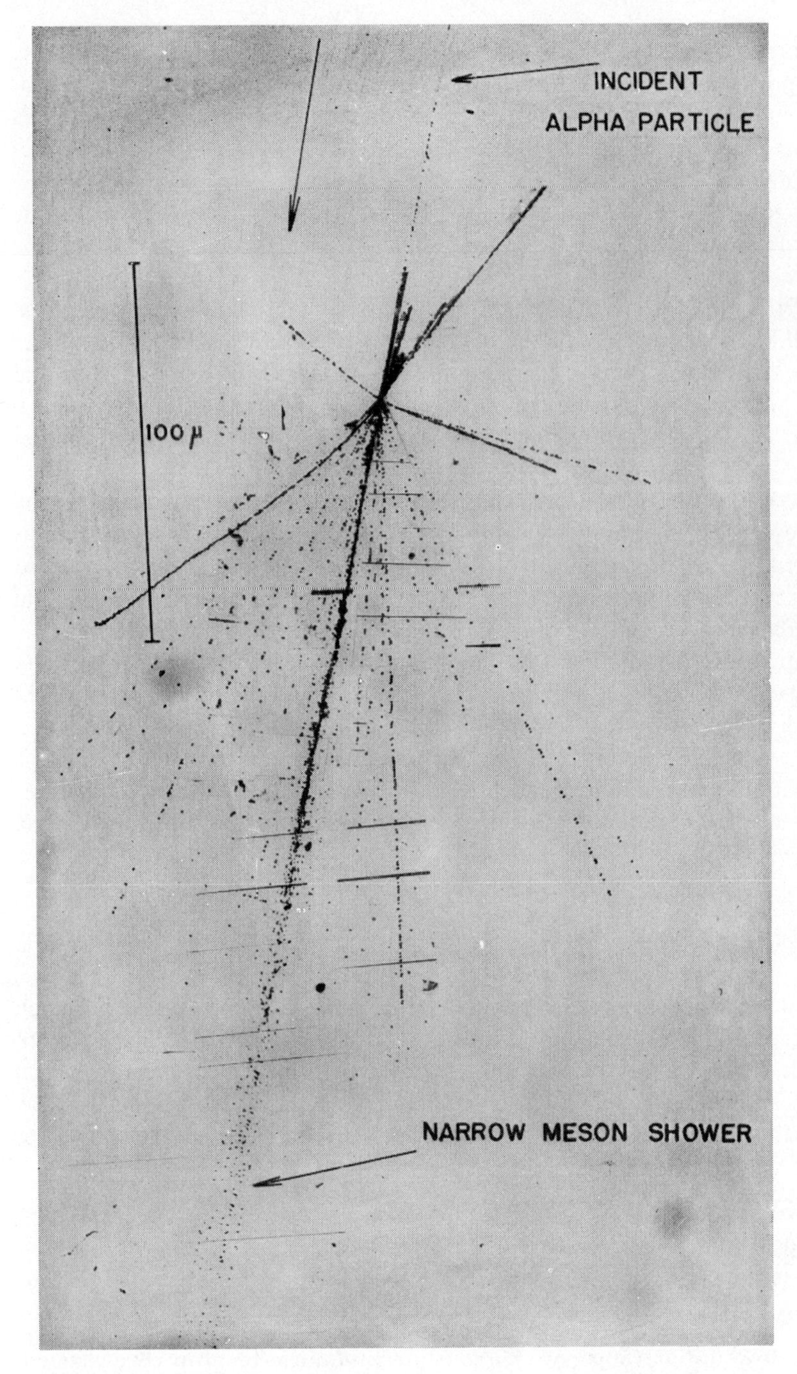

Figure 20-4. A meson shower produced by a high-energy (10^6 MeV) alpha particle. (From M. F. Kaplon, B. Peters, and H. L. Bradt, *Phys. Rev.* **76**, 1735, 1949.)

been degraded to energies which give more uniform tissue doses. Cosmic-ray intensities vary with altitude and geomagnetic latitude by a factor of 2 or more.

Although rich deposits are rare, low concentrations of radioactive nuclides are very widespread. Exposure from these sources will be least over the oceans. On land, sedimentary rocks will have a lower content of active materials than will igneous formations. In a few areas, notably in Brazil and India, substantial numbers of people have lived for many generations in areas of exceptionally high local radiation.

Radium, and other naturally occurring radioactive nuclides, are so wide-spread that they are found in very low concentrations in many water supplies and foodstuffs. Much that is ingested will be eliminated, but the chemical similarities between radium and calcium lead to some deposition, principally in bone. Over a normal lifetime, the accumulated body burden of radium may amount to as much as 0.1 microcurie.

Two naturally occurring radioactive nuclides contribute to the radiation dose from internal emitters. We have seen that ^{14}C is produced in the upper atmosphere by neutron reactions with nitrogen. Under long-established equilibrium conditions, the ^{14}C entered into the total carbon pool of the earth to produce a specific activity of about 15 disintegrations per minute per gram of carbon. Living tissues, as part of the general carbon pool, will also contain ^{14}C at that level. Atmospheric testing of nuclear devices released sufficient neutrons to increase the previously existing specific activity by about 2 per cent. With the mutual agreement to cease atmospheric testing by the U.S. and the U.S.S.R., the enhanced activity is returning toward its former level.

Most of the radiation dose from internal emitters comes from ^{40}K. Natural potassium contains 0.0119 per cent ^{40}K, which has two modes of decay:

$$^{40}\mathrm{K} \longrightarrow {}^{40}\mathrm{Ca} + \beta^- \qquad 89\% \tag{20-1}$$

$$^{40}\mathrm{K} + e^- \longrightarrow {}^{40}\mathrm{Ar} + \gamma \qquad 11\% \tag{20-2}$$

Potassium is an essential constituent of living tissues, particularly of lean muscle. All of the mean beta-particle energy, 0.46 MeV, will be absorbed locally. The 1.32-MeV gamma ray from the reaction of Eq. (20-2) will add only a small fraction of the beta-particle dose.

The gamma rays from reaction (20-2) that escape from the body have been put to good use in the human whole-body counter. In this counter, the subject is partially or completely surrounded with gamma-sensitive detectors such as sodium iodide or a liquid scintillator. The efficient counting geometry permits a determination of the body burden of potassium with only a few minutes' counting time. Whole-body counters are finding increasing uses in diagnostic medicine for potassium counting and for determining the uptakes of a variety of administered radionuclides.

REFERENCES

Anderson, K. A., "Energetic Particles in the Earth's Magnetic Field." *Ann. Rev. Nuc. Sci.* **16,** 291, 1966.

Bierman, L., "Origins and Propagation of Cosmic Rays." *Ann. Rev. Nuc. Sci.* **2,** 335, 1953.

Friedman, H., "X-rays from Stars and Nebulae." *Ann. Rev. Nuc. Sci.* **17,** 317, 1967.

Holcomb, R. W., "Galaxies and Quasars: Puzzling Observations and Bizarre Theories." *Science* **167,** 1601, 1970.

PROBLEMS

20-1. About 1300 primary protons reach each cm^2 of the earth every hour. To what current flow is this equivalent? Assume that this flow is the only charge reaching or leaving the earth and calculate the rate of change of the earth potential.

20-2. What is the cosmic-ray power delivered to the earth, assuming that the average proton energy is 6 GeV?

20-3. Astronauts leaving earth on a mission to the moon observe an intense solar flare. How much later can they expect to encounter the increased proton flux released in the flare?

20-4. Calculate the annual radiation dose to the human body from its natural content of ^{14}C and ^{40}K, and compare these values to the dose received from cosmic-ray and general background.

20-5. Calculate the total energy released when a ^{40}Ca nucleus in bone is evaporated into its constituent neutrons and protons by a 3-GeV-proton primary cosmic ray.

21

Transuranium Elements

21.01 Neptunium and Plutonium

The possibility of producing elements with atomic numbers beyond that of uranium was recognized soon after the neutron and its ability to produce nuclear reactions were discovered. Fermi suspected that he had produced transuranium elements by the neutron bombardment of uranium, when in fact he was observing the radioactivity of fission products.

In 1940, McMillan and Abelson produced, and correctly identified, the first manmade transuranium elements. They bombarded a uranium target with slow neutrons and observed two negatron activities with half-lives of 23 minutes and 2.3 days. The initial reaction formed a compound nucleus with ^{238}U with a good yield because of the strong low-energy resonances, Fig. 18-1.

$$^{238}_{92}U + ^{1}_{0}n \longrightarrow ^{239}_{92}U \tag{21-1}$$

The 23-m half-life was identified with the negatron decay of the compound nucleus:

$$^{239}_{92}U \xrightarrow{23\text{ m}} ^{239}_{93}Np + \beta^- \tag{21-2}$$

Neptunium was appropriately chosen as the name of the new element, since in our solar system the planet Neptune lies beyond Uranus.

Neptunium-239 decays in turn by negatron emission with a half-life of 2.3 days:

$$^{239}_{93}Np \xrightarrow{2.3\text{ d}} ^{239}_{93}Pu + \beta^- \tag{21-3}$$

Plutonium was chosen as the name of the new daughter product, continuing the association with the planetary sequence Uranus, Neptune, Pluto.

Plutonium is an important stepping-stone to still heavier elements because it can be produced in large quantities and can be stockpiled. ^{239}Pu decays by alpha-particle emission and spontaneous fission, but the half-lives are long, 24,000 and 5.5×10^{15} y, respectively.

21.02 Elements Beyond Plutonium

The potentialities of ^{239}Pu as a nuclear fuel and a bomb material stimulated intense research activity on elements at the upper end of the periodic classification. This activity has continued, and several countries have research teams working on the transuranium elements. Any compilation of manmade nuclides may soon be outdated. Table 21-1 shows the transuranium census as of September, 1970.

TABLE 21-1
TRANSURANIUM ELEMENTS

Name	*Symbol*	*Discovered*	*Z*	*Mass range*	*Estimated production in 1975 (grams*)*
Neptunium	Np	1940	93	231–241	4×10^5
Plutonium	Pu	1940	94	232–246	1×10^7
Americium	Am	1945	95	237–246	8×10^4
Curium	Cm	1944	96	238–250	2×10^3
Berkelium	Bk	1950	97	243–250	10^{-1}
Californium	Cf	1950	98	244–254	1
Einsteinium	Es	1955	99	245–256	10^{-2}
Fermium	Fm	1955	100	248–258	10^{-11}
Mendelevium	Md	1955	101	255–257	
Nobelium	No	1958	102	251–257	
Lawrencium	Lw	1961	103	256–260	

*Estimated for the principal isotope with perhaps small quantities of other mass numbers. From G. T. Seaborg, "Elements Beyond 100, Present Status and Future Prospects," *Ann. Rev. Nuc. Sci.*, Annual Reviews, Inc., Palo Alto, Cal., 1968.

By tradition, newly identified elements are named by the discoverer. The names listed in Table 21-1 reflect the intense activity in this field centered at the University of California, Berkeley. A study of the production dates and the estimated production figures will indicate the fantastic rate at which experimental difficulties increase as one goes up the atomic-number scale. Two production methods are available for the creation of transuranium elements: neutron capture followed by negatron decay, and the bombardment of heavy nuclei by heavy charged particles such as ^{10}Ne, ^{12}C, or perhaps even ^{238}U.

21.03 Neutron-production Methods

Reactions (21-1) and (21-2) are examples of a general scheme used to create new elements of higher atomic number: neutron capture followed by negatron emission. Neutron capture brings its binding energy at least, and perhaps some kinetic energy as well, to the compound nucleus. The neutron capture has increased the N/Z ratio, which can be nearly restored by the emission of a negative beta particle.

Elements below $Z = 82$ tend to decay by beta emission, but at $Z = 83$, alpha-particle emission becomes important and predominates at higher atomic numbers. Alpha decay in turn is superseded by fission at still higher atomic numbers. The (n, β^-) reaction needed to produce a higher-Z nuclei is in sharp competition with (n, γ), (n, p), (n, α), and (n, f); production of the desired element is, in general, very small.

The ultimate concentration of any species produced by neutron bombardment depends linearly on the neutron flux, Eq. (16-22). Interest in transuranium production has led to the construction of specially designed high-flux reactors. These reactors consume nuclear fuel at a rapid rate, but thermal fluxes approaching 10^{16} cm^2 sec^{-1} have been attained.

Every transuranium nucleus thus far produced is radioactive and most have relatively short half-lives. The most attractive stepping-stones upward after ^{239}Pu are ^{242}Pu and ^{252}Cf. Intense neutron irradiation of ^{239}Pu produces 3.8×10^5 y ^{242}Pu by the $3n$ reaction of Eq. (21-4):

$$^{239}\text{Pu} + n \begin{cases} \nearrow 73\%\ \text{fission} \\ \searrow {}^{240}\text{Pu} + n \longrightarrow {}^{241}\text{Pu} + n \begin{cases} \nearrow 70\%\ \text{fission} \\ \searrow {}^{242}\text{Pu} \end{cases} \end{cases} \tag{21-4}$$

Large production reactors can produce enough ^{242}Pu so that it is not a bottleneck to subsequent reactions. Ten nucleons, four of them protons, must be added to ^{242}Pu to reach 2.6 y ^{252}Cf. Several combinations of neutron absorption and beta decays turn the trick, Fig. 21-1. There are competing reactions at every step, and a special reactor was constructed for this irradiation, in the hope of ultimately attaining an annual production of 1 gram of ^{252}Cf.

With a supply of ^{252}Cf in hand, further neutron irradiation leads to the production of ^{257}Fm, Fig. 21-1. Here the neutron–beta sequence runs into trouble. The half-life of ^{258}Fm that is produced by the next neutron capture is apparently too short to permit any appreciable accumulation. The neutron-production chain appears to be broken at this point unless some way can be found to circumvent the roadblock, perhaps through ^{257}Es.

Nuclear detonations have been used successfully as neutron sources for transuranium-isotope production. Einsteinium and fermium were first isolated from the debris of a thermonuclear explosion.

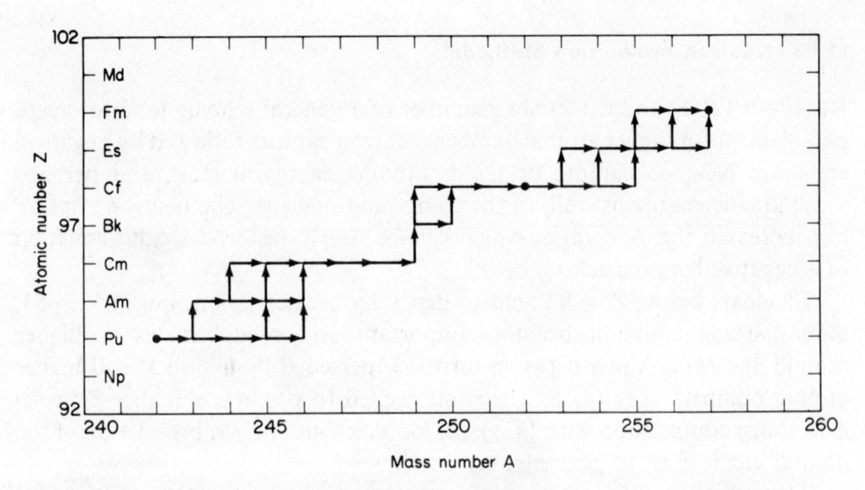

Figure 21-1. Production pathways from ^{242}Pu to the heavier elements.

21.04 Charged-particle Production

Bombardment with high-speed charged particles will produce nuclei with relatively large jumps in both mass and charge. Targets may be any of the materials available from the neutron–beta sequence. Bombarding particles can range upward from protons, deuterons, and alpha particles to the most massive ions. A typical charged-particle reaction that has been used successfully is

$$^{253}_{99}\text{Es} + ^{4}_{2}\text{He} \longrightarrow ^{256}_{101}\text{Md} + ^{1}_{0}n \tag{21-5}$$

Product nuclei from charged-particle capture will be neutron-deficient, rather than neutron-rich, and so they are apt to decay by alpha-particle emission, or by positron emission or its equivalent, electron capture. The latter is the case for the ^{256}Md produced by the reaction of Eq. 21-5:

$$^{256}_{101}\text{Md} + e^{-} \xrightarrow{77\text{m}} ^{256}_{100}\text{Fm} \tag{21-6}$$

Still heavier ions were used to produce lawrencium:

$$^{252}_{98}\text{Cf} + ^{10}_{5}\text{B} \longrightarrow ^{256}_{103}\text{Lw} + 6n \tag{21-7}$$

Several accelerators in use or under construction have characteristics that are especially suitable for heavy-ion bombardments. Magnetic-resonance machines lack some of the versatility of nonresonant accelerators, since they are restricted to certain charge–mass ratios. In the Soviet Union, however, the cyclotron has been the device of choice, and other resonance machines have been used successfully.

A double Van de Graaff arrangement will accelerate any ions, even up to uranium, to energies in excess of that needed to penetrate the potential barrier of a uranium target. Cascaded linear accelerators are also capable of accelerating any particles to energies needed for effective barrier penetration.

21.05 Product Identification

The amount of material available for a target, and the amount of product formed, decrease rapidly as the atomic number increases. Mendelevium, first produced by the reaction of Eq. (21-5), was formed by the bombardment of an einsteinium target of less than 10^{-12} g, or 10^9 atoms. Product identification was made on a yield calculated to be less than one atom for every hour of bombardment time.

Obviously, special methods must be used to separate and identify the product nuclei. Fortunately, the chemical properties of the new elements can be calculated with a high degree of certainty from lower analogs.

Product nuclei, recoiling from the capture of a charged particle, are ejected from the thin foil target and are caught on a second thin gold foil. The latter is dissolved and subjected to ion-exchange chromatography. Unknown activities found in the various eluted fractions can be assigned to specific elements. Mass assignments can be made by considering the details of the reaction, and by obtaining a given nuclide by a variety of reactions. Descriptions of the ingenious techniques devised for production and identification make fascinating reading, for which the original literature should be consulted.

21.06 Chemical Properties of the New Elements

The transuranium elements fall in the part of the periodic classification where the $5f$ subshell is filling up successively, Fig. 21-2. Each member of the $5f$, or actinide, series may then be compared to its analog in the lanthanide series, $Z = 58$–71, where the $4f$ subshell is filling. For example, terbium, $Z = 65$, has a xenon core plus 11 outer electrons whose configuration is [Xe] $4f^8 5d6s^2$. The corresponding element in the actinide series is einsteinium, with a radon core and 11 outer electrons. There is some uncertainty in the calculations, but the most probable configuration here is [Rn] $5f^7 6d^2 7s^2$. This differs from its analog by only one electron between the f and d states.

Series similarities can be transferred to still higher atomic numbers and chemical properties predicted with some confidence. Thus element 129, the terbium analog in the next or hyperactinide series, will have a rare-gas core and an outer configuration very nearly $6f^7 7d^2 8s^2$. The capability of calculating chemical and physical properties before the fact is an invaluable aid in locating the minute products from the series of eluants, and in making positive chemical identifications.

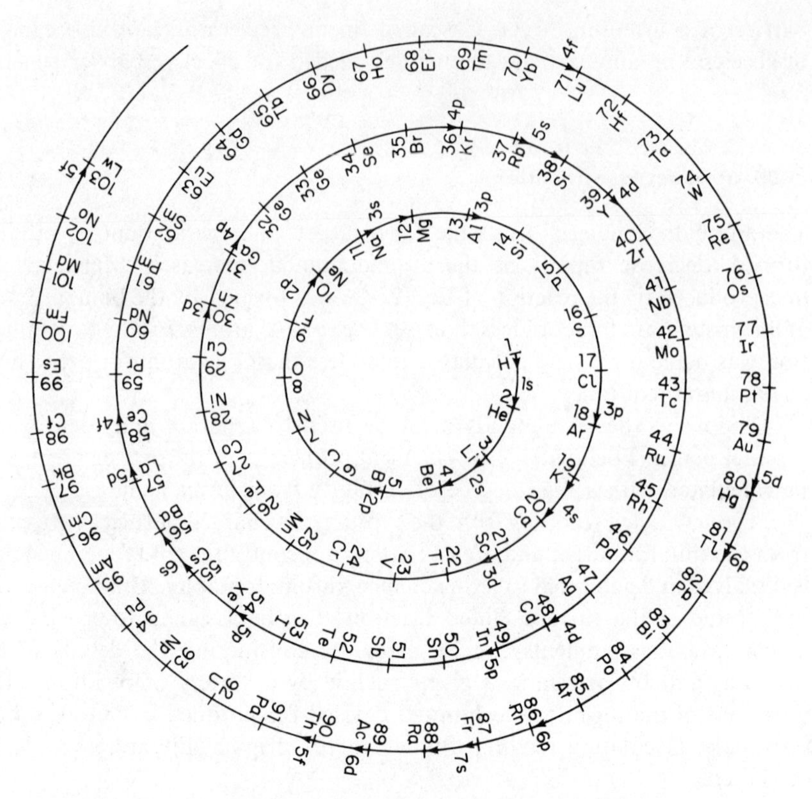

Figure 21-2. Filling sequence of electron shells.

Some of the transuranium nuclides have properties of more than academic interest. ^{242}Cm and ^{244}Cm are outstanding examples of heat sources in which high specific output is combined with a long life. The alpha-particle emissions from 1 gram of ^{242}Cm release 30 calories of energy per second. All of this energy is absorbed locally to produce a high temperature that can be used in many applications. This rate of energy release will be reduced to one-half in 163 days. One gram of gasoline, releasing energy at the same rate, will be completely consumed in 6.4 min. ^{244}Cm releases 0.68 cal sec^{-1} g^{-1}. The half-life of this nuclide is 17.6 y; the whole-life for gasoline at the same rate is 4.7 h.

21.07 Islands of Stability

It is evident that a nucleus has many of the attributes of the independent-particle model, with a series of shell-like structures. At the same time, it behaves in many ways like a liquid drop. Calculations based on the shell model have been particularly useful at or near the magic numbers, and it is obviously desirable to extend these calculations to the transuranium region.

As more nuclear structures have become known in detail, it has become possible to account for the magic numbers as evidence of shell closures, and to predict the numbers with some success.

Extrapolation from known structures can be made up into the transuranium region, and the order of filling neutron and proton states predicted with some degree of confidence. When this is done, shell closures at $Z = 114$ or 126, and at $N = 184$, are to be expected. Nuclei $^{298}_{114}X$ and $^{301}_{126}Y$ should be doubly magic, and as such should be relatively stable. If these predictions are fulfilled, these structures may stand out as islands of relative (but not absolute) stability surrounded by a sea of very unstable, short-lived nuclides.

Potential barrier heights can be calculated from the assumed shell structures, and these results can be used to estimate half-lives for each of the decay processes. For $^{301}_{126}Y$, the half-life for spontaneous fission may well be over 1 year, and for alpha decay more than 1 day. These are indeed long half-lives when compared to the increasing instability seen as one progresses painfully upward from ^{252}Cf.

The postulated X and Y elements can only be produced by hurling massive projectiles ($A = 60$–80) at suitable heavy-element targets. This is obviously a job for a large charged-particle accelerator. If these reactions do go, the entire region between the superelements and the present limit will become populated through successive decays from X and Y. The decay sequence will probably be complete unless interrupted by a spontaneous fission that happens to have a large cross-section. With this step taken, where will the next one lead?

REFERENCES

Seaborg, G. T., "Elements Beyond 100. Present Status and Future Prospects." *Ann. Rev. Nuc. Sci.* **18**, 53, 1968.

Seaborg, G. T., *The Transuranium Elements*. Addison-Wesley, Reading, Mass., 1958.

Seaborg, G. T., et al., eds., *The Transuranium Elements*. Nat. Nuc. Engin. Series. McGraw-Hill, New York, 1949.

Appendix

TABLE 1
VALUES OF PHYSICAL CONSTANTS

Symbol	*Quantity*	*Value*
c	Velocity of light in vacuum	2.9979×10^{8} m sec^{-1}
e	Elementary charge	1.6021×10^{-19} coulomb 1.6021×10^{-20} emu 4.8029×10^{-10} esu
h	Planck's constant	6.6256×10^{-34} joule-sec
$\hbar$	$h/2\pi$	1.0545×10^{-34} joule-sec
N_a	Avogadro's number	6.02252×10^{23} mole^{-1}
V_0	Volume of 1 mole of ideal gas	2.241×10^{-2} m^{3} mole^{-1}
k	Boltzmann's constant	1.380×10^{-23} joule °K^{-1}
g	Gravitational constant	6.670×10^{-11} newton m^{2} Kg^{-2}
R	Rydberg constant	1.09737×10^{7} m^{-1}
a_0	Bohr radius	5.2917×10^{-11} m
λ_c	Compton wavelength of electron	2.426×10^{-12} m
α	Fine-structure constant	7.2972×10^{-3}
	$1/\alpha$	1.37039×10^{2}
m_e	Electron mass	9.109×10^{-31} Kg 5.4859×10^{-4} u 5.1098×10^{5} eV
m_p	Proton mass	1.6725×10^{-27} Kg 1.00727 u 9.3855×10^{8} eV
m_n	Neutron mass	1.6748×10^{-27} Kg 1.00866 u 9.39767×10^{8} eV
μ_B	Bohr magneton	9.273×10^{-24} joule tesla^{-1}
μ_N	Nuclear magneton	5.050×10^{-27} joule tesla^{-1}

TABLE 2
GENERAL CONVERSIONS

Quantity	*MKSA*	*CGS*	*Other*
Length	1 meter equals	100 cm	10^6 microns (μ) 10^8 angstroms (Å)
Mass	1 Kg	10^3 g	6.0225×10^{26} umu
Force	1 newton	10^5 dynes	
Energy	1 joule	10^7 ergs	0.4799 calories
Power	1 watt	10^7 ergs sec^{-1}	

TABLE 3
TIME CONVERSIONS

Sec	*Min*	*Hours*	*Days*	*Years*
60	1.0			
3.60×10^3	60	1.0		
8.64×10^4	1.44×10^3	24	1.0	
3.15×10^7	5.26×10^5	8.76×10^3	365	1.0

TABLE 4
ENERGY CONVERSIONS

ergs	*eV*	*grams*	*mass units*
1.0	6.24×10^{11}	1.1126×10^{-21}	6.701×10^2
1.602×10^{-12}	1.0	1.7826×10^{-33}	1.0736×10^{-9}
8.9875×10^{20}	5.610×10^{32}	1.0	6.0225×10^{23}
1.4923×10^{-3}	9.3148×10^8	1.6604×10^{-24}	1.0

TABLE 5
ELECTRICAL CONVERSIONS

Quantity	*MKSA*	*CGS-ESU*	*CGS-EMU*
Charge	1 coulomb	3×10^9	10^{-1}
Current	1 ampere	3×10^9	10^{-1}
Potential	1 volt	1/300	10^8
Electric field strength	1 volt meter^{-1}	$1/3 \times 10^4$ cm^{-1}	10^4 cm^{-1}
Capacitance	1 farad	9×10^{11} cm	10^{-9}
Magnetic flux	1 weber	1/300	10^8 maxwells
Mag. flux density	1 weber meter^{-2}	$1/3 \times 10^6$ cm^{-2}	10^4 gauss
Vacuum polarization	8.85×10^{-12} farads meter^{-1} = 1.0 ESU		

TABLE 6
COMMON PREFIXES

giga-	G-	10^9	milli-	m-	10^{-3}
mega-	M-	10^6	micro-	μ-	10^{-6}
kilo-	K-	10^3	nano-	n-	10^{-9}
			pico-	p-	10^{-12}

TABLE 7
PROPERTIES OF SOME RADIOACTIVE NUCLIDES

Z	*Element*	*A*	*Half-life*	λ	$\bar{E}_\beta(MeV)$	$\Gamma(Rhm)$
1	H	1	12.6 y	0.055	0.0057	—
6	C	14	5680 y	1.22×10^{-4}	0.049	—
11	Na	22	2.6 y	0.265	0.191	0.50
		24	15.0 h	0.0462	0.55	1.84
15	P	32	14.3 d	0.0485	0.69	—
16	S	35	87.9 d	7.9×10^{-3}	0.046	—
17	Cl	36	3×10^5 y	2.3×10^{-6}	0.27	—
19	K	40	1.3×10^9 y	5.3×10^{-10}	1.46	—
		42	12.4 h	0.055	1.49	0.15
20	Ca	45	165 d	4.2×10^{-3}	0.074	—
		47	4.5 d	0.154	0.816	0.44
24	Cr	51	27.5 d	0.0252	—	0.018
26	Fe	55	2.6 y	0.267	—	—
		59	45 d	0.0154	0.12	0.6
27	Co	57	270 d	2.57×10^{-3}	—	0.059
		58	71 d	9.8×10^{-3}	0.35	0.035
		60	5.26 y	0.123	0.095	1.28
36	Kr	85	10.7 y	0.0649	0.250	—
38	Sr	89 }	53 d	0.0131	0.56	—
39	Y	89 }				
53	I	131	8.05 d	0.0862	0.20	0.22
55	Cs	137	30 y	0.0232	0.23	0.32
77	Ir	192	74 d	9.4×10^{-3}	0.17	0.4
79	Au	198	2.7 d	0.0257	0.32	0.23
80	Hg	203	46.9 d	0.0148	0.058	0.16
88	Ra	226	1602 y	4.34×10^{-4}	In 0.5-mm Pt	0.84

TABLE 8

X-RAY CRITICAL-ABSORPTION AND EMISSION ENERGIES

Calculated from the conversion KeV = $12.398/\lambda$ (angstroms)

		KeV			
Z	*Element*	K_{abs}	K_{emm}*	L_{abs}	L_{emm}*
13	Aluminum	1.56	1.49	0.087	—
22	Titanium	4.96	4.51	0.53	0.45
23	Vanadium	5.46	4.95	0.60	0.51
24	Chromium	5.99	5.41	0.68	0.57
25	Manganese	6.54	5.90	0.76	0.64
26	Iron	7.11	6.40	0.85	0.71
27	Cobalt	7.71	6.93	0.93	0.78
28	Nickel	8.33	7.48	1.02	0.85
29	Copper	8.98	8.05	1.10	0.93
30	Zinc	9.66	8.64	1.20	1.02
42	Molybdenum	20.00	17.48	2.88	2.29
45	Rhodium	23.22	22.21	3.42	2.69
46	Palladium	24.35	21.18	3.62	2.84
47	Silver	25.52	22.16	3.81	2.98
48	Cadmium	26.71	23.17	4.02	3.13
49	Indium	27.93	24.21	4.25	3.29
50	Tin	29.19	25.27	4.46	3.44
53	Iodine	33.17	28.60	5.19	3.94
73	Tantalum	67.40	57.52	11.67	8.15
74	Tungsten	69.51	59.31	12.09	8.40
79	Gold	80.71	68.79	14.35	9.71
80	Mercury	83.11	70.82	14.81	9.99
82	Lead	88.00	74.96	15.87	10.55
92	Uranium	115.60	98.43	21.75	13.61

* The most prominent emission line in each group.

TABLE 9
SOME ATOMIC AND NUCLEAR PROPERTIES

Element	*Z*	*A*	*Chem. atm. wgt.*	*Atm. mass (MeV)*	*Percent abundance*
Hydrogen	1	1	1.008	938.767	99.98
		2		1,876.092	0.02
		3		2,809.384	
Helium	2	3	4.0026	2,809.365	10^{-4}
		4		3,728.337	100
Lithium	3	6	6.939	5,602.956	7.5
		7		6,535.253	92.5
		8		7,472.770	
Beryllium	4	7	9.013	6,536.115	
		9		8,394.653	100
Boron	5	8	10.82	7,474.747	
		10		9,326.832	18.4
		11		10,254.876	81.6
		12		11,191.106	
Carbon	6	11	12.011	10,256.856	
		12		11,177.736	98.892
		13		12,112.339	1.108
		14		13,043.712	
Nitrogen	7	14	14.008	13,043.556	99.64
		15		13,972.270	0.36
Oxygen	8	16	16	14,898.911	99.759
		17		15,834.318	0.037
		18		16,765.822	
Fluorine	9	19	19.00	17,696.596	100
Neon	10	20	20.183	18,662.518	90.9
		21		19,555.308	0.27
		22		20,484.491	8.83
Sodium	11	22	22.990	20,487.334	
		23		21,414.466	100
		24		22,347.054	
Magnesium	12	24	24.32	22,341.539	78.6
		25		23,273.759	10.2
		26		24,202.214	11.2
Aluminum	13	27	26.98	25,132.710	100
Silicon	14	28	28.09	26,059.894	92.27
		29		26,990.968	4.68
		30		27,919.901	3.05
Phosphorous	15	31	30.974	28,851.380	100
		32		29,782.993	
Sulphur	16	32	32.066	29,781.283	95.02
		33		30,712.191	0.75
		34		31,640.318	4.12
		35		32,572.883	

TABLE 9
CONTINUED

Element	*Z*	*A*	*Chem. atm. wgt.*	*Atm. mass (MeV)*	*Percent abundance*
Chlorine	17	35	35.457	32,572.715	75.4
		37		34,432.921	24.6
Argon	18	36	39.944	33,502.976	0.337
		38		35,361.446	0.063
		40		37,224.082	99.600
Potassium	19	39	39.100	36,293.839	93.08
		40		37,225.587	0.0119
		41		38,155.046	6.91
		42		39,087.056	
Calcium	20	40	40.08	37,224.272	96.97
		44		40,943.572	2.06
Vanadium	23	51	50.95	47,453.179	99.75
Chromium	24	50	52.01	46,523.651	4.31
		51		47,453.931	
		52		48,381.445	83.76
Iron	26	54	55.85	50,243.566	5.84
		55		51,176.816	
		56		52,102.163	91.68
		57		53,034.070	2.17
		58		53,963.577	0.31
		59		54,896.542	
Cobalt	27	59	58.94	54,894.969	100
		60		55,827.029	
Nickel	28	58	58.69	53,965.49	67.76
		60		55,824.21	26.16
Copper	29	63	63.54	58,617.531	69.09
		64		59,549.164	
		65		60,578.80	30.91
Zinc	30	64	65.37	59,548.592	48.87
		65		60,480.15	
		66		61,406.67	27.62
Silver	47	107	107.88	99,579.7	51.35
		109		101,442.3	48.65
Lead	82	204	207.21	189,996.4	1.48
		206		191,860.7	23.6
		207		192,793.6	22.6
		208		193,725.8	52.3
Bismuth	83	209	209.0	194,661.4	100
Uranium	92	234	238.07	218,004.0	0.0058
		235		218,938.2	0.715
		236		219,186.3	
		238		221,739.0	99.28
		239		222,673.8	

TABLE 10
MASS, VELOCITY, AND ENERGY RELATIONS FOR ELECTRONS

Energy (KeV)	$\beta = \frac{v}{c}$	*Velocity* (cm sec^{-1})	m/m_0
1	0.06245	1.872×10^9	1.002
2	0.08832	2.648	1.004
3	0.1078	3.231	1.006
4	0.1245	3.732	1.008
5	0.1389	4.165	1.010
6	0.1519	4.554	1.012
7	0.1639	4.912	1.014
8	0.1749	5.244	1.016
9	0.1850	5.546	1.018
10	0.1950	5.847	1.020
20	0.2719	8.152	1.039
30	0.3284	9.846	1.059
40	0.3742	1.121×10^{10}	1.078
50	0.4128	1.237	1.098
60	0.4463	1.338	1.118
70	0.4759	1.427	1.137
80	0.5025	1.506	1.157
90	0.5265	1.578	1.176
100	0.5483	1.644	1.196
200	0.6954	2.085	1.392
300	0.7766	2.328	1.587
400	0.8278	2.482	1.783
500	0.8629	2.587	1.979
600	0.8880	2.662	2.175
700	0.9066	2.718	2.371
800	0.9210	2.761	2.566
900	0.9322	2.794	2.762
1,000	0.9411	2.821	2.957
2,000	0.9791	2.935	4.916
3,000	0.9893	2.966	6.873
4,000	0.9936	2.979	8.831
5,000	0.9957	2.985	10.79
6,000	0.9969	2.989	12.75
7,000	0.9976	2.991	14.38
8,000	0.9982	2.992	16.66
9,000	0.9985	2.993	18.62
10,000	0.9988	2.994	20.58

TABLE 11
RANGE OF CHARGED PARTICLES IN AIR

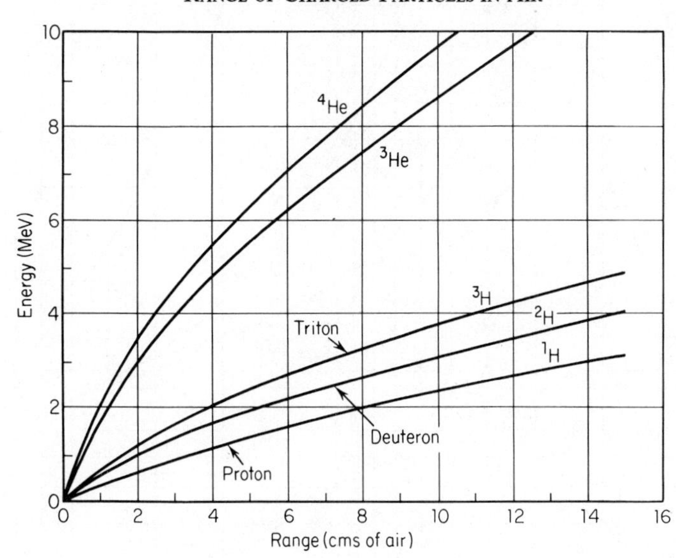

Table 12
Total-absorption Coefficients

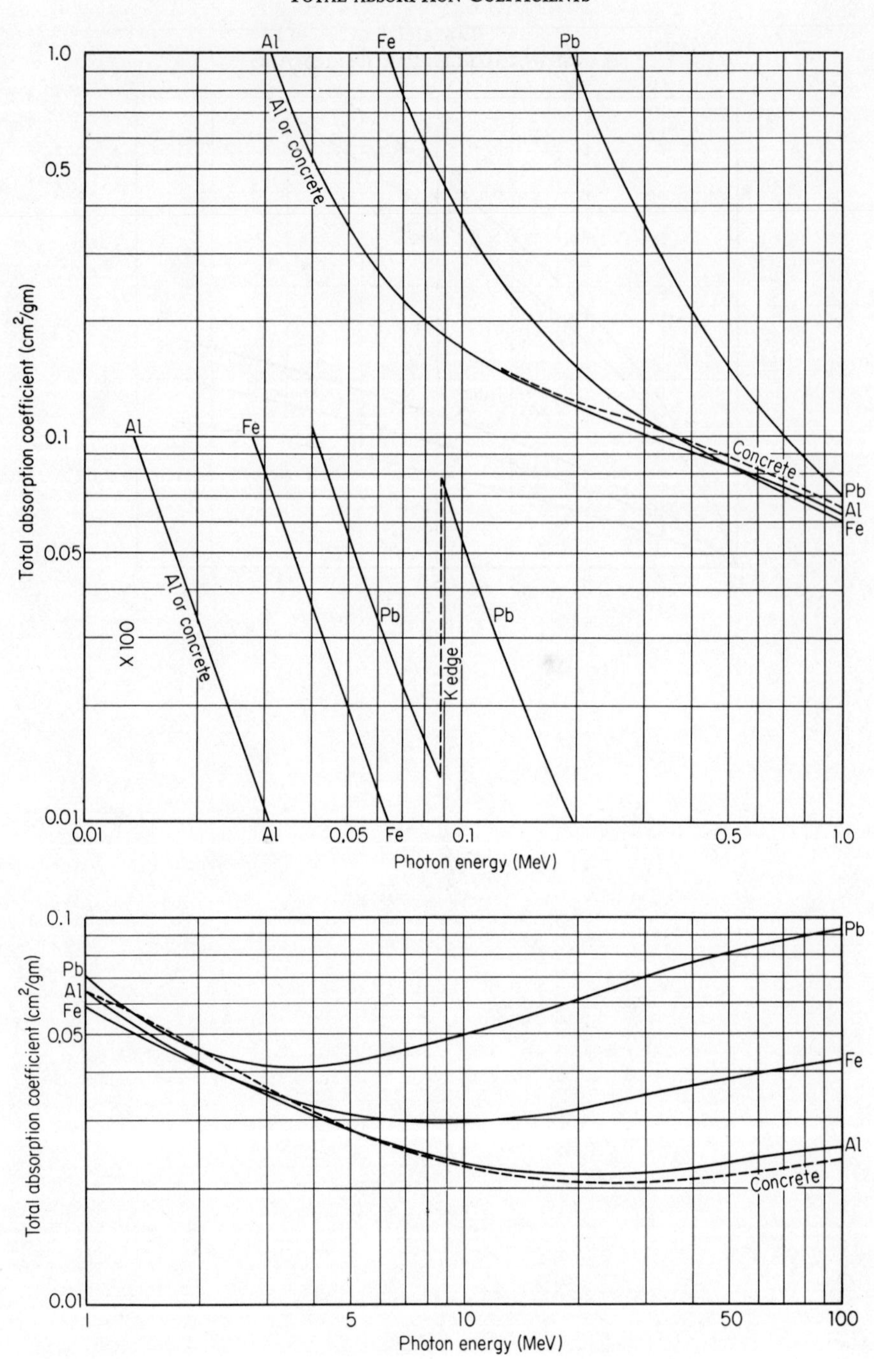

TABLE 13
IONIZATION BY PHOTONS

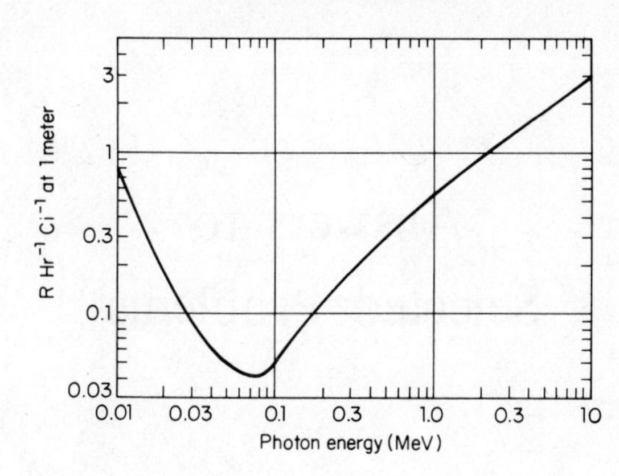

Answers to Selected Problems

1-1. $3.48 \times 10^3\ sec^{-1}$.
1-3. 1.16×10^4°K, 1.16×10^5°K, 1.16×10^6°K.
1-4. 15.5, 11.8 KV.
1-6. 4.04 pF.
1-7. 0.412 pF.
1-8. 67.2 volts
1-9. 91%, 50%, 213 volts, 672 volts.
1-10. 2.68 mV, 12 megohms.
1-11. $t = 0.693$ RC.
1-12. 4.48×10^{-5} volt peak-to-peak.
1-13. 1.44×10^{18} ohms.
1-14. 3.5×10^4 cm sec^{-1}, 3.9×10^4 cm sec^{-1}.
1-15. 24 ft, 1.8×10^6 ohms.
1-16. Add a 294-pF capacitor.
1-17. 21.6 cm^3, 2.5×10^8 ohms.
1-18. 2.26 cm, 1.27×10^{-4} sec.
1-19. 12.8 cm, 1.1×10^{-3} sec.
1-20. 406 and 34.8 volts cm^{-1}, 621 and 124 volts cm^{-1}.

2-1. $1.31 \times 10^4\ sec^{-1}$.
2-2. 65 sec^{-1}.
2-3. 1.1×10^4 ion pairs, or 0.0055%.
2-4. 1180, 1020 with argon.
2-5. 775 volts.
2-6. 0.087 cm from the wire.
2-7. ± 39.4 cpm, operation satisfactory.

2-8. 4.6×10^{-3} sec, 72 sec^{-1}.
2-10. Equal counts $40.80 each, equal time $44.00, minimum $46.50.
2-11. Save $1250 by doubling the amount.
2-12. 64 days.
2-14. $\pm 1.32\%$, ± 1.26 cpm, 470 min.
2-15. $\pm 1.32\%$, ± 12.6 cpm, 47 min.
2-16. $\pm 1.32\%$, ± 5.9 cpm, 296 min.
2-17. 1385 cpm.
2-18. $\tau = \dfrac{kn_1 - n_2 - B(k-1)}{(k-1)(n_1 n_2 - Bn_1 - Bn_2)}$ if $n_1\tau$ and $n_2\tau \gg 1$.
2-19. 100 μsec, 0.4 μsec, 40 pF.
2-20. 5.5×10^{-6} min.

3-2. ± 31 KeV, 4.68%.
3-4. 318 pF, 6.5×10^{-4} volt.
3-5. 8.7×10^{-3} sec.
3-6. 2.2×10^{-10} sec, 6.6×10^{-10} sec.
3-7. 530 pF, 5.3×10^{-2} sec, 1.4×10^{-9} sec, 3.3×10^{-4} volt.
3-8. 374 pF, 3.7×10^{-2} sec, 1.4×10^{-9} sec, 4.6×10^{-4} volt.
3-9. 0.25%, 0.36%, 0.093%, and 2.2%, 2.6%, 0.70%.

4-1. 6.95 KeV.
4-2. 8,170Å, 7.4×10^{5} m sec^{-1}.
4-3. 0.021 Kg.
4-4. 0.312Å, 39.8 KeV, X-ray region.
4-5. 96.5 μm, 0.0129 eV, far infrared, bolometer.
4-9. 1.76×10^{-31} Kg, 4.36×10^{7} m sec^{-1}, 1.76×10^{-31} Kg, 2.0×10^{4} m sec^{-1}.
4-10. 0.0080 m.
4-11. 39.978 u.
4-12. 2.71×10^{6} sec.
4-13. 73 years.
4-14. 1.004, 24.
4-16. 1.06×10^{-10} newton, yes, 2.3×10^{-7} newton.
4-17. 8.27×10^{2} newtons m^{-1}, 0.134 Å.
4-18. 4, 3.4×10^{-9} m.

5-1. 2.8 Å, 15°49′.
5-2. 0.562 Å.
5-3. 58.1 KeV, 3.32 KeV, 3.42×10^{7} m sec^{-1}.
5-4. 5.05×10^{-3} Å, $4.9m_e$, 2.46 MeV.
5-5. 6.4×10^{-6} Å, 6.2×10^{-6} Å.
5-6. 8.06×10^{-12} m.
5-7. 6.6×10^{-13} sec.
5-8. 35.2 Å.
5-9. 3.9×10^{-28} m.
5-10. 1.31×10^{-21} Kg-m sec^{-1}, 0.0545 m, 1.066×10^{-21} Kg-m sec^{-1}.

6-1. 12.81 eV, none.
6-2. 2.81×10^{-10} newton.
6-3. The first Lyman line.
6-4. 1.46×10^{-8} u.
6-5. $18 = 3s^2 3p^6 3d^{10}$.
6-6. 1.05×10^{-3} Å, 9.2×10^{-24} joule tesla^{-1}.
6-7. 1.58×10^{5}°K.
6-8. 2420, 2060, 1940Å, 5.1, 6.0, 6.4 eV.
6-9. 2.87×10^{-13} m.
6-10. 6.22×10^{-15} m.
6-11. 1.20×10^{-38} eV.
6-12. 8.79 MHz.
6-13. 4.80 MHz.
6-14. Too high by factor of 1.000545, 2.88×10^{-14} m.
6-15. 9.25×10^{-8} eV, 3.26 msec^{-1}.
6-16. 6.893, 1.279, 0.446 KeV.
6-17. 2.36.

7-1. 32.000, 58.165, 76.259, 91.166 MeV.
 5.33 6.46 6.93 7.68
7-2. 0.0000, −0.0136, −0.021, −0.0448, +0.0657 u.
 0.0000 −4.37 −5.28 −7.10 +3.14
7-3. 1.3×10^{17} Kg m^{-3}.
7-4. 17.1 MeV.
7-5. 28.297, 23.847, 19.823 MeV.
7-6. 457.7 MeV.
7-7. 1.66×10^{10}°C.
7-8. 2.88×10^{-11} m.
7-9. 1.11, 5.50, 11.8, 1.84, 9.85, 21.9 MeV.
7-11. 1.65×10^{-13}, 9.2×10^{-19} m.
7-12. ^{39}K 327.3 MeV, ^{39}Ca 320.5 MeV, latter unstable and radioactive.
7-13. Ni 547.2, Cu 544.3, Zn 544.4 MeV: two modes of ^{64}Cu decay.
7-14. 1.48×10^{4} m, 8.93×10^{12} joules.
7-15. 1.4×10^{9} Km.

8-4. 4.9 KW, 130 watts.
8-6. 73.9543 u, positron emission and EC.
8-7. Defect −0.07609 u, decrement 0.68987 u, 642.92 MeV.
8-8. 226.12, +0.0174 u.
8-9. 1.821 MeV.
8-10. 2.29 MeV.
8-11. 4.95-KeV and 0.5-KeV X rays, 747, 427 KeV.
8-12. 2.224 KeV.
8-14. Ni anode, 8.33 KeV.
8-15. 101 KeV.
8-16. 8.3 eV.
8-17. 0.79 m sec^{-1}.
8-18. 2.08×10^{2}, 1.09 m sec^{-1}.

9-1. 16.31 mCi.
9-2. 1.7×10^{-10} m^3.
9-3. 81, 73 Ci g^{-1}.
9-4. ^{14}C 0.104, ^{40}K 0.13 μCi.
9-5. 538 mCi, 15.7 days.
9-6. 6.95×10^8 y.
9-7. 2280 watts, 1.72×10^6 g.
9-8. 0.379 m, 3.8 μCi.
9-9. ^{235}U series 0.011, ^{238}U series 0.30, ^{40}K 14.4 dps liter^{-1}.
9-10. 38.4, 5.95×10^{-4} cm^3.
9-12. 12.6 μCi.
9-13. 11,800 yr.
9-14. 1.67×10^3, 1.91×10^4 min.

10-1. 0.014%, 210 MeV.
10-2. 22.8, 12.3 MeV.
10-3. 4.87–24.3, 21.8–67, 34.5–87 MeV.
10-4. $\log R(\text{cm}) = 0.790 + 0.0198 \log \lambda(y^{-1})$.
10-5. 1270, 6.8, 212.
10-6. 58.4, −1.7.
10-7. ^{144}Nd 3.8×10^8, ^{147}Sm 2.6×10^3, ^{209}Bi 1.5×10^8
0, no 0, no ??
10-8. About at $^{190}_{79}$Au.

11-4. 1.75 MeV.
11-5. 0.37 MeV allowed, 1.48 MeV first forbidden (4, +), (2, +), (0, +), fourth forbidden.
11-6. 5.515 MeV, ground-state transition fourth forbidden.
11-8. 0.318 MeV, 0.274.
11-9. 78, 49 eV.
11-10. 135, 300, 481 KeV.

12-1. 511–171, 0–340, 2750–235, 0–2515 KeV.
12-2. 1800, 887, 399, 258, 224 KeV.
12-3. 0.93 cm, 0.745 cm^{-1}, 0.0660 cm^2 g^{-1}, 1.34 cm, 1.1 or 25 MeV.
12-4. 2.08 cm, 0.321 cm^{-1}, 0.119 cm^2 g^{-1}, 3.0 cm, 250 KeV.
12-5. Best choices Pd and Mo; economics dictates Mo and Zn.
12-6. 108, 4.1, 0.75; —, —, 2.2×10^3 barns.
12-9. 172 calories, 2.15×10^{-3}°C.
12-11. 1.35 R hr^{-1} Ci^{-1} at 1 meter.
12-12. 1.87 ergs min^{-1} cm^{-2}.
12-13. 7.2×10^5 Ci.
12-14. 260, 73; 112, 43 rads min^{-1}.
12-15. 73 m.

13-1. 2.94 cm, 0.752.
13-2. 48.2°, 472 MeV.
13-3. 650 mg cm^{-2}, 1.4 MeV.

13-4. 6200 cpm, 0.040 cm, 6.38 $cm^2\ g^{-1}$.
13-5. 33.6 rads beta and 2.4 rads gamma.
13-6. 9.75 mrads beta and 16.9 mrads gamma, 67%.
13-7. 0.38 mrad, 68% in 50 years.
13-8. 1.31 mCi, 7540 rads.
13-9. 3 mrads.
13-10. 440 rads.
13-11. 2 hr first day, 2 hr 30 min second, 3 hr 20 min third, 4 hr 20 min fourth.

14-1. 2.05 MeV.
14-2. 7.94 MeV.
14-3. +8.072 MeV, 0.00867 u; +7.289 MeV, 0.00783 u.
14-4. 189.8 MeV.
14-5. 0.156, 0.019 MeV.
14-6. 0.546 MeV.
14-7. 0.374 mCi.
14-8. 0.64 MeV required.
14-9. 8.397 MeV exoergic.
14-10. 138.8, 140.4 MeV.
14-11. ^{52m}Mn.
14-12. Beta decay 105.1 MeV, proton reaction 104.8 MeV.
14-13. $^{24}Mg(n, p)$ 4.33 MeV endoergic, $^{27}Al(n, \alpha)$ 3.13 MeV endoergic, $^{21}Ne(\alpha, p)$ 2.86 MeV exoergic but poor target.
14-14. 1877.534 MeV.

15-1. 1100 g on a number basis; 3000 g on an energy basis.
15-2. 51.6 MeV, 3.22 MeV/A.
15-3. 198, 222, 243 cm.
15-4. 0.645, 2.148 u.
15-5. 1.2, 0.424 ft.
15-6. 2.08, 11.74, 108.4, 1075 u; 1 to 0.484, 0.0798, 0.00865, 0.000872.
15-7. 14.5 MeV, 18.4 MHz.
15-8. 42.6% decrease from f to $0.573f$.
15-9. 1.35 mm at the outer radius.
15-10. 4.57×10^7, 1.23×10^6 cm, 2.4×10^{-4}, 8.6×10^{-8} ampere, 3.6×10^{-4}.
15-11. 183 volts per revolution.
15-12. 230 eV.
15-13. 6.06×10^6, 546 watts, 1.6×10^6 rads min^{-1}.

16-1. 1.009030 u.
16-2. 1.008805 u.
16-3. 0.797 Km sec^{-1}, 3.32×10^{-3} eV, 9.1×10^{-7} m.
16-4. 1.008982 ± 0.000014 u.
16-6. 30.
16-7. 0.058 eV.
16-8. 3.5%.

16-9. 1.33:1.
16-10. 14.2 MeV.
16-11. 5.55×10^{14} atoms yr^{-1}.
16-12. 0.0108.
16-13. 3.54 yr.
16-14. 1.0:6.3.
16-15. 26.7 hr, 10.7 mCi.
16-16. 21.6 Ci g^{-1}, 2.2 yr.
16-17. 89.7 R hr^{-1} at 1 m.
16-18. 8.1×10^4.

17-1. 6.4, 12.7, 9.1, 10.7, 0.511 MeV.
17-2. 187.3 MeV.
17-3. 2.61×10^{-8}, 6.87×10^{-5}; 7.0, 15.2 μg.
17-6. 29.5%.
17-7. 1.84×10^{20} cm^{-2}.
17-8. 17.39 MeV required.
17-9. 5.14×10^4, 2.01 Kg, 0.717 g.
17-10. 35 g, 4.85×10^3 Ci; 128 g, 1.12×10^4 Ci.
17-12. 95, 26 R.
17-13. 8.4×10^5 Ci ^{90}Sr and 1.94×10^6 Ci ^{137}Cs.

18-1. 24, 2.3×10^{-7} sec.
18-2. 66.
18-3. 11 hr, 16 times.
18-4. 810 mCi g^{-1}, 30 hr.
18-5. 906 g day^{-1}.
18-6. 0.058 sec.
18-7. 11.1 g, 1.6×10^3 Ci.
18-8. 1.18 KT.
18-9. 656 joules.
18-10. 750 mCi g^{-1}.

19-1. 8.27×10^{21} g yr^{-1}, 4.13×10^{-12}.
19-2. 2.29×10^{48}, 2.5×10^{47} ergs.
19-3. 2.9×10^{47} ergs, 3.24×10^{26} g.
19-4. 1.08×10^5 joules, 10 sec^{-1}.
19-6. 1.14 MeV, 1.32×10^{10}°K.

20-1. 294 mA.
20-2. 1.76×10^4 watts.
20-3. About 8 min.
20-4. 6, 12, 100 mrads.
20-5. 342 MeV.

Index